Law of Sines

$$\frac{a}{\sin A} = \frac{b}{\sin B} = \frac{c}{\sin C}.$$

Law of Cosines

$$a^2 = b^2 + c^2 - 2bc \cos A,$$
$$b^2 = c^2 + a^2 - 2ca \cos B,$$
$$c^2 = a^2 + b^2 - 2ab \cos C.$$

Law of Tangents

$$\frac{a-b}{a+b} = \frac{\tan\frac{1}{2}(A-B)}{\tan\frac{1}{2}(A+B)}, \qquad \frac{b-c}{b+c} = \frac{\tan\frac{1}{2}(B-C)}{\tan\frac{1}{2}(B+C)},$$

$$\frac{c-a}{c+a} = \frac{\tan\frac{1}{2}(C-A)}{\tan\frac{1}{2}(C+A)}.$$

The Half-angle Formulas in Terms of the Sides of a Triangle

$$\tan\tfrac{1}{2}A = \frac{r}{s-a}, \qquad \tan\tfrac{1}{2}B = \frac{r}{s-b}, \qquad \tan\tfrac{1}{2}C = \frac{r}{s-c},$$

$$\sin\tfrac{1}{2}A = \sqrt{\frac{(s-b)(s-c)}{bc}}, \qquad \sin\tfrac{1}{2}B = \sqrt{\frac{(s-a)(s-c)}{ac}},$$

$$\sin\tfrac{1}{2}C = \sqrt{\frac{(s-a)(s-b)}{ab}},$$

where $\quad r = \sqrt{\dfrac{(s-a)(s-b)(s-c)}{s}} \quad$ and $\quad s = \tfrac{1}{2}(a+b+c).$

The Area of a Triangle

$$K = \tfrac{1}{2}ab \sin C.$$

$$K = \tfrac{1}{2}a^2 \frac{\sin B \sin C}{\sin A}.$$

$$K = \sqrt{s(s-a)(s-b)(s-c)},$$

where $\quad s = \dfrac{a+b+c}{2}.$

PLANE TRIGONOMETRY
A New Approach

PLANE TRIGONOMETRY
A New Approach

Second Edition

C. L. JOHNSTON

East Los Angeles College

PRENTICE-HALL, INC., *Englewood Cliffs, N.J.* 07632

Library of Congress Cataloging in Publication Data

Johnston, Carol Lee.
   Plane trigonometry, a new approach.

   Includes index.
   1. Trigonometry, Plane. I. Title.
QA533.J65 1978   516'.22   77-16841
ISBN 0-13-677666-3

Printed in the United States of America

10 9 8 7

PRENTICE-HALL INTERNATIONAL, INC., *London*
PRENTICE-HALL OF AUSTRALIA PTY. LIMITED, *Sydney*
PRENTICE-HALL OF CANADA, LTD., *Toronto*
PRENTICE-HALL OF INDIA PRIVATE LIMITED, *New Delhi*
PRENTICE-HALL OF JAPAN, INC., *Tokyo*
PRENTICE-HALL OF SOUTHEAST ASIA PTE. LTD., *Singapore*
WHITEHALL BOOKS LIMITED, *Wellington, New Zealand*

Contents

2 THE SINE FUNCTION   20

3 THE COSINE FUNCTION   55

4 THE TANGENT FUNCTION   79

5 THE RECIPROCAL FUNCTIONS   99

6 RADIAN MEASURE OF ANGLES   113

7 THE FUNDAMENTAL RELATIONS BETWEEN THE TRIGONOMETRIC FUNCTIONS   128

8 COMBINATION OF ANGLES   146

9 GRAPHICAL METHODS   175

10 LOGARITHMS   190

11 SOLUTION OF OBLIQUE TRIANGLES   213

Contents

12 INVERSE TRIGONOMETRIC FUNCTIONS   231

13 COMPLEX NUMBERS   246

TABLES

ANSWERS TO ODD-NUMBERED EXERCISES   297

INDEX   349

Preface

In the second edition we have (1) added a set of review exercises at the end of each chapter, (2) added a diagnostic test at the end of each chapter, (3) introduced basic instructions on use of the hand calculator as an aid in solving trigonometric problems, and (4) made minor changes to improve the clarity of the material.

We have retained the four-place tables. Thus the instructor has the option of having his students (1) do all their calculations with the aid of tables without using a calculator, (2) combine use of a minimal calculator and tables, or (3) do all the work using a more complete calculator.

Students of plane trigonometry often encounter more difficulty in study of the subject than is justified by the essential simplicity of the material. It is for this reason that in this book we avoid what appears to be one of the major causes of their confusion: discussion of the six trigonometric functions at one time.

In traditional textbooks on trigonometry, the student is introduced to the six basic trigonometric functions each immediately following the other. Then, very soon, he is required to give from memory, or to find from tables, the value of each of the six functions for an angle of any magnitude. For us who have lived with this subject for a long time, it requires no effort to determine both the correct algebraic sign and the numerical value, regardless of the value of the angle or real number on which the function depends. But experience shows that this is not such an easy matter for many students.

One of the principal innovations of this textbook is that we take just one trigonometric function at a time and carry it through evaluations, trigonometric equations, applications to the right and the general triangles, and graphs. For example, by the time the student has lived with only the sine function through these applications, it has become thoroughly established in his mind and will not easily be forgotten. This is done with the sine function in Chapter 2, the cosine function in Chapter 3, and the tangent function in Chapter 4. Then, having this thorough knowledge of the sine, cosine, and tangent functions, the student has no difficulty extending his knowledge to include the definitions and applications of the reciprocal functions, which are treated in Chapter 5.

After studying the first five chapters a student should be able (1) to solve both right and general triangles and many basic problems involving vectors, (2) to draw the graphs of all the trigonometric functions as well as composite functions, and (3) to solve many trigonometric equations. In fact, for many students who are enrolled in a two-year curriculum such as drafting, mechanical engineering, or electronics, or a four-year curriculum such as medicine, dentistry, pharmacy, or optometry, the first five chapters will give the necessary trigonometry. This does not mean that this textbook presents a "watered down" version of trigonometry. In fact I have used the first five chapters, experimentally, in classes for several years and found that one can ultimately cover the subject matter in less time by using this approach than can be done by the traditional method. This leaves more time for the subjects of the later chapters, such as inverse functions and complex numbers, which are often slighted at the end of the course. The entire book offers a complete course in plane trigonometry and is especially adaptable to courses of various lengths and purposes.

The unit circle has been used extensively as an aid in defining functions and developing formulas and identities.

An abundance of illustrative examples with solutions, about three hundred in number, anticipates the student's difficulties and at the same time sets before him applications of basic principles and orderly solutions of exercises. The discussion of trigonometric equations and graphs is more complete than in most books on the subject. One hundred thirty-seven illustrative figures help to clarify the proofs, definitions, and examples. The over thirteen hundred exercises allow a student to obtain practice on all parts of the theory by working either the odd-numbered or the even-numbered exercises. Answers to odd-numbered exercises and the diagnostic tests appear at the end of the text. Answers to the even-numbered exercises are available in a separate pamphlet.

Basic trigonometric formulas are listed inside the front cover and the basic trigonometric identities inside the back cover so as to be immediately available when needed.

I am deeply indebted to Professor Raymond W. Brink for his many suggestions and painstaking attention to my manuscript. The elegant proof of the Law of Cosines, the set of tables, and more than a score of other passages in the text of this book were taken by permission from the *Third Edition* of Dr. Brink's *Plane Trigonometry*.

Whittier, California C. L. JOHNSTON

1

Introduction

1-1 TRIGONOMETRY

Trigonometry—a branch of mathematics that deals with the relationships between the angles and sides of triangles and the theory of the periodic functions connected with them—is a basic tool used in the development of mathematics and many sciences such as physics, engineering, astronomy, and the like. The name "trigonometry" is derived from two Greek words, combining to mean measurement of triangles.

Early astronomers used trigonometry. Two Greek astronomers Hipparchus (about 140 B.C.) and Ptolemy (about 150 A.D.) used trigonometry in solving spherical triangles.

In this revised edition we show how the hand calculator can be used to find values of functions and do calculations. The instructions in this book are written for both the student who does not have a calculator and for the student who does have a calculator.

Before giving instructions in the use of the calculator and beginning the study of trigonometry we review **order of operations.**

1-2 ORDER OF OPERATIONS

When addition, subtraction, multiplication, division, powers, and roots are written in the same expression, and there are no parentheses to change the normal order, we perform the operations in the following order:

Order of operations:

First, powers and roots are done in any order.

Next, multiplication and division are done in order from left to right.

Finally, addition and subtraction are done in order from left to right.

In the following examples we show the correct order of operations and what happens when an incorrect order of operations is used.

Example 1. $16 \div 2 \times 4$.

Correct order (division is done first because it is written first)

$$16 \div 2 \times 4 = 8 \times 4 = 32.$$

Incorrect order

$$16 \div 2 \times 4 \neq 16 \div 8 = 2.$$

(Division was written first and should, therefore, have been done before multiplication.)

Example 2. $9 + 4 \times 5$.

Correct order (multiplication must be done before addition)

$$9 + 4 \times 5 = 9 + 20 = 29.$$

Incorrect order

$$9 + 4 \times 5 \neq 13 \times 5 = 65.$$

(Incorrect because addition was done before multiplication.)

Example 3. $6 \times 2 + 4 - 10 \div 2$.

Solution:

$$
\begin{aligned}
& 6 \times 2 + 4 - 10 \div 2 \\
=\ & 12 + 4 - 5 \\
=\ & 16 - 5 = 11.
\end{aligned}
$$

Example 4. $5 \times 3^2 \div 5 + \sqrt{16} \times 2$.

Solution:

$$5 \times 3^2 \div 5 + \sqrt{16} \times 2$$
$$= 5 \times 9 \div 5 + \quad 4 \times 2$$
$$= \quad 45 \quad \div 5 + \quad 4 \times 2$$
$$= \qquad 9 \quad + \qquad 8 = 17.$$

Example 5. $\dfrac{2 + 6}{2}$.

Solution:

$$\frac{2 + 6}{2} = \frac{2}{2} + \frac{6}{2} = 1 + 3 = 4.$$

Alternate solution:

$$\frac{2 + 6}{2} = \frac{8}{2} = 4.$$

Verify the following answers:

	ANSWERS
1. $25 \div 5 \times 2$	1. 10
2. $4 \times 6 \div 3$	2. 8
3. $2 + 5 \times 6$	3. 32
4. $4 \times 6 - 2$	4. 22
5. $10 - 2 \times 4$	5. 2
6. $20 - 8 \div 4$	6. 18
7. $10 - 3 + 4 \times 5$	7. 27
8. $18 + 2 \times 10 - 6$	8. 32
9. $6^2 \div 3 \times 4 - \sqrt{25} \times 2$	9. 38
10. $\sqrt{100} + 2 \times 3^2 - 8$	10. 20
11. $\dfrac{9 + 3}{3}$.	11. 4

1-3   THE HAND CALCULATOR

From time to time in addition to solving problems with the aid of tables we shall show how the calculation can be done on a calculator. We shall only show the key operations for calculators that use the **algebraic mode of entry.** Even with this restriction, however, there are many different kinds of calculators. Only standard procedures that apply to most of these calculators are given here.

Rules for entering numbers and operations on the calculator

1. Press $\boxed{\text{C}}$ before starting a calculation.

2. Enter numbers and operations by pressing only one key at a time.

3. Press $\boxed{\cdot}$ for the decimal point.

4. Your calculation ends when you press $\boxed{=}$. After completing a calculation by pressing $\boxed{=}$, it is not necessary to press $\boxed{\text{C}}$ before doing the next calculation.

5. If after you start a calculation you press a wrong number, press $\boxed{\text{CE}}$, then press the correct number, and continue with the calculation.

6. If you press $\boxed{\text{C}}$, it clears the current calculation in progress and the display.

We show the key operations used on the calculator in solving the following examples.

Example 1. Find $25.5 + 13 - 17.3$.

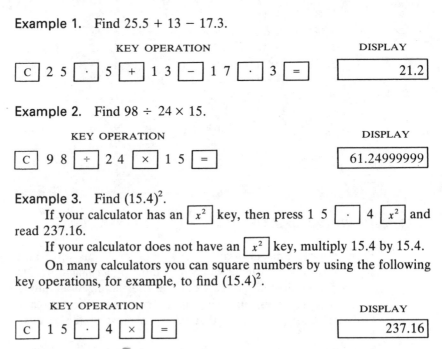

KEY OPERATION

$\boxed{\text{C}}$ 2 5 $\boxed{\cdot}$ 5 $\boxed{+}$ 1 3 $\boxed{-}$ 1 7 $\boxed{\cdot}$ 3 $\boxed{=}$

DISPLAY

$\boxed{21.2}$

Example 2. Find $98 \div 24 \times 15$.

KEY OPERATION

$\boxed{\text{C}}$ 9 8 $\boxed{\div}$ 2 4 $\boxed{\times}$ 1 5 $\boxed{=}$

DISPLAY

$\boxed{61.24999999}$

Example 3. Find $(15.4)^2$.

If your calculator has an $\boxed{x^2}$ key, then press 1 5 $\boxed{\cdot}$ 4 $\boxed{x^2}$ and read 237.16.

If your calculator does not have an $\boxed{x^2}$ key, multiply 15.4 by 15.4.

On many calculators you can square numbers by using the following key operations, for example, to find $(15.4)^2$.

KEY OPERATION

$\boxed{\text{C}}$ 1 5 $\boxed{\cdot}$ 4 $\boxed{\times}$ $\boxed{=}$

DISPLAY

$\boxed{237.16}$

1-4 RECTANGULAR COORDINATES

Although it is assumed that the student has had experience with the rectangular coordinate system, a brief review of the subject is given.

Two perpendicular lines are drawn meeting at O (Fig. 1-1). The point O is called the **origin**, the line OX the **x-axis**, and the line OY the **y-axis**. A convenient unit of length is used to mark off distances to the right and left and up and down from the origin O. Distances to the right are taken as positive values of x, and distances to the left are taken as negative values of x. Positive values of y are measured upward, and negative values of y are measured downward.

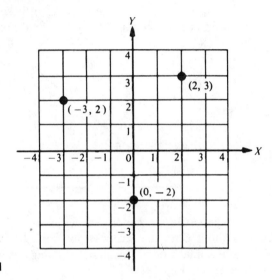

Figure 1-1

The position of any point P on the xy-plane is determined by an ordered pair of numbers called the **coordinates** of the point. The distance of P to the right or left of the origin is called the **abscissa**, or **x-coordinate**, of point P, and y, the vertical distance of P from the x-axis, is the **ordinate**, or **y-coordinate**, of point P. Point P is said to have the coordinates (x, y) and may be referred to as *the point* (x, y).

Example 1. Locate the point $(2, 3)$.

Start at the origin and move two units to the right, then up three units (Fig. 1-1).

Example 2. Locate the point $(0, -2)$.

Start at the origin, but, because the first number is zero, do not move right or left. The second number being negative directs us down two units (Fig. 1-1).

5

Example 3.   Locate the point $(-3, 2)$.

Start at the origin and move left three units and then up two units (Fig. 1-1).

1-5 THE FORMATION OF ANGLES

A plane angle is formed if two rays have the same end-point. This end-point is called the **vertex** of the angle, and the two rays are the **sides of the angle** (Fig. 1-2). We can think of the angle as being **generated** when a ray whose end-point is the vertex of the angle rotates in the plane about the vertex from the position of one side of the angle until it coincides with the other side. Its initial position is the **initial side** and its final position is the **terminal side** of the angle.

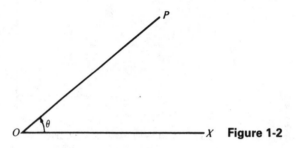

Figure 1-2

Angle *XOP* is generated when a ray with the initial position *OX* rotates about *O* to the terminal position *OP*. We call *OX* the initial side and *OP* the terminal side of the angle. An angle is often denoted by a single Greek or italic letter (Fig. 1-2).

An angle may be positive or negative and of any magnitude. Angles are said to be positive when the generating ray rotates counterclockwise, and negative when the generating ray rotates clockwise (Fig. 1-3).

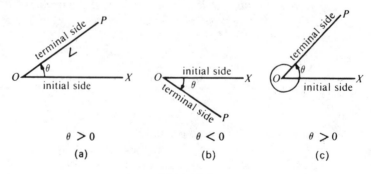

Figure 1-3

1-6 ANGLES IN STANDARD POSITION

An angle is said to be in **standard position**, with respect to the rectangular coordinate system, when its vertex is at the origin and its initial side extends along the positive x-axis.

The xy-plane is divided by the coordinate axes into four quadrants as shown in Figure 1-4. On some occasions we shall use the abbreviation Q I, Q II, Q III, and Q IV to represent the respective quadrants.

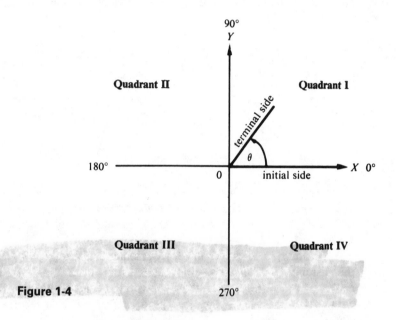

Figure 1-4

Angles in standard position are often spoken of as being in the quadrant that contains the terminal side (Fig. 1-5).

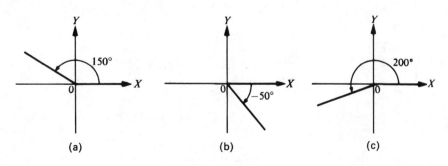

Figure 1-5

Example 1.   Second, fourth, and third quadrant angles.

(a) 150° is an angle in the second quadrant;
(b) −50° is an angle in the fourth quadrant;
(c) 200° is an angle in the third quadrant.

Coterminal Angles   *Two or more angles are said to be* **coterminal** *if their terminal sides coincide when the angles are in standard position.* Examples of coterminal angles: 90° and −270°; 30° and 390°; −30° and 330°; etc. (Fig. 1-6). Any angle has an unlimited number of coterminal angles.

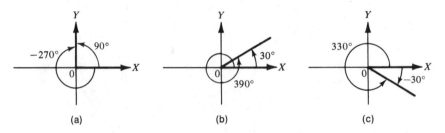

(a)   (b)   (c)

Figure 1-6

Quadrantal Angles   *Angles in standard position and having their terminal sides along a coordinate axis are called* **quadrantal angles.** The quadrantal angles include 0°, 90°, 180°, 270°, and their coterminal angles.

In addition to the x- and y-coordinates of the point P, a third number, r, the distance of P from the origin, is often used. This distance is called the **radius** of P. In this course the radius, r, will always be considered positive or zero regardless of the signs of x and y. Referring to the right triangle OAP (Fig. 1-7), we have:

$$r^2 = x^2 + y^2,$$

$$r = \sqrt{x^2 + y^2}.$$

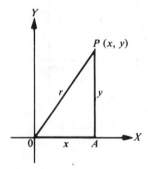

Figure 1-7

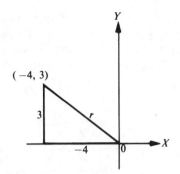

Figure 1-8

Example 2. Find the distance from the origin to the point $(-4, 3)$ (Fig. 1-8).

Solution: Plot the point. Then

$$r^2 = (-4)^2 + 3^2$$
$$= 16 + 9$$
$$= 25,$$
$$r = \sqrt{25} = 5.$$

1-7 DISTANCE BETWEEN TWO POINTS

Let two points, P_1 and P_2, have coordinates (x_1, y_1) and (x_2, y_2), and let d be the length of the line segment joining them (Fig. 1-9). Draw a horizontal line

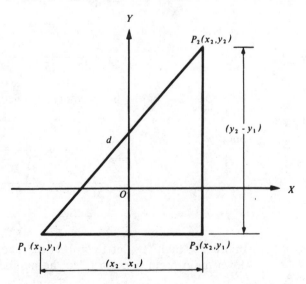

Figure 1-9

through P_1 and a vertical line through P_2. These two lines intersect, forming a 90° angle at point $P_3(x_2, y_1)$. Draw the line segment P_1P_2 which represents the distance d between points P_1 and P_2.

Then by the theorem of Pythagoras,

$$d^2 = (x_2 - x_1)^2 + (y_2 - y_1)^2,$$

$$d = \sqrt{(x_2 - x_1)^2 + (y_2 - y_1)^2}.$$

(1)

> *Distance between two points $P_1(x_1, y_1)$ and $P_2(x_2, y_2)$:*
>
> $$d = \sqrt{(x_2 - x_1)^2 + (y_2 - y_1)^2}$$

If the points P_1 and P_2 were interchanged, the results would be the same. That is:

$$d = \sqrt{(x_2 - x_1)^2 + (y_2 - y_1)^2} = \sqrt{(x_1 - x_2)^2 + (y_1 - y_2)^2}.$$

Example 1. Find the distance between the points $A(-3, -2)$ and $B(9, 3)$.

Solution: We use Formula (1) and Table 5 to find the required distance.

$$d = \sqrt{(x_2 - x_1)^2 + (y_2 - y_1)^2}$$

$$= \sqrt{[9 - (-3)]^2 + [3 - (-2)]^2}$$

$$= \sqrt{(9 + 3)^2 + (3 + 2)^2}$$

$$= \sqrt{144 + 25} = \sqrt{169} = 13.$$

Example 2. Find the distance between the points $A(-3, 1)$ and $B(0, -2)$.

Solution:

$$d = \sqrt{(x_2 - x_1)^2 + (y_2 - y_1)^2}$$

$$= \sqrt{[0 - (-3)]^2 + [(-2) - (+1)]^2}$$

$$= \sqrt{3^2 + (-3)^2} = \sqrt{18}.$$

In order to find $\sqrt{18}$, let $18 = 10(1.8)$. Then in Table 5, page 288, follow down the column under "n" to the number 1.80. Go across the line to the column under "$\sqrt{10n}$" and read $\sqrt{18} = 4.24264$ (to five decimal places). Therefore, the required distance is 4.24264.

If you have a calculator with a square root key ($\sqrt{}$ or $\sqrt{x}$), you can find $\sqrt{18}$ as follows:

KEY OPERATION	DISPLAY
C 18 $\boxed{\sqrt{x}}$	4.242640687

When rounded off to five decimal places this number becomes the same as the number found by using the table.

Example 3. Use Table 5 to find the value of each of the following expressions:

(a) $\sqrt{235}$.

Solution: Set $235 = 100(2.35)$. Then

$$\sqrt{235} = \sqrt{100(2.35)}$$

$$= 10\sqrt{2.35}.$$

To find $\sqrt{2.35}$, see Table 5, page 289. Follow down the column under "n" to the number 2.35. Go across the line to the column under "$\sqrt{n}$" and read $\sqrt{2.35} = 1.53297$. Therefore, $10\sqrt{2.35} = 15.3297$ (to four decimal places).

Finding $\sqrt{235}$ on the calculator.

KEY OPERATION	DISPLAY
C 235 $\boxed{\sqrt{}}$	15.32970972

(b) $\sqrt{.0785}$.

Solution: Set $.0785 = .01(7.85)$. Then

$$\sqrt{.0785} = \sqrt{.01(7.85)}$$

$$= .1\sqrt{7.85} = .1(2.80179) = .280179.$$

(to six decimal places).

Finding $\sqrt{.0785}$ on the calculator.

KEY OPERATION	DISPLAY
C · 0785 $\boxed{\sqrt{}}$	.2801785145

1-8 RELATED ANGLES

Angles are usually designated by Greek letters such as α (alpha), β (beta), γ (gamma), θ (theta), ϕ (phi), ω (omega), etc.

Let θ be any angle in standard position. *We define the **related angle** of θ as the angle, in the interval from 0° to 90° inclusive, formed by the terminal side of θ and the horizontal axis.*

Examples: (See Fig. 1-10.)

1. The related angle of 150° is 30°.
2. The related angle of 210° is 30°.
3. The related angle of −150° is 30°.

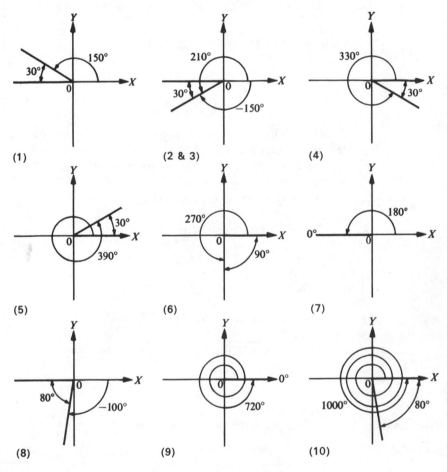

Figure 1-10

   4. The related angle of 330° is 30°.
   5. The related angle of 390° is 30°.
   6. The related angle of 270° is 90°.
   7. The related angle of 180° is 0°.
   8. The related angle of −100° is 80°.
   9. The related angle of 720° is 0°.
10. The related angle of 1000° is 80°.

Example 11. Find the related angle of 127° 34.2'.

Solution: We subtract 127° 34.2' from 180° in the following manner:

$$\begin{array}{ll} 179°\ 60' & (=180°) \\ 127°\ 34.2' & (-) \\ \hline \end{array}$$

Related angle = 52° 25.8'

Example 12. Find the related angle of 223° 17.8'.

Solution: We subtract 180° from 223° 17.8' in the following manner:

$$\begin{array}{ll} 223°\ 17.8' & \\ 180°\ 00.0' & (-) \\ \hline \end{array}$$

Related angle = 43° 17.8'

Exercises 1-8

In Exercises 1–12 name the quadrant in which each of the given angles terminate.

1. 100°.	2. 170°.	3. 200°.	4. 260°.
5. −10°.	6. −70°.	7. −200°.	8. −250°.
9. 400°.	10. 440°.	11. 500°.	12. 530°.

In Exercises 13–48 find the related angle of each of the given angles.

13. 100°.	14. 110°.	15. 210°.	16. 200°.
17. 265°.	18. 269°.	19. 275°.	20. 280°.
21. −140°.	22. 140°.	23. −160°.	24. −200°.
25. −210°.	26. −300°.	27. −280°.	28. 540°.
29. 360°.	30. −270°.	31. 450°.	32. 180°.
33. 720°.	34. 900°.	35. 530°.	36. 630°.
37. 10° 15'.	38. 20° 50'.	39. −40° 10'.	40. −60° 50'.
41. 140° 25'.	42. 150° 35'.	43. 225° 10.3'.	44. 235° 27.3'.
45. 300° 10.7'.	46. 350° 50.6'.	47. 151° 15.4'.	48. 179° 25.9'.

In Exercises 49–54 find the radius, r, for each of the points.

49. $(3, 4)$.   **50.** $(5, 12)$.   **51.** $(-5, 12)$.   **52.** $(-3, 4)$.

53. $(5, 0)$.   **54.** $(0, -4)$.

In Exercises 55–60 use Formula (1) and Table 5 to find the distance between the pairs of points.

55. $(2, 1)$ and $(6, 4)$.   **56.** $(-3, 0)$ and $(0, 4)$.

57. $(2, -3)$ and $(5, 8)$.   **58.** $(2, -3)$ and $(-5, 2)$.

59. $(2, 3)$ and $(-6, 1)$.   **60.** $(0, -5)$ and $(-5, 0)$.

Find the perimeters of the triangles whose vertices are at the points given in Exercises 61 and 62.

61. $(0, 1)$, $(4, 4)$, $(6, 2)$ **62.** $(10, 10)$, $(-1, -2)$, $(-4, 5)$.

1-9   THE 30°– 60° RIGHT TRIANGLE

Each angle of an equilateral triangle is 60°. The bisector of any angle of this triangle is the perpendicular bisector of the side opposite the angle. (Consider, for example, the equilateral triangle of Figure 1-11 in which each side is two

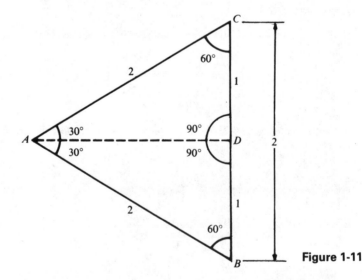

Figure 1-11

units in length.) It follows that in any 30°–60° right triangle the side opposite the 30° angle is half the length of the hypotenuse (Fig. 1-12).

We use the Pythagorean Theorem to find the side opposite the 60° angle. The following solutions relate to parts *a*, *b*, and *c* of Figure 1-12.

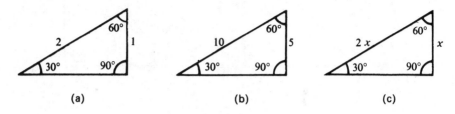

(a)   (b)   (c)

Figure 1-12

$a^2 + 1^2 = 2^2,$	$b^2 + 5^2 = 10^2,$	$c^2 + x^2 = (2x)^2,$
$a^2 + 1 = 4,$	$b^2 + 25 = 100,$	$c^2 + x^2 = 4x^2,$
$a^2 = 3,$	$b^2 = 75,$	$c^2 = 3x^2,$
$a = \sqrt{3}.$	$b = \sqrt{(25)(3)} = 5\sqrt{3}.$	$c = \sqrt{3}x.$

The lengths of the respective sides of the 30–60° right triangle are in the ratio of 1 to 2 to $\sqrt{3}$, and the student should memorize this. Notice that the shortest side is opposite the smallest angle and the longest side is opposite the largest angle. We will have occasion to refer frequently to this triangle and to the ratios of the lengths of the respective sides (Fig. 1-13).

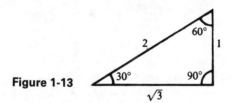

Figure 1-13

1-10   THE 45° RIGHT TRIANGLE

*An **acute** angle is a positive angle less than* 90°.

The sum of the two acute angles of a right triangle is 90°. If one of the two acute angles is 45°, the other must also be 45°, making an isosceles right triangle.

We use the Pythagorean Theorem to find the hypotenuse of a 45° right triangle. We denote the length of the equal sides by *s* and that of the hypotenuse by *h*. First consider the 45° right triangle having equal sides one unit long (Fig. 1-14).

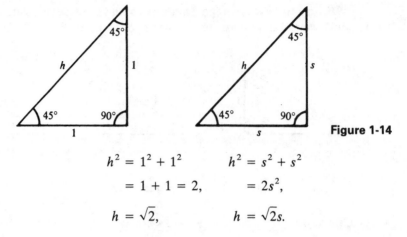

Figure 1-14

$$h^2 = 1^2 + 1^2 \qquad h^2 = s^2 + s^2$$
$$= 1 + 1 = 2, \qquad = 2s^2,$$
$$h = \sqrt{2}, \qquad h = \sqrt{2}s.$$

In the 45° right triangle the hypotenuse is always $\sqrt{2}$ times the length of one of the equal sides. The student should memorize the 1 to 1 to $\sqrt{2}$ ratio of the respective sides. These will be used many times in this course (Fig. 1-15).

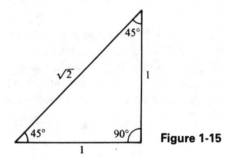

Figure 1-15

1-11 PYTHAGOREAN RIGHT TRIANGLES

Right triangles whose three sides are integers are often referred to as **Pythagorean right triangles;** for example, the 3, 4, 5 right triangle and the 5, 12, 13 right triangle are Pythagorean triangles (Fig. 1-16).

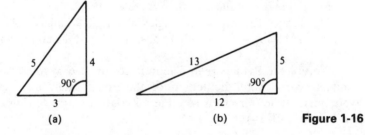

(a)   (b)   Figure 1-16

These triangles will be used on a number of occasions in this course. Note:

$$5^2 = 4^2 + 3^2, \qquad 13^2 = 12^2 + 5^2,$$
$$25 = 16 + 9 = 25. \qquad 169 = 144 + 25 = 169.$$

Exercises 1-11

1. Find the length of the sides of a 30°–60° right triangle whose hypotenuse is 5. Express the answer in simplified radical form.

2. Find the length of the hypotenuse of a 45° right triangle whose sides are each 4. Express answer in simplified radical form.

3. The lengths of the sides of a certain right triangle are in the ratio of 3 to 4 to 5. If the hypotenuse of this triangle is 60 centimeters long, what are the lengths of the other two sides?

4. Find the y-coordinate of the point in the fourth quadrant having a x-coordinate of 4 and a radius of 5.

Solution:

$$4^2 + y^2 = 5^2,$$
$$16 + y^2 = 25,$$
$$y^2 = 9,$$
$$y = \pm\sqrt{9} = \pm 3.$$

Because P is in Q IV, y must be negative. Therefore, the y-coordinate of P is -3.

In Exercises 5–12 find the missing coordinate for each of the points. As was explained in Section 1-6, we have three important numbers associated with every point P in the xy-plane. These are the x- and y-coordinates of P, and r, its distance from the origin.

5. $x = 3$, $r = 5$, point is in Q I.

6. $x = 3$, $r = 5$, point is in Q IV.

7. $y = 5$, $r = 13$, point is in Q II.

8. $y = -5$, $r = 13$, point is in Q III.

9. $x = 1$, $r = 2$, point is in Q IV.

10. $x = 1$, $r = \sqrt{2}$, point is in Q I.
11. $y = -1$, $r = \sqrt{2}$, point is in Q III.
12. $y = 1$, $r = 2$, point is in Q II.

REVIEW EXERCISES

In Exercises 1–6 perform the indicated calculations.

1. $20 \div 2 \times 5$.
2. $3 + 4 \times 2$.
3. $10 - 3 \times 5 + 4^2$.
4. $4\sqrt{16} + 7 - 3$.
5. $\dfrac{7 + 21}{7}$.
6. $\dfrac{15 + 5}{5}$.

In Exercises 7–14 name the quadrant in which the given angles terminate.

7. $260°$.
8. $300°$.
9. $-260°$.
10. $-200°$.
11. $370°$.
12. $210°$.
13. $-75°$.
14. $-280°$.

In Exercises 15–20 find the related angle of each of the given angles.

15. $190°$.
16. $281° \, 45.6'$.
17. $-60°$.
18. $-260°$.
19. $270°$.
20. $380°$.

In Exercises 21–24 find the radius, r, for each of the given points.

21. $(12, 5)$.
22. $(-4, 3)$.
23. $(-5, -12)$.
24. $(0, 6)$.

In Exercises 25–28 find the distance between the pairs of points.

25. $(-2, 0)$ and $(2, 3)$.
26. $(-3, 2)$ and $(4, -1)$.
27. $(0, 4)$ and $(-3, -2)$.
28. $(-5, 6)$ and $(7, 1)$.

29. The length of the side opposite the $30°$ angle of a $30°$–$60°$ right triangle is 4. Find the length of the side opposite the $60°$ angle.

30. Find the x-coordinate of the point in the second quadrant having a y-coordinate of $\sqrt{5}$ and a radius of $\sqrt{10}$.

31. The lengths of the sides of a certain right triangle are in the ratio of 3 to 4 to 5. If the length of the hypotenuse of this triangle is 12 inches, what is the length of the shortest side of the triangle?

CHAPTER 1: DIAGNOSTIC TEST

The purpose of this test is to see how well you understand the work covered in this introductory chapter. We recommend that you work this test before your instructor tests you on this chapter. Allow yourself approximately 50 minutes to do this test.

Solutions to the problems, together with section references, are given in the Answer Section at the end of this book. You should study the sections referred to for the problems you do incorrectly.

---

1. Calculate each of the following.

    (a) $18 \div 2 \times 3$.   (b) $\dfrac{12 + 3}{3}$.   (c) $4\sqrt{9} + 3^2 - 3$.

2. Are $90°$ and $-270°$ coterminal angles? (yes or no)

3. Is $180°$ a quadrantal angle? (yes or no)

4. Name the related angle of each of the following angles.
    (a) $170°$.   (b) $-30°$.   (c) $270°$.   (d) $-150°$.
    (e) $180°$.   (f) $390°$.   (g) $130° \, 15.6'$.

5. Name the quadrant in which each of the following angles terminate.
    (a) $210°$.   (b) $-45°$.   (c) $120°$.   (d) $400°$.

6. Find the radius, r, for the point $(4, -3)$.

7. Find the distance between the points $(-1, -1)$ and $(2, 3)$.

8. Find the hypotenuse of the $45°$ right triangle whose sides are each three units long. Express your answer in simplied radical form.

9. Find the x-coordinate of the point in the third quadrant having a y-coordinate of -3 and a radius of 5.

10. The length of the hypotenuse of a $30°$–$60°$ right triangle is 4. Find the length of the side opposite the $60°$ angle. Express your answer in simplified radical form.

2

The Sine Function

By means of the relation

$$A = \pi r^2,$$

we can find the area of a circle when its radius is known. For example, when the radius is 3 inches the area is 9π, or approximately 28.3 square inches. The area, A, is said to be a **function** of the radius, r, because the value of A can be found when the value of r is known.

> Definition.   *A variable y is said to be a **function** of a variable x if x and y are so related that when a value is assigned to x, one and only one corresponding value is determined for y.*

2-2 THE SINE FUNCTION

Let θ be any angle. Place it in standard position (Fig. 2-1), and let P be a point on the terminal side of θ, other than the origin. Associated with P are three numbers; these are its x- and y-coordinates and its radius, r, the distance of P from the origin.

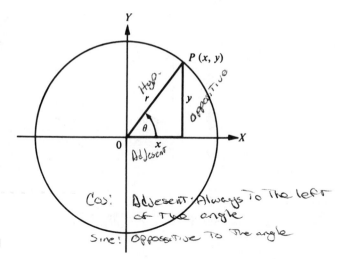

Figure 2-1

Six different fractions, or ratios, can be formed from the numbers x, y, and r. These ratios are called **trigonometric functions** of θ. In this chapter we shall learn a great deal about one of these ratios, namely, y/r, which is called the **sine function.**

2-3 DEFINITION OF THE SINE OF AN ANGLE

1. Place an angle θ in standard position.
2. Let point P with coordinates (x, y) be any point, other than the origin, on the terminal side of angle θ.
3. Let r be the distance of P from the origin. This distance r will always be considered as a positive number no matter what position P may occupy on the coordinate plane.

> Then the sine of θ (written sin θ) is defined as the ratio of y to r, or
>
> $$\sin \theta = \frac{y}{r}. \qquad r \neq 0$$

Special case. Let θ be an acute angle of a right triangle. Let this triangle be placed in the first quadrant with θ in standard position. Let the point (x, y) be the end of the hypotenuse in the first quadrant. With the triangle in this position, y becomes the side of the triangle opposite θ, and r becomes the hypotenuse of the triangle. Then, from the definition of the sine of an angle,

$$\sin \theta = \frac{y}{r} = \frac{\text{side opposite } \theta}{\text{hypotenuse}}.$$

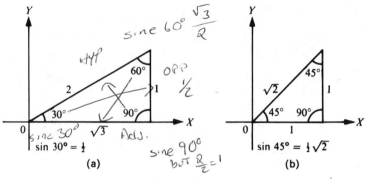

Figure 2-2

Examples.   (See Fig. 2-2.)

1. $\sin 30° = \frac{1}{2}.$

2. $\sin 60° = \dfrac{\sqrt{3}}{2} = \frac{1}{2}\sqrt{3}.$

3. $\sin 45° = \dfrac{1}{\sqrt{2}} = \frac{1}{2}\sqrt{2}.$

2-4 SIMILAR TRIANGLES

If two triangles have the three angles of one equal respectively to the three angles of the other, the triangles are similar. In plane geometry we learned that the ratios of corresponding sides of similar triangles are the same regardless of the size of the triangle. Therefore, corresponding angles of any two similar right triangles have equal sines. This is illustrated in Figure 2-3.

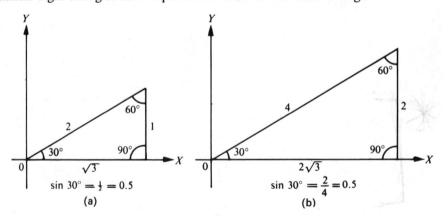

Figure 2-3

2-5 THE UNIT CIRCLE

*The circle with radius r = 1 unit and with its center at the origin is called **the unit circle**.* Because of the fact that corresponding sides of similar triangles have equal ratios, we can simplify the task of finding the sine of a given angle θ as follows (Fig. 2-4);

Unit circle

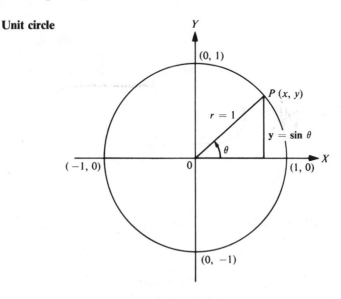

Figure 2-4

1. Place θ in standard position.
2. Use the origin as a center, and construct a circle with radius equal to 1 unit.
3. Let $P(x, y)$ be the point where the terminal side of θ meets the unit circle.

Then $r = 1$, and

$$\sin\theta = \frac{y}{1} = y$$

(in the unit circle).

Important relationship. *In the unit circle sin θ = y.*

It follows that the sine of an angle is positive when y is positive and negative when y is negative. Therefore, *the sines of first and second quadrant angles are positive, and the sines of third and fourth quadrant angles are negative. That is, the sine of an angle is positive when a point on the terminal side lies above the x-axis and negative when a point on the terminal side lies below the x-axis* (Fig. 2-5).

23

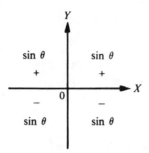

Figure 2-5

From Figure 2-4 we observe the sines of the quadrantal angles to be as follows:

$$\sin 0° = 0, \qquad \sin 180° = 0,$$
$$\sin 90° = 1, \qquad \sin 270° = -1.$$

Again, by observing Figure 2-4, note that the sine of an angle is never greater than 1 nor less than -1.

2-6 THE ABSOLUTE VALUE OF A NUMBER

Symbols that will be used in this section are:

$<$, read *is less than,*
$\leqq$, read *is less than or equal to,*
$>$, read *is greater than,*
$\geqq$, read *is greater than or equal to.*

The **absolute value** of a real number a is written $|a|$, and is equal to a if $a \geqq 0$ and to $-a$ if $a < 0$.

Thus, for example, $|5| = 5$, since $5 > 0$. But $|-3| = 3$, for $-3 < 0$, and, consequently, $|-3| = -(-3) = 3$. These examples illustrate the fact that *the absolute value of a number is never negative.*

Example. $|-5| = |5| = 5$.

2-7 VALUES OF THE SINES
OF CERTAIN SPECIAL ANGLES

Let θ be any angle in standard position, and let $P(x, y)$ be the point of intersection of the terminal side of θ with the unit circle. Recall that under these conditions the sine of θ has the same value as the y-coordinate of point

P. Therefore, as θ varies from 0° to 90°, to 180°, to 270°, and finally to 360°, the sine of θ correspondingly varies from 0 to 1, to 0, to −1, and back to 0 (Fig. 2-6).

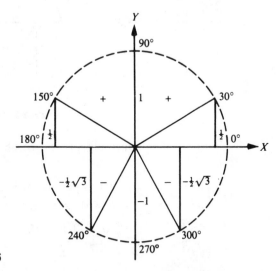

Figure 2-6

From the unit circle, Figure 2-6, we can find the values of certain special angles. For example, $\sin 30° = \frac{1}{2}$, $\sin 90° = 1$, $\sin 150° = \frac{1}{2}$, $\sin 180° = 0$, $\sin 240° = -\frac{1}{2}\sqrt{3}$, $\sin 270° = -1$, $\sin 300° = -\frac{1}{2}\sqrt{3}$, $\sin 360° = 0$.

The sine of any angle has the same absolute value as the sine of its related angle. The sine is positive when a point on the terminal side lies above the x-axis and negative when a point on the terminal side lies below the x-axis (Fig. 2-5).

To find the sine of an angle, place the angle in standard position and then determine its quadrant and the sine of its related angle.

Example 1. Find $\sin 150°$. (See Fig. 2-7a.)
 Since 150° is in the second quadrant, $\sin 150°$ is positive. The related angle is 30°. Therefore, $\sin 150° = \sin 30° = \frac{1}{2}$.

Example 2. Find $\sin 300°$. (See Fig. 2-7b.)
 Because 300° is in the fourth quadrant, $\sin 300°$ is negative. The related angle is 60°. Therefore, $\sin 300° = -\sin 60° = -\frac{1}{2}\sqrt{3}$.

Example 3. Find $\sin(-225°)$. (See Fig. 2-7c.)
 Since −225° is in the second quadrant, $\sin(-225°)$ is positive. The related angle is 45°. Therefore, $\sin(-225°) = \sin 45° = \frac{1}{2}\sqrt{2}$.

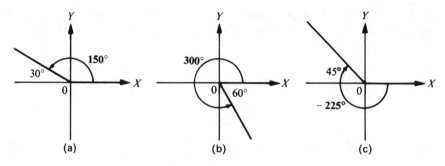

Figure 2-7

From elementary algebra we recall certain facts about *powers* and *roots*:

(1) $\left(\dfrac{a}{b}\right)^n = \dfrac{a^n}{b^n};$ n is a positive whole number.
    $b \neq 0$ ($\neq$ is read *is not equal to*).

(2) $\sqrt{a}\sqrt{b} = \sqrt{ab};$ $a > 0, b > 0,$

(3) $(\sqrt{a})^2 = \sqrt{a}\sqrt{a} = \sqrt{a^2} = a.$

(4) $\left(\dfrac{3}{2}\right)^2 = \dfrac{9}{4}.$

(5) $\dfrac{1}{\sqrt{2}} = \dfrac{1}{\sqrt{2}} \cdot \boxed{\dfrac{\sqrt{2}}{\sqrt{2}}} = \tfrac{1}{2}\sqrt{2}.$

Powers of $\sin \theta$ are written:

$$(\sin \theta)^3 = \sin^3 \theta, \qquad (\sin \theta)^2 = \sin^2 \theta, \qquad (\sin \theta)^n = \sin^n \theta.$$

Example 4. $\sin^2 60° = (\sin 60°)^2 = (\tfrac{1}{2}\sqrt{3})^2 = \tfrac{3}{4}.$

Example 5. $\sin^3 210° = (\sin 210°)^3 = (-\tfrac{1}{2})^3 = -\tfrac{1}{8}.$

Example 6.   Let θ be an angle in standard position with its terminal side passing through the point $(3, -4)$. Find $\sin \theta.$

Solution: Sketch the angle θ (Fig. 2-8).
    Solve for r.

$$r = \sqrt{3^2 + (-4)^2}$$

$$= \sqrt{9 + 16}$$

$$= \sqrt{25} = 5.$$

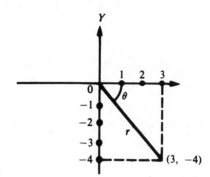

Figure 2-8

Hence,

$$\sin \theta = \frac{y}{r} = \frac{-4}{5} = -\frac{4}{5}.$$

Exercises 2-7

In Exercises 1–20 find the value of each term, and then combine the values.

1. $\sin 30° + \sin 180° + \sin 60°$.

2. $\sin 90° + \sin 150° + \sin 0°$.

3. $\sin (-90°) + \sin (-150°) + \sin (-30°)$.

4. $\sin 210° + \sin 270° + \sin 360°$.

5. $\sin (-30°) + \sin 720° + \sin 45°$.

6. $\sin (-90°) + \sin (-180°) + \sin (135°)$.

7. $\sin^2 45° + \sin^2 135° + \sin^2 180°$.

8. $\sin^2 60° + \sin^2 90° + \sin^2 150°$.

9. $2 \sin 150° + \sin^2 (-60°) + \sin^2 90°$.

10. $\sin^2 (-60°) + \sin 180° + \sin 210°$.

11. $\sin 330° + \sin (-150°) + \sin 150°$.

12. $\sin^3 30° + \sin^4 45° + \sin^2 135°$.

13. $\sin^3 210° + \sin^4 60° + \sin^2 225°$.

14. $\sin 390° + \sin 450° + \sin 540°$.

15. $\sin 720° + \sin (-270°) - \sin (-150°)$.

16. $\sin (-135°) - \sin^2 (-60°) + \sin 360°$.

17. $\sin 0° + \sin 180° + \sin 360°$.

18. $\sin 720° + \sin (-180°) + \sin 540°$.

19. $2 \sin^2 60° + 4 \sin^2 45° + 6 \sin^2 135°$.

20. $4 \sin^2 120° + 2 \sin^2 150° + 6 \sin^2 210°$.

In each of the Exercises 21–28 let θ be an angle in standard position with its terminal side passing through the point given. Find $\sin \theta$.

21. $(3, 4)$.   22. $(5, 12)$ 23. $(-3, 4)$.   24. $(\sqrt{3}, 1)$.

25. $(-1, -1)$.   26. $(0, 3)$.   27. $(-2, 0)$.   28. $(1, -\sqrt{3})$.

2-8   PERIODIC FUNCTIONS

We have seen that if an angle is increased by 360°, the value of the sine of the angle is unchanged.

Examples.

1. $\sin 0° = \sin 360°$.
2. $\sin 30° = \sin 390°$.
3. $\sin 60° = \sin 420°$.

That is, if θ is any angle, then

$$\sin (\theta + 360°) = \sin \theta, \text{ or } \sin (\theta + n \cdot 360°) = \sin \theta \ (n \text{ an integer}).$$

The values of the sine of an angle thus recur periodically, the period of recurrence being 360°.

It will be seen later that all the trigonometric functions repeat their values in cycles and are therefore called **periodic functions.**

The periodic nature of the trigonometric functions makes them of great importance in the study of natural science. Many natural phenomena are of a periodic nature, as for example a vibrating string, a swinging pendulum, the rotation of the earth, the movements of the planets, and the wave motions in the transmission of sound, light, and electricity.

2-9   A TABLE OF SINES

To find $\sin \theta$ when θ is known, and to find θ when $\sin \theta$ is known, refer to Table 1.

For angles in the interval 0° to 45°, use the degree and sine columns that read down from the top of the page. For angles in the interval 45° to 90°, read up from the bottom of the page.

Example 1.   Evaluate $\sin 20° 30'$.

Solution: Locate 20° 30′ in the degrees column of Table 1; then read value of sin 20° 30′ = .3502 to the right of 20° 30′ in the sine value column.

Example 2. Evaluate sin 74° 20′.

Solution: Locate 74° 20′ in the degrees column of Table 1; then read the value of sin 74° 20′ = .9628 to the left of 70° 20′ in the sine value column that reads up from the bottom of the page.

Example 3. Evaluate sin 200°.

Solution: The related angle of 200° is 20°. 200° is a third quadrant angle. The sine of a third quadrant angle is negative. Therefore,

$$\sin 200° = -\sin 20° = -.3420.$$

Example 4. Find the acute angle θ if sin θ = .3283

Solution: Look in the sine column of Table 1 until you find .3283. Then read θ = 19° 10′.

Example 5. Find the second quadrant angle θ such that sin θ = .8988.

Solution: Search the sine column of Table 1 until you find .8988. This number corresponds to sin 64° 00′. 64° 00′ is the related angle of the desired angle. (See Fig. 2-9.)

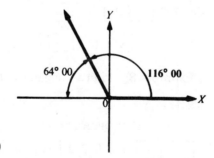

Figure 2-9

One value of θ is 180° − 64° 00′ = 116° 00′. Other values of θ can be found as follows: θ = 116° 00′ + n360° where n is any integer.

Example 6. Find an angle θ in the third quadrant such that sin θ = −.3420.

Solution: Search the sine columns of Table 1 until you find .3420 in the column marked "Sine" at the top of the page. To the left in the degree column find 20° 00′. Therefore, the related angle is 20° 00′. (See Fig. 2-10.) Thus, $\theta = 180° + 20° = 200°$. A more general answer would be $\theta = 200° + n360°$, where n is an integer.

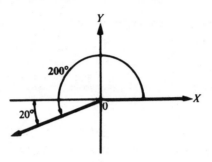

Figure 2-10

We now show how to use the calculator to evaluate trigonometric functions. If your calculator has a Degree–Radian switch, move the switch to degrees. Display answers will sometimes vary in the final right-hand numbers, depending on the calculator used.

Example 7.   Evaluate each of the following functions:

KEY OPERATIONS   DISPLAY

(a) sin 15°. $\boxed{\text{C}}$ 1 5 $\boxed{\text{sin}}$ $\boxed{.2588190451}$

(b) sin 132°. $\boxed{\text{C}}$ 1 3 2 $\boxed{\text{sin}}$ $\boxed{.7431448255}$

(c) sin 214°. $\boxed{\text{C}}$ 2 1 4 $\boxed{\text{sin}}$ $\boxed{-.5591929035}$

(d) sin 285° 17′. $\boxed{\text{C}}$ 1 7 $\boxed{\div}$ 6 0 $\boxed{+}$ 2 8 5 $\boxed{=}$ $\boxed{\text{sin}}$ $\boxed{-.9646341352}$

Example 8.   Find an angle θ in the first quadrant such that

$$\sin \theta = .48022$$

KEY OPERATION   DISPLAY

$\boxed{\text{C}}$ $\boxed{\cdot}$ 4 8 0 2 2 $\boxed{\text{inv}}$ $\boxed{\text{sin}}$ $\boxed{28.69977154}$

The answer shown is in degrees. 28.69977154° is equivalent to 28° + .69977154°. To change .69977154° to minutes, multiply by 60. An easy way to accomplish this when 28.69977154 is already in the display is to continue the above key operation as follows:

$\boxed{-}$ 2 8 $\boxed{=}$ $\boxed{\times}$ 6 0 $\boxed{=}$ $\boxed{41.98629269}$

Therefore, .69977154° = 41.98629269′ ≈ 42′ and $\theta \approx 28° 42′$ (*rounded off to the nearest minute*).

Exercises 2-9

In Exercises 1–21 find the value of the functions listed. Use Table 1.

1.	sin 10°.	**2.**	sin 40°.	**3.**	sin 160°.
4.	sin 170°.	**5.**	sin 220°.	**6.**	sin 230°.
7.	sin 280°.	**8.**	sin 290°.	**9.**	sin (−10°).
10.	sin (−15°).	**11.**	sin (−100°).	**12.**	sin (−110°).
13.	sin 20° 40′.	**14.**	sin 30° 50′.	**15.**	sin 100° 50′.
16.	sin 100° 10′.	**17.**	sin 265° 40′.	**18.**	sin 255° 10′.
19.	sin 760°.	**20.**	sin 1000°.	**21.**	sin 45°.

In Exercises 22–27 find the acute angle, θ of the functions listed. Use Table 1.

22. $\sin \theta = .1219$.   **23.** $\sin \theta = .3907$.   **24.** $\sin \theta = .6157$.

25. $\sin \theta = .6905$.   **26.** $\sin \theta = .7660$.   **27.** $\sin \theta = .9827$.

In Exercises 28–31, find the angle θ where $90° < \theta < 180°$ and $\sin \theta$ has the given value. Use Table 1.

28. $\sin \theta = .1045$.   **29.** $\sin \theta = .1736$.

30. $\sin \theta = .9890$.   **31.** $\sin \theta = .8307$.

In Exercises 32–35 find the angle θ where $180° < \theta < 270°$ and $\sin \theta$ has the given values. Use Table 1.

32. $\sin \theta = -.4067$.   **33.** $\sin \theta = -.5446$.

34. $\sin \theta = -.9283$.   **35.** $\sin \theta = -.8675$.

In Exercises 36–39 find the angle θ where $270° < \theta < 360°$ and $\sin \theta$ has the given value. Use Table 1.

36. $\sin \theta = -.4226$.   **37.** $\sin \theta = -.4384$.

38. $\sin \theta = -.9474$.   **39.** $\sin \theta = -.8897$.

2-10 SIGNIFICANT DIGITS

Applied problems are usually stated in terms of data obtained by *measurement* or *observation*. The accuracy of such data is subject to the mechanical limitations of the instruments used.

When an astronomer states that it is 93,000,000 miles from the earth to the sun, or a machinist states that the diameter of a bar of metal is 1.287 inches, we appreciate that both numbers are only approximate. The entries in the tables of trigonometric functions are usually approximations of never-ending decimals.

Digit defined.   *A digit is any one of the ten Arabic numerals*, 0, 1, 2, 3, 4, 5, 6, 7, 8, 9, *so named from counting upon the fingers* (Latin *digitus*, finger). Thus, 785 is a three-digit number; 80 is a two-digit number; etc.

In most computations we necessarily deal with approximations. Thus, if a four-place table gives sin 34° 00′ = .5592, we know merely that the exact value of sin 34° 00′ is at least as near to .5592 as it is to either .5591 or .5593, and therefore it lies in the range from .55915 to .55925. In this case, we say that 2 is the **last significant digit.** *In general, the position of the last significant digit is determined such that the error made by using the approximate value instead of the exact value is not more than 5 units in the next place to the right of the last significant digit.*

The *significant digits* of a number written in decimal form are those digits that begin with the first digit not zero and end with the last digit which is definitely known. Thus, in the number 27.43 the significant digits are 2, 7, 4, 3; in the number .00412 the significant digits are 4, 1, 2; in the number .315 the significant digits are 3, 1, 5. In the number recorded as 3.510 the significant digits are 3, 5, 1, 0, the final 0 being significant since it is used to indicate a degree of accuracy. Zeros used to fix the position of the decimal point are not significant.

In a measurement recorded as 180,000, for example, it is impossible to tell which of the zeros are significant until further information is given concerning the known accuracy of the measurement. By expressing such numbers in a form called *scientific notation* (Section 2-12), or *standard notation*, the significant digits can be indicated. In the absence of this additional information concerning the position of the last significant digit, we shall assume that it is the last nonzero digit in a number. Thus, in the absence of other information, the last significant digit of 1800 is 8, that of 1492 is 2, and that of .00326 is 6.

2-11   APPROXIMATIONS

It is often desirable to express a number with fewer digits than were originally given. This process is called **rounding off the number.** We shall use the following rules whenever rounding off takes place:

(1) *If the part to be dropped amounts to less than 5 units in the first discarded place, keep the last retained digit unchanged.* (See Example 1.)

(2) *If the part to be dropped amounts to more than 5 units in the first discarded place, increase the last retained digit by one unit.* (See Example 2.)

(3) *If the part to be dropped is exactly 5, round off the preceding digit to an even number.* (See Examples 3 and 4.) Hand calculators, with round off built in, do not obey rule (3). They round up.

Examples.

1. 3.1416 = 3.14 (to three significant digits).
2. 3.1416 = 3.142 (to four significant digits).
3. 47.45 = 47.4 (to three significant digits).
4. 47.35 = 47.4 (to three significant digits).
5. 9138 = 9000 (to one significant digit).

The symbol $\approx$ is put between two expressions to indicate that the right-hand expression is an **approximation** to the left-hand expression. This symbol is read *is approximately equal to*. Thus, $\sqrt{3} \approx 1.73$; $\pi \approx 3.1416$; sin 21° $\approx$ .3584, etc.

In general, the result of a calculation is only as accurate as the least accurate number used in the calculation.

Unless otherwise stated we shall use the following relationship between the accuracy of sides and angles of triangles.

ACCURACY IN SIDES	ACCURACY IN ANGLES
2 significant digits	Nearest degree.
3 significant digits	Nearest multiple of 10 minutes.
4 significant digits	Nearest minute.
5 significant digits	Nearest tenth of a minute.

If a problem comes in a form involving *seconds* in angular measure, immediately round off to the nearest tenth of a minute, remembering that $0.1' = 6''$. In using tables, one should carry out accuracy as far as the tables in use permit and *round off figures only after all computation has been completed*. If one rounds off earlier, there is often a serious accumulation of error that goes much beyond that of rounding off a single entry.

2-12 SCIENTIFIC NOTATION

Scientific notation (sometimes called *standard notation*) gives us a convenient way of writing numbers so as to display their degree of accuracy. Assuming that 40950 and .04095 have four-place accuracy, in scientific notation we should write them as

$$4.095 \times 10^4 \quad \text{and} \quad 4.095 \times 10^{-2}, \text{respectively.}$$

It will be noticed that the factor 4.095 is the same for both numbers, and that it consists of the sequence of significant digits with the decimal point placed immediately after the first such digit.

> A number is said to be written in scientific notation when it is in the form
>
> $$a \times 10^n$$
>
> where $1 \leqq a < 10$ and n is an integer.

To change a number to scientific notation, always place the first nonzero digit immediately to the left of the decimal point and then multiply the number so written by the appropriate power of ten.

Examples.

1. $830 = 8.3 \times 100 = $ . . . . 8.3×10^2.

2. $83 = 8.3 \times 10 = $ . . . . 8.3×10^1.

3. $8.3 = 8.3 \times 1 = $ . . . . 8.3×10^0.
    (By definition, $10^0 = 1$)

4. $.83 = \dfrac{83}{100} = \dfrac{8.3}{10} = $ . . . . 8.3×10^{-1}.

5. $.083 = \dfrac{83}{1000} = \dfrac{8.3}{100} = \dfrac{8.3}{10^2} = $ . . . . 8.3×10^{-2}.

The number 1800, by our assumption in Section 2-10, is understood to be accurate to two significant digits, that is, to hundreds. If we wish to indicate that this number is accurate to units, then scientific notation can be used as follows: $1800 = 1.800 \times 10^3$.

2-13 APPLICATION OF THE SINE FUNCTION TO RIGHT TRIANGLES

In studying triangles it is common practice to denote the vertices and also the values of the corresponding angles by capital letters such as A, B, C. When this is done we denote the lengths of the opposite sides by the corresponding small letters such as a, b, c (Fig. 2-11).

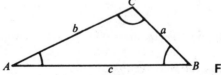

Figure 2-11

In general, we shall use a capital letter, such as *A*, to denote the point or vertex as well as the angle. When there is any possibility of confusing two such meanings we shall denote the angle whose vertex is *A* by ∠*A* or ∠*CAB*.

As you know, a triangle has six *parts*, three sides and three angles. To *solve* a triangle, when we are given some of its parts, means to find the remaining parts. However, in some of the problems the student will be asked to solve for only specified parts.

We can solve for the sides of a right triangle when the hypotenuse and an acute angle are known.

Example 1. Solve the right triangle *ABC*, in which *c* = 1.243, *A* = 32° 50′, and *C* = 90° 00′ (Fig. 2-12). Here we have used the letter *A* to denote a *point* (namely, one of the vertices of the triangle) and also an angle whose vertex is at that point.

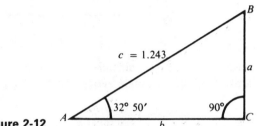

Figure 2-12

Solution: We use Table 1 and the sine function.

(1) $\quad \sin 32° 50′ = \dfrac{a}{1.243}$,

(2) $\qquad a = (1.243)(\sin 32° 50′)$

$\qquad\qquad = (1.243)(.5422) = .674$ (to three significant digits; see Sec. 2.11

$\quad B = 90° - 32° 50′ = 57° 10′.$

The calculator can be used to find the value of (2) in the solution of Example 1, as follows:

$$a = (1.243)(\sin 32° 50′)$$

KEY OPERATION   DISPLAY

| C | 5 0 | ÷ | 6 0 | + | 3 2 | = | sin | × | 1 | · | 2 4 3 | = | .6739510455 |

Rounding off to three significant digits, we have .674, which is the same number found by using tables.

$$\sin 57° \, 10' = \frac{b}{1.243},$$

$$b = (1.243)(sin\ 57° \, 10')$$

$$= (1.243)(.8403) = 1.04 \text{ (to three significant digits)}.$$

Example 2. Solve the oblique triangle ABC for the altitude drawn from vertex C to side AB, given $A = 41° \, 20'$ and $AC = 1.065$.

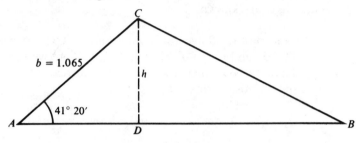

Figure 2-13

Solution: Let the altitude h meet side AB at D (Fig. 2-13). The altitude divides triangle ABC into two right triangles. In triangle ADC,

$$\sin 41° \, 20' = \frac{h}{1.065},$$

$$h = (1.065)(\sin 41° \, 20')$$

$$= (1.065)(.6604) = (.703) \text{ (to three significant digits)}.$$

We can use the sine function to solve for the angles of a right triangle when one side and the hypotenuse are known.

Example 3. Solve the right triangle ABC, in which $c = 10.00$, $a = .6428$ and $C = 90° \, 00'$.

Solution:

(1) $$\sin A = \frac{.6428}{10.00} = .6428$$

$$A = 40° \, 00' \text{ (see Table 1)}.$$

(2) $$B = 90° \, 00' - 40° \, 00' = 50° \, 00'.$$

We can use the calculator to find angle A from step (1) of Example 3 as follows:

KEY OPERATION	DISPLAY
	40.00092673

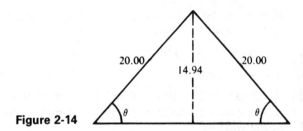

Figure 2-14

Example 4. The equal sides of an isosceles triangle are 20.00 units long, and the altitude is 14.94. Find the base angles of the triangle (Fig. 2-14).

Solution:

$$\sin \theta = \frac{14.94}{20.00} = .7470$$

$$\theta = 48° \ 20' \text{ (see Table 1)}.$$

Exercises 2-13

State the number of significant digits in each of the numbers in Exercises 1–6.

1. 2001.   **2.** .002.   **3.** 200.5.

4. .050.   **5.** 40.050.   **6.** .00505.

In Exercises 7–12 round off each of the approximate numbers to one less significant digit.

7. 12.24.   **8.** 3.95.   **9.** 4.05.

10. 8.666.   **11.** 1600.   **12.** 24,000.

In Exercises 13–18 write each of the numbers in scientific notation.

13. 93,000,000.   **14.** .000006.   **15.** 31.47.

16. 184.50.   **17.** .015.   **18.** .108

In Exercises 19–24 write each of the numbers in a form to indicate definitely that the number is known to three significant figures.

19. 4000.   **20.** 20,000.   **21.** 93,000,000.

22. .03200.   **23.** .00150.   **24.** 21.5.

In Exercises 25–28 use Table 1 and solve each of the right triangles for side a. Angle $C = 90°$. Assume the given angles accurate to minutes.

25. $A = 38° \ 10'$, $c = 4.551$.   **26.** $A = 44° \ 40'$, $c = .1852$.

27. $A = 51° \ 10'$, $c = 2485$.   **28.** $A = 3° \ 40'$, $c = 5.085$.

In Exercises 29–32 use Table 1 and solve each of the right triangles for angle *A*. Angle $C = 90°$. Express answer accurate to the nearest 10 minutes.

29. $a = 5.640, c = 30.00.$ **30.** $a = 2.060, c = 4.000.$

31. $a = 41.45, c = 50.00.$ **32.** $a = 22.47, c = 30.00.$

33. A 15-foot ladder just reaches the base of a second-story window. The ladder makes an angle of 73° with the level paving below. Find the height of the window.

34. A surveyor measures the distance down a sloping surface from point *A* to point *C* to be 97.98 feet. The sloping surface makes an angle of 25° 30′ with the horizontal. Determine the difference in elevation of points *A* and *C*.

2-14 THE LAW OF SINES

*In **any** triangle the sides are proportional to the sines of the opposite angles.*

> *Law of sines*:
>
> $$\frac{a}{\sin A} = \frac{b}{\sin B} = \frac{c}{\sin C}.$$

Proof. Let *A*, *B*, and *C* be the angles of any triangle, and let *a*, *b*, and *c* be the opposite sides. We consider two triangles, one in which all angles are acute (Fig. 2-15a), and the other in which one angle, angle *A*, is obtuse (Fig. 2-15b).

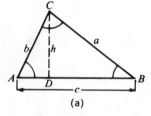   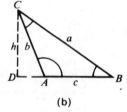

(a)   (b)   **Figure 2-15**

Draw the altitude *h* from vertex *C* to side *AB* or *AB* extended. In both right triangles thus formed,

$$\sin B = \frac{h}{a} \quad \text{and} \quad \sin A = \frac{h}{b}.$$

Therefore, $h = a \sin B \quad \text{and} \quad h = b \sin A.$

Then by equating the two values of h we have

$$a \sin B = b \sin A.$$

Divide both sides of the equation by $\sin A \sin B$.

$$\frac{a \sin B}{\sin A \sin B} = \frac{b \sin A}{\sin A \sin B}.$$

This reduces the equation to

(1)
$$\frac{a}{\sin A} = \frac{b}{\sin B}.$$

By drawing the altitude from angle B we can show that

$$\sin C = \frac{h}{a} \quad \text{and} \quad \sin A = \frac{h}{c}.$$

Then
$$h = a \sin C \quad \text{and} \quad h = c \sin A,$$

and
$$a \sin C = c \sin A,$$

which yields

(2)
$$\frac{a}{\sin A} = \frac{c}{\sin C}.$$

Combining Equations (1) and (2) yields the *Law of Sines*:

$$\frac{a}{\sin A} = \frac{b}{\sin B} = \frac{c}{\sin C}.$$

The Law of Sines is used in solving any triangle when two opposite parts (angle A and side a, or angle B and side b, or angle C and side c) are given along with another known side or angle.

Example. Find side b of triangle ABC when $a = 7.05$, $A = 40°\,10'$, and $B = 35°\,10'$.

Solution:

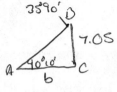

$$\frac{b}{\sin B} = \frac{a}{\sin A},$$

$$\frac{b}{\sin 35°\,10'} = \frac{7.05}{\sin 40°\,10'},$$

(1)
$$b = \frac{(7.05)\sin 35°\,10'}{\sin 40°\,10'}$$

(2)
$$= \frac{(7.05)(.5760)}{.6450} = 6.30 \text{ (to three significant digits).}$$

If your calculator only performs the fundamental operations, use it to calculate step (2) above. If your calculator has a memory and sine key, solve step (1) above as follows. (We use key 1 as a memory.)

KEY OPERATION   DISPLAY

| C | 1 0 | ÷ | 60 | + | 4 0 | = | sin | STO | 1 | 10 | ÷ | 60 |

| + | 3 5 | = | sin | × | 7 | . | 0 5 | ÷ | RCL | 1 | = | 6.295212992 |

Rounding off to three significant digits, we have 6.30 which is the same number found by using tables.

Exercises 2-14

In Exercises 1–4 use Table 1 and solve the oblique triangles for the missing sides.

1. $A = 52° 00'$, $B = 63° 00'$, $b = 1.15$.

2. $B = 43° 00'$, $C = 77° 10'$, $c = 5.61$.

3. $A = 38° 10'$, $C = 110° 20'$, $a = 2.10$.

4. $A = 41° 20'$, $B = 35° 10'$, $c = 1.20$.

5. A building along a line AB makes direct measurement of the line AB impossible. An offset point C is established, and, by measurement, it is found that $CB = 70.5$ feet, angle $CAB = 54°$, and angle $ABC = 47°$. Find the length of AB.

6. Points A and B are on opposite banks of a river. Point C is 235 feet from A and on the same side of the river as A. By measurement, $\angle CAB = 105° 30'$ and $\angle ACB = 50° 40'$. Find the distance across the river between points A and B.

7. Points A and B are on the same side of a river and are 155 feet apart. Point C is located across the river in such a way that $\angle ABC = 77° 10'$ and $\angle BAC = 69° 20'$. Find the distance across the river between points B and C.

2-15 THE AMBIGUOUS CASE

To solve a triangle when given two sides and an angle opposite one of them, the following possibilities exist:

1. Only one triangle is possible.
2. Two different triangles are possible.
3. A triangle cannot be constructed using the given parts. Let the given parts be a, b, and A. (Fig. 2-16.)

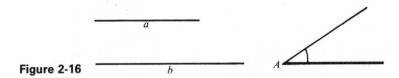

Figure 2-16

First, draw a freehand figure to help you visualize the problem (Fig. 2-17).

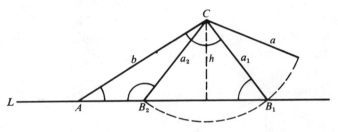

Figure 2-17

Lay out the figure by always placing one side of the given angle (in this example, A) along a horizontal working line, L, as shown (Fig. 2-17). Of the two given sides the one which is not opposite the given angle (b in this example) should be placed along the oblique side of the given angle. The other end of b determines the point of vertex C. Side a must begin at C. Next, compute the perpendicular distance from C to line L. This perpendicular distance ($b \sin A$) is the altitude h of the triangle as drawn from vertex C.

If A is an acute angle, there are the following possibilities:

1. If $a < h$ ($h = b \sin A$), then no triangle is possible (Fig. 2-18a).
2. If $a = h$, then there is just one triangle—a right triangle (Fig. 2-18b).
3. If a is greater than h but less than b (that is, $h < a < b$), then there are two triangles—triangles AB_1C and AB_2C (Fig. 2-17).
4. If a is greater than h and also greater than b (that is, $h < b < a$), then there is only one triangle (Fig. 2-19).

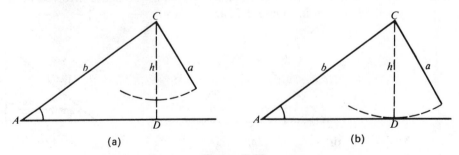

(a)   (b)

Figure 2-18

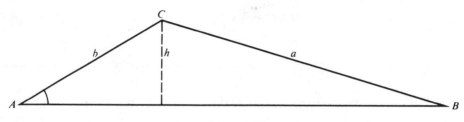

Figure 2-19

If *A* is a right angle or an obtuse angle, and if a triangle is possible, then one and only one triangle exists (Fig. 2-20).

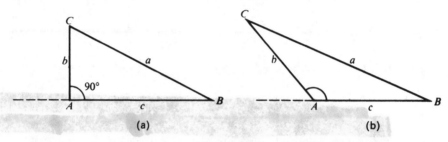

Figure 2-20

After we have determined which of the possibilities is applicable, we can complete the solution using the Law of Sines.

If there are two triangles (see Fig. 2-17), then $\angle AB_1C$ and $\angle AB_2C$ are supplementary angles; that is, the sum of these angles is 180° ($\angle B_1B_2C = \angle AB_1C$, being base angles of the isoscles triangle B_2B_1C. $\angle B_1B_2C$ is a supplement of $\angle AB_2C$; therefore, its equal, $\angle AB_1C$, is also supplementary to $\angle AB_2C$).

Example 1. Solve the triangle *ABC*, given $a = 6.45$, $b = 10.0$, $A = 40°\ 10'$.

Solution: Make a freehand sketch of the figure involved in the problem, as shown in Figure 2-21a.

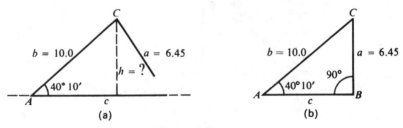

Figure 2-21

Compute the altitude h.

$$\sin A = \frac{h}{b},$$

$$h = b \sin A$$

$$= 10 \sin 40° \, 10'$$

$$= 10 \, (.6450) = 6.45 \text{ (to three significant figures).}$$

Hence $h = a$, and therefore triangle ABC is a right triangle (Fig. 2-21b). Solve for angle C and side c.

$$C = 90° - 40° \, 10' = 49° \, 50'.$$

$$\sin C = \frac{c}{b},$$

$$c = b \sin C$$

$$= 10 \sin 49° \, 50'$$

$$= 10 \, (.7642) = 7.64 \text{ (to three significant figures).}$$

Example 2. Solve the triangle ABC, given $a = 6.00$, $b = 10.0$, $A = 30° \, 00'$.

Solution: Make a freehand sketch of the figure involved in the problem, as shown in Figure 2-22a.

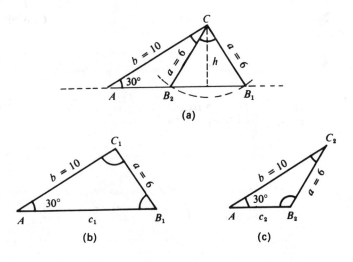

Figure 2-22

First compute the altitude h.

$$\sin A = \frac{h}{b},$$

$$h = b \sin A$$

$$= 10 \sin 30°$$

$$= 10 \,(.5000) = 5.$$

Hence $h < a < b$; therefore, two different triangles are possible.

First, we use the Law of Sines to solve the larger triangle AB_1C_1. (See Fig. 2-22b.)

$$\frac{\sin B_1}{b} = \frac{\sin A}{a},$$

$$\sin B_1 = \frac{b \sin A}{a}$$

$$= \frac{10 \sin 30° \; 00'}{6} = .8333;$$

$$B_1 = 56° \; 30' \qquad \text{(to the nearest 10')}.$$

$$A + B_1 + C_1 = 180°,$$

$$30° \; 00' + 56° \; 30' + C_1 = 180°,$$

$$C_1 = 93° \; 30'.$$

$$\frac{c_1}{\sin C_1} = \frac{a}{\sin A},$$

$$c_1 = \frac{a \sin C_1}{\sin A}$$

$$= \frac{6 \sin 93° \; 30'}{\sin 30° \; 30'}$$

$$= \frac{6 \,(.9981)}{.5000} = 12.0 \;\text{(to three significant figures)}.$$

Next, we solve the smaller triangle AB_2C_2. (See Fig. 2-22c.)

$$B_2 = 180° - B_1 \qquad \text{(B_2 and B_1 are supplemetary}$$
$$\text{angles)}$$

$$= 180° - 56° \; 30' = 123° \; 30'.$$

$$A + B_2 + C_2 = 180°,$$

$$30° \; 00' + 123° \; 30' + C_2 = 180°,$$

$$C_2 = 26° \; 30'.$$

$$\frac{c_2}{\sin C_2} = \frac{a}{\sin A},$$

$$c_2 = \frac{a \sin C_2}{\sin A}$$

$$= \frac{6 \sin 26° 30'}{\sin 30° 00'}$$

$$= \frac{6\,(.4462)}{.5000} = 5.35 \text{ (to three significant figures).}$$

Exercises 2-15

In the following exercises find the number of possible triangles for the data, and complete the solution of all possible triangles. Draw the figure approximately to scale.

1. $A = 41° 00', a = 11.0, b = 10.0.$
2. $A = 51° 00', a = 11.0, b = 10.0.$
3. $C = 25° 00', c = 4.11, b = 10.0.$
4. $C = 28° 00', c = 4.51, b = 10.0.$
5. $A = 43° 00', a = 8.00, b = 10.0.$
6. $A = 37° 00', a = 7.00, b = 10.0.$
7. $B = 28° 10', a = 10.0, b = 4.72.$
8. $B = 35° 10', a = 10.0, b = 5.76.$
9. $C = 63° 00', a = 10.0, c = 9.00.$
10. $C = 67° 40', a = 10.0, c = 9.50.$

2-16 TRIGONOMETRIC EQUATIONS

A **trigonometric equation** is an equality that involves one or more trigonometric functions of an unknown angle, and which is true for certain particular values of the unknown angle but not for all values. Any value of the unknown angle for which the equality is true is called a **solution** of the trigonometric equation and is said to **satisfy** the equation. Since the same functions of coterminal angles are equal (and there are an unlimited number of angles coterminal with a given angle), a trigonometric equation, in general, has an unlimited number of solutions, provided that it does have a solution. In most cases we will limit our solutions to angles from 0° to 360°—we shall include 0° but not 360°. Our first examples are not of trigonometric equations but of algebraic equations whose methods of solution are often useful in solving trigonometric equations.

Example 1. Solve $x^2 + x = 0$.

Solution: Factor the left-hand member of the equation.

$$x(x + 1) = 0.$$

Set each factor equal to zero. Thus,

$$x = 0 \quad \text{or} \quad x + 1 = 0.$$

Therefore,

$$x = 0 \quad \text{and} \quad x = -1$$

are the solutions or roots of the given equation.

Example 2. Solve $4x^2 - 1 = 0$.

Solution:

FIRST METHOD	SECOND METHOD
$4x^2 - 1 = 0,$	$4x^2 - 1 = 0,$
$4x^2 = 1,$	$(2x + 1)(2x - 1) = 0,$
$x^2 = \frac{1}{4},$	$2x + 1 = 0 \quad \text{or} \quad 2x - 1 = 0,$
$x = \pm \frac{1}{2}.$	$2x = -1 \quad \text{or} \quad 2x = 1,$
	$x = -\frac{1}{2} \quad \text{and} \quad x = \frac{1}{2}.$

Example 3.   Solve the equation $\sin \theta = \frac{1}{2}$, for θ where $0° \leqq \theta < 360°$.

Solution: First, determine the related angle. The related angle is 30°, because $\sin 30° = \frac{1}{2}$. Second, determine the quadrants which apply. They are quadrants I and II because $\sin \theta$ is positive. Third, place the related angle in each of these quadrants. Thus, 30° is the related angle of itself and of 150°. (Fig. 2-23). Therefore, the solutions are 30° and 150°.

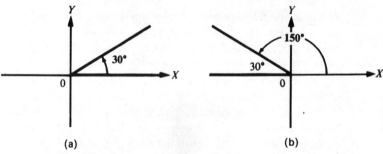

(a)   (b)

Figure 2-23

Example 4. Solve the equation $\sin \theta = -\frac{1}{2}$, for θ where $0° \leqq \theta < 360°$.

Solution: The related angle is 30°, and $\sin \theta$ is negative. Therefore, θ is in quadrant III or IV, and the solutions are 210° and 330°. (Fig. 2-24.)

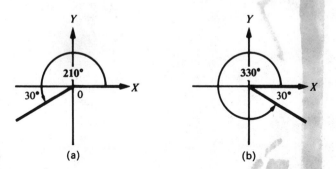

Figure 2-24   (a)   (b)

Example 5. Solve the equation $\sin 2\theta = 0$, for θ where $0° \leqq \theta < 360°$.

Solution: Let $\alpha = 2\theta$. Then

$$\sin \alpha = 0,$$

and $\qquad \alpha = 0°, 180°, 360°, 540°, 720°,$ etc.

But $\alpha = 2\theta$; therefore,

$$2\theta = 0°, 180°, 360°, 540°, 720°, \text{ etc.}$$

We were required to find θ where $0° \leqq \theta < 360°$; therefore,

$$\theta = 0°, 90°, 180°, 270°.$$

Example 6. Solve the equation $\sin \frac{1}{2}\theta = 1$, for θ where $0° \leqq \theta < 360°$.

Solution: Let $\alpha = \frac{1}{2}\theta$. Then

$$\sin \alpha = 1,$$

and $\qquad \alpha = 90°, 450°,$ etc.

But $\alpha = \frac{1}{2}\theta$; therefore,

$$\frac{1}{2}\theta = 90°, 450°, \text{ etc.}$$

Then $\qquad\qquad \theta = 180°, 950°,$ etc.

But we were required to find θ where $0° \leqq \theta < 360°$. Therefore,

$$\theta = 180°$$

is the only solution.

Example 7. Solve the equation $2 \sin^3 \theta + \sin^2 \theta - \sin \theta = 0$, for θ where $0° \leqq \theta < 360°$.

Solution: Let $\sin \theta = x$. Then the equation

$$2 \sin^3 \theta + \sin^2 \theta - \sin \theta \doteq 0$$

takes the form

$$2x^3 + x^2 - x = 0,$$

which factors into

$$x(2x^2 + x - 1) = 0,$$

or $\qquad\qquad\qquad\qquad x(2x - 1)(x + 1) = 0.$

Set each factor equal to zero. Thus,

$$x = 0 \quad\text{or}\quad 2x - 1 = 0 \quad\text{or}\quad x + 1 = 0.$$

Then

$$x = 0, \qquad x = \frac{1}{2}, \qquad x = -1.$$

But $\qquad\qquad\qquad\qquad x = \sin \theta.$

Therefore,

$$\sin \theta = 0, \qquad \sin \theta = \frac{1}{2}, \qquad \sin \theta = -1.$$

Hence,

$$\theta = 0°, 180°, \qquad \theta = 30°, 150°, \qquad \theta = 270°$$

are the solutions of the equation.

Exercises 2-16

Solve each of the following equations for all the values of θ such that $0° \leqq \theta < 360°$.

1. $\sin \theta = 1$.

2. $\sin \theta = -1$.

3. $\sin \theta = 0$.

4. $\sin \theta = \frac{1}{2}$.

5. $\sin \theta = \dfrac{\sqrt{3}}{2} = \frac{1}{2}\sqrt{3}$.

6. $\sin \theta = -\dfrac{\sqrt{3}}{2} = -\frac{1}{2}\sqrt{3}$.

7. $\sin \theta = -\dfrac{1}{\sqrt{2}} = -\frac{1}{2}\sqrt{2}$.

8. $\sin \theta = 3$.

9. $\sin \theta = 2$.

10. $2 \sin \theta = 1$.

11. $2 \sin \theta = -1.$ 12. $\sin^2 \theta = 1.$

13. $\sin^2 \theta + \sin \theta = 0.$ 14. $\sin^2 \theta - \sin \theta = 0.$

15. $4 \sin^2 \theta - 1 = 0.$ 16. $4 \sin^2 \theta - 3 = 0.$

17. $2 \sin^2 \theta + \sin \theta - 1 = 0.$ 18. $\sin^2 \theta - 2 \sin \theta + 1 = 0.$

19. $\sin^2 \theta + 2 \sin \theta + 1 = 0.$ 20. $\sin^3 \theta - \sin \theta = 0.$

21. $2 \sin^3 \theta - \sin^2 \theta - \sin \theta = 0.$ 22. $\sin^2 \theta + \sin \theta - 2 = 0.$

23. $\sin 2\theta = 1.$ 24. $\sin 2\theta = -1.$

25. $\sin^2 2\theta + \sin 2\theta = 0.$ 26. $\sin^2 2\theta - \sin 2\theta = 0.$

27. $2 \sin^2 2\theta + \sin 2\theta - 1 = 0.$ 28. $\sin^2 \tfrac{1}{2}\theta - 2 \sin \tfrac{1}{2}\theta + 1 = 0.$

2-17   GRAPH OF THE SINE FUNCTION

In this section we discuss the graph of the sine function, $y = \sin \theta$. Graphs of the more general sine function, $y = a \sin b\theta$, will be discussed in Chapter 9. The unit circle can be used to show the increasing and decreasing values of the sine function as the angle changes from 0° to 360°. Let the variable angle θ be in standard position with respect to a coordinate system. It was shown in Section 2-5 that the value of $\sin \theta$ is the same as the value of y at the point where the terminal side of θ crosses the unit circle. We shall show the relationship of θ and $\sin \theta$ by means of a graph.

Example 1.   Graph $y = \sin \theta$.

We take values of θ along the horizontal axis and the corresponding values of y on the vertical axis. On the θ-axis we choose a convenient length to represent the equal distances between 0°, 30°, 60°, etc. Then by plotting many points of $y = \sin \theta$, we determine the shape of the curve. The coordinates of some of these points are given in the following table:

θ	0°	30°	60°	90°	120°	150°	180°	210°	240°	270°	300°	330°	360°
y	0	$\tfrac{1}{2}$	$\tfrac{1}{2}\sqrt{3}$	1	$\tfrac{1}{2}\sqrt{3}$	$\tfrac{1}{2}$	0	$-\tfrac{1}{2}$	$-\tfrac{1}{2}\sqrt{3}$	-1	$-\tfrac{1}{2}\sqrt{3}$	$-\tfrac{1}{2}$	0

We use Figure 2-25 to emphasize that on the unit circle the value of $\sin \theta$ is the same as the value of y. In Figure 2-26 the sine function is carried through one complete cycle.

The curve shown in Figures 2-26 and 2-27 is called **the graph of the sine function,** or **the sine wave curve.** Because of its wave form the sine curve is very important in the study of wave motion in electrical engineering and physics. The maximum distance of the curve from the θ-axis is called the **amplitude** of the curve. The amplitude of this curve is said to

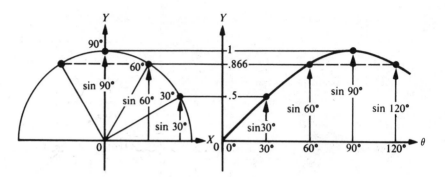

Figure 2-25

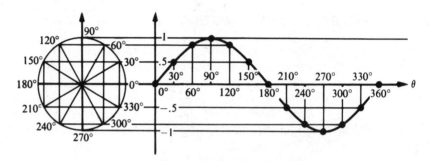

Figure 2-26

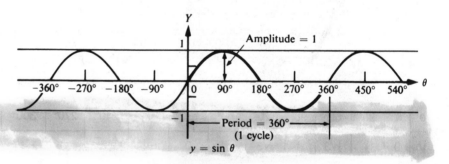

Figure 2-27

be 1, not ±1. This curve repeats its cycle in an interval of 360° and is therefore said to have a **period** of 360°. (Fig. 2-27.)

Example 2. Graph $y = 2 \sin \theta$. (Fig. 2-28.)
    The y values on the $y = 2 \sin \theta$ curve are double those of the $y = \sin \theta$ curve. Therefore, the $y = 2 \sin \theta$ curve has an amplitude of 2

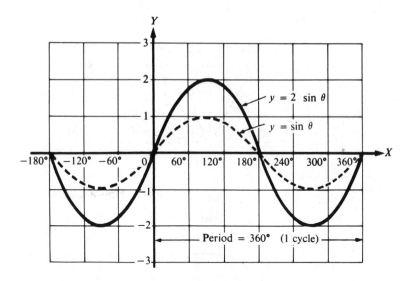

Figure 2-28

and oscillates back and forth between $y = 2$ and $y = -2$. Both curves have a period of 360°.

Example 3. Graph $y = 2 + \sin \theta$. (Fig. 2-29.)

For each value of θ the corresponding value of y for this curve is 2 greater than the value of y for the standard sine curve, $y = \sin \theta$. Thus, by the addition of the constant 2, the curve was **translated** upward 2 units. The period and amplitude remain unchanged from that of $y = \sin \theta$.

Adding a nonzero constant c to the $\sin \theta$ function causes a vertical translation of c units but does not change the shape of the curve. The translation is upward $|c|$ units if $c > 0$, and, the translation is downward $|c|$ units if $c < 0$.

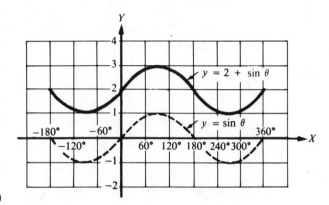

Figure 2-29

Exercises 2-17

1.   Graph $y = \frac{1}{2}\sin\theta$ from $\theta = 0°$ to $\theta = 360°$.
2.   Graph $y = 3\sin\theta$ from $\theta = 0°$ to $\theta = 360°$.
3.   Graph $y = -\sin\theta$ from $\theta = 0°$ to $\theta = 360°$.
4.   Graph $y = 1 + \sin\theta$ from $\theta = 0°$ to $\theta = 360°$.
5.   Graph $y = 1 - \sin\theta$ from $\theta = 0°$ to $\theta = 360°$.
6.   Graph $y = -1 + \sin\theta$ from $\theta = 0°$ to $\theta = 360°$.
7.   Graph $y = -2 - \sin\theta$ from $\theta = 0°$ to $\theta = 360°$.
8.   Graph $y = 1 + 2\sin\theta$ from $\theta = -180°$ to $\theta = 360°$.
9.   Graph $y = 1 - 2\sin\theta$ from $\theta = -180°$ to $\theta = 360°$.
10.   Graph $y = -2 + 3\sin\theta$ from $\theta = 0°$ to $\theta = 360°$.

REVIEW EXERCISES

In Exercises 1–6 find the value of each term and then combine their values.

1. $\sin 90° + \sin 210° + \sin(-30°)$.
2. $\sin 45° + \sin 135° + \sin 180°$.
3. $\sin^2 225° + \sin 330° + \sin 90°$.
4. $\sin^2 120° + \sin 270° + \sin 390°$.
5. $\sin(-90°) + \sin(-270°) + \sin 540°$.
6. $\sin(-45°) + \sin 405° - \sin 150°$.

In Exercises 7 and 8 find $\sin\theta$ where θ is an angle in standard position with its terminal side passing through the given point.

7. $(4, 3)$.   8. $(1, \sqrt{3})$.

In Exercises 9–12 use Table 1 to find the value of the given function.

9. $\sin 15° 20'$.   10. $\sin 67° 20'$.
11. $\sin 130° 10'$.   12. $\sin(-200° 10')$.

In Exercises 13–16 use Table 1 to find the angle that satisfies the given conditions.

13. $\sin\theta = .6841$ $0° < \theta < 90°$.
14. $\sin\theta = .7679$ $90° < \theta < 180°$.
15. $\sin\theta = -.1074$ $270° < \theta < 360°$.
16. $\sin\theta = -.9988$ $180° < \theta < 270°$.

In Exercises 17–19 state the number of significant digits in each number.

17. .0600. **18.** 18,030. **19.** 5.000.

In Exercises 20 and 21 round off to the indicated number of significant digits.

20. 4.85 (two significant digits).

21. 0.36791 (three significant digits).

22. A 24-foot ladder leans against a building. The ladder makes an angle of 74° with the level paving below. Find the height of the point where the top of the ladder touches the building. Round off your answer to the nearest foot.

23. Points A and B are on the same side of a river and are 93.4 feet apart. Point C is located across the river in such a way that $\angle ABC = 44° 20'$ and $\angle BAC = 66° 40'$. Find the distance across the river between points A and C.

24. How many different triangles (if any) can be constructed with $A = 43° 00'$, $a = 7.35$, and $b = 10.0$? Do not solve for all the unknown parts of the triangle(s).

In Exercises 25–28 solve each equation for all the values of θ such that $0° \leqq \theta < 360°$.

25. $\sin \theta = -\frac{1}{2}$. **26.** $\sin \theta = \frac{\sqrt{2}}{2}$.

27. $2 \sin^2 \theta - \sin \theta = 0$. **28.** $\sin^2 2\theta - 2 \sin 2\theta + 1 = 0$.

29. Graph $y = 4 \sin \theta$ from $\theta = 0$ to $\theta = 360°$.

30. Graph $y = -2 + \sin \theta$ from $\theta = 0$ to $\theta = 360°$.

CHAPTER 2: DIAGNOSTIC TEST

The purpose of this test is to see how well you understand the work covered in Chapter 2. We recommend that you work this test before your instructor tests you on this chapter. Allow yourself approximately 50 minutes to do the test.

Solutions to the problems, together with section references, are given in the Answer Section at the end of this book. You should study the sections referred to for the problems you do incorrectly.

---

1. Find the value of $\sin \theta$ where θ is in standard position and its terminal side passes through the point $(-5, 12)$.

2. Find the value of each of the following functions. Do not use tables.
(a) $\sin(-60°)$. (b) $\sin 270°$. (c) $\sin^2 315°$.

3. Use Table 1 as an aid and find the value of each of the following functions.
   (a) sin 123° 20'.   (b)   sin 228° 10'.   (c)   sin (−41° 50').

4. Use Table 1 as an aid and find the value of θ that satisfies each of the following conditions.
   (a) sin θ = .6967   90° < θ < 180°.
   (b) sin θ = −.6041   180° < θ < 270°.
   (c) sin θ = −.0785   270° < θ < 360°.

5. Round off each of the following numbers to the indicated number of significant digits.
   (a) 70.505 (four significant digits).
   (b) 0.03617 (three significant digits).
   (c) 186,300 (two significant digits).

6. Write each of the following numbers in scientific notation.
   (a) 3500.   (b) .0615.   (c) 85.6.

7. Given triangle ABC with A = 38° 10', C = 90° 00', and side c = 10.00, solve for side a.

8. Given triangle ABC with B = 57° 30', C = 84° 10', and side b = 84.4, solve for side c.

9. How many different triangles (if any) can be constructed with A = 53° 50', a = 7.45, and b = 10.0? Do not solve for all the unknown parts of the triangle(s).

10. Solve the equation, $\sin^2 \theta - 2 \sin \theta = 0$, for all the values of θ where $0° \leqq \theta < 360°$.

11. Graph $y = 2 - \sin \theta$ from $\theta = 0°$ to $\theta = 360°$.

3

The Cosine Function

3-1 *THE SINE FUNCTION REDEFINED*

The student will recall (Section 1-7) that the distance of a point $P(x, y)$ to the right or left of the origin is called the *abscissa*, or x-coordinate, of point P, and that y, the vertical distance of P from the x-axis, is called the *ordinate*, or y-coordinate, of point P. The radius of P is the distance of P from the origin.

The definition of the sine of an angle, given in Section 2-3, can now be phrased as follows: Let θ be an angle placed in standard position with respect to a system of coordinates. Choose any point on the terminal side of θ other than the origin. Then *the sine of θ is the ratio of the ordinate to the radius of that point.* (*See* Fig. 3-1.)

$$\sin \theta = \frac{\text{ordinate}}{\text{radius}} = \frac{y}{r}.$$

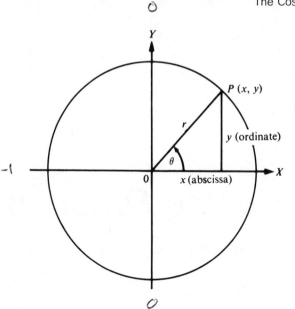

Figure 3-1

3-2 DEFINITION OF THE COSINE OF AN ANGLE

1. *Place an angle θ in standard position.*
2. *Let point P, whose coordinates are* (x, y), *be any point on the terminal side of angle θ, other than the origin.*
3. *Let r be the distance of P from the origin. This distance r will always be considered as a positive number.*

Then the cosine of θ (written cos θ) *is defined as the ratio of x to r:*

$$\cos \theta = \frac{x}{r},$$

or *the cosine of θ is the ratio of the abscissa to the radius of P:*

$$\boxed{\cos \theta = \frac{\text{abscissa}}{\text{radius}} = \frac{x}{r}.}$$

Special case. Let θ be an acute angle of a right triangle. Let this triangle be placed in the first quadrant with θ in standard position. Let the point (x, y) be the end of the hypotenuse in the first quadrant. With the triangle in this position, x becomes the side of the triangle adjacent to θ, and r is the hypotenuse of the triangle. Then, from the definition of the cosine of an angle,

$$\cos \theta = \frac{x}{r} = \frac{\text{side adjacent to } \theta}{\text{hypotenuse}}.$$

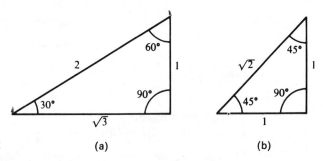

Figure 3-2   (a)   (b)

Examples: (See Fig. 3-2.)

1. $\cos 30° = \dfrac{\sqrt{3}}{2} = \tfrac{1}{2}\sqrt{3}.$

2. $\cos 60° = \tfrac{1}{2}.$

3. $\cos 45° = \dfrac{1}{\sqrt{2}} = \tfrac{1}{2}\sqrt{2}.$

3-3 COMPLEMENTARY ANGLES

Two angles, A and B, are complementary, and each is the complement of the other, if their sum is equal to 90°. We shall adopt this definition even when one of the angles is negative or 0°. Thus, 30° and 60°, 0° and 90°, 120° and −30°, are pairs of complementary angles.

   Observe that the cosine of an angle is equal to the sine of its complement:

$$\cos 60° = \frac{1}{2} \quad \text{and} \quad \sin 30° = \frac{1}{2}.$$

Therefore, $\cos 60° = \sin 30°.$

   Note in Figure 3-3 that

$$\cos A = \frac{b}{c} \quad \text{and} \quad \sin B = \frac{b}{c}.$$

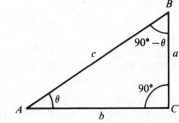

Figure 3-3

Therefore, $\cos \theta = \sin (90° - \theta).$

Also $\sin \theta = \cos (90° - \theta).$

Examples.

1. $\cos 10° = \sin 80°.$
2. $\sin 40° = \cos 50°.$
3. $\sin 120° = \cos (-30°).$

The sine and the cosine are said to be **cofunctions**, each of the other.

3-4 THE UNIT CIRCLE AND THE COSINE FUNCTION

Let θ be any angle in standard position, and let $P(x, y)$ be the point of intersection of the terminal side of θ with the unit circle. Then, by definition,

$$\cos \theta = \frac{x}{1} = x$$

(in the unit circle).

Unit circle

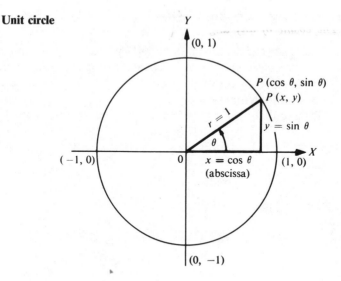

Figure 3-4

Clearly, the sign of $\cos \theta$ is the same as the sign of x when θ is in standard position and (x, y) is a point on the terminal side of θ. Therefore, *the cosine is positive in quadrants I and IV, negative in quadrants II and III.* That is, the cosine of an angle is positive when a point on the terminal side lies to the right of the y-axis and negative when a point on the terminal side lies on the left side

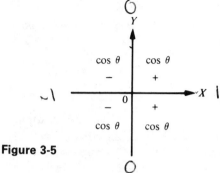

Figure 3-5

of the y-axis. (Fig. 3-5.) From Figure 3-4 we find the values of the cosines of quadrantal angles.

$$\cos 0° = 1, \qquad \cos 180° = -1,$$
$$\cos 90° = 0, \qquad \cos 270° = 0.$$

Again by observing Figure 3-4, note that the cosine of an angle is never greater than 1 or less than -1.

From Figure 3-6 we can find the values of certain special angles. For example, $\cos 60° = \frac{1}{2}$, $\cos 150° = -\frac{1}{2}\sqrt{3}$, $\cos 240° = -\frac{1}{2}$, $\cos 300° = \frac{1}{2}$.

*The cosine of any angle has the same **absolute value** as the cosine of its related angle. The cosine is positive when a point on the terminal side lies to the*

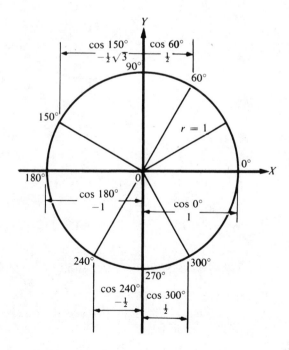

Figure 3-6

*right of the y-axis and negative when a point on the terminal side lies to the left
of the y-axis.*

3-5 $\sin^2 \theta + \cos^2 \theta = 1$

Let θ by any angle in standard position, and let $P(x, y)$ be the point of
intersection of the terminal side of θ with the unit circle. Then the coordinates
of P are $(\cos \theta, \sin \theta)$. From these coordinates of P we find **$\sin^2 \theta +$
$\cos^2 \theta = 1$**, by the Pythagorean Theorem. (Fig. 3-7.) The expression, $\sin^2 \theta +$
$\cos^2 \theta = 1$, is called a **trigonometric identity** because it is true for any value of
θ.

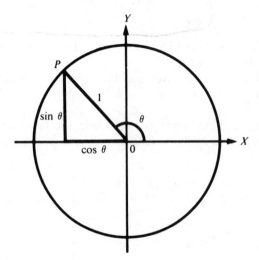

Figure 3-7

Examples.

1. $\sin^2 45° + \cos^2 45° = (\tfrac{1}{2}\sqrt{2})^2 + (\tfrac{1}{2}\sqrt{2})^2 = \tfrac{1}{2} + \tfrac{1}{2} = 1.$
2. $\sin^2 120° + \cos^2 120° = (\tfrac{1}{2}\sqrt{3})^2 + (-\tfrac{1}{2})^2 = \tfrac{3}{4} + \tfrac{1}{4} = 1.$
3. $\sin^2 180° + \cos^2 180° = (0)^2 + (-1)^2 = 1.$
4. $\sin^2 315° + \cos^2 315° = (-\tfrac{1}{2}\sqrt{2})^2 + (\tfrac{1}{2}\sqrt{2})^2 = \tfrac{1}{2} + \tfrac{1}{2} = 1.$

To find the cosine of an angle, place the angle in standard position,
and then determine its quadrant and related angle.

Example 5. Find cos 120°.

Since 120° is in the second quadrant, cos 120° is negative. The
related angle is 60°. Therefore, $\cos 120° = -\cos 60° = -\tfrac{1}{2}.$

Example 6. Find cos 300°.

Because 300° is in the fourth quadrant, cos 300° is positive. The related angle is 60°. Therefore, $\cos 300° = \cos 60° = \frac{1}{2}$.

Example 7. Find cos (−450°).

−450° is a quadrantal angle. The related angle is 90°. Therefore, $\cos (-450°) = \cos 90° = 0$.

If you have a calculator with sin and cos keys, and a memory key, we think that you will find it interesting to verify the identity, $\sin^2 \theta + \cos^2 \theta = 1$, for various values of θ.

Example 8. We use the calculator to show that $\sin^2 \theta + \cos^2 \theta = 1$ when $\theta = 215°$. We use key 1 as a memory.

$$\sin^2 215° + \cos^2 215° = 1.$$

KEY OPERATION .... DISPLAY

| C | 2 1 5 | sin | x^2 | STO | 1 |

| 2 1 5 | cos | x^2 | + | RCL | 1 | = | .... | 1 |

Exercises 3-5

In Exercises 1–20 find the value of each term and then combine the values.

1. cos 0° + cos 90° + cos 180° + cos 270°.

2. $\cos 60° + \cos^2 45° + \cos^2 30° + \cos 90°$.

3. $\cos 120° + \cos^2 180° + \cos 300°$.

4. cos 135° + cos 225° + cos 330°.

5. $\sin^2 30° + \cos^2 30°$.

6. $\sin^2 60° + \cos^2 60$.

7. $\sin^2 135° + \cos^2 135°$.

8. sin (−60°) + cos (−60°) + cos (−270°).

9. sin 180° + cos 180° + sin 60° + cos 30°.

10. cos (−30°) + sin (−30°) + cos (−300°).

11. cos 30° + sin 60°.

12. cos 60° + sin 30°.

13. cos 0° + sin 90°.

14. sin 0° + cos 90°.

15. cos 720° + cos (−720°) − cos (−150°).

16. $\cos(-135°) - \sin(-60°) + \cos(-45°)$.
17. $\cos(-60°) + \sin(-60°) + \cos 120°$.
18. $\cos 540° + \cos 0° + \cos 90°$.
19. $2 \cos^2 30° + 4 \cos^2 45° + 6 \cos^2 135°$.
20. $4 \cos^2 120° + 2 \cos^2 150° + 6 \cos^2 225°$.

In each of the Exercises 21–26, let θ be an angle in standard position with its terminal side passing through the given point. Find both the sine and the cosine of θ.

21. $(4, 3)$.   22. $(12, 5)$.   23. $(3, -4)$.
24. $(1, \sqrt{3})$.   25. $(-1, \sqrt{3})$.   26. $(3, 0)$.

In Exercises 27–44 use Table 1 and the given or related angle to find the value of each function.

27. $\cos 10°$.   28. $\cos 40°$.   29. $\cos 160°$.
30. $\cos 170°$.   31. $\cos 220°$.   32. $\cos 230°$.
33. $\cos 280°$.   34. $\cos 290°$.   35. $\cos(-10°)$.
36. $\cos(-15°)$.   37. $\cos(-100°)$.   38. $\cos(-110°)$.
39. $\cos 100° 50'$.   40. $\cos 100° 10'$.   41. $\cos 265° 10'$.
42. $\cos 255° 10'$.   43. $\cos 760°$.   44. $\cos 1000°$.

Use Table 1 to find the acute angle θ for each of the Exercises 45–50.

45. $\cos \theta = .9483$.   46. $\cos \theta = .8732$.
47. $\cos \theta = .3557$.   48. $\cos \theta = .5257$.
49. $\cos \theta = .0262$.   50. $\cos \theta = .0407$.

3-6 INTERPOLATION

Interpolation is the process of approximating a number that, although not an entry in the table, lies between two consecutive entries. Several methods of interpolation are used. The method used here is known as **linear interpolation**. In interpolation, fractions of table differences are rounded off to the same number of decimal places as are used in the table. In rounding off a number that is *exactly halfway* between two entries, it is conventional to choose the number that makes the *final result even* rather than odd. This procedure will be followed throughout this book.

Example 1. Find sin 45°.

Of course, sin 45° can be read directly from the table, but we use it here in conjunction with the $y = \sin \theta$ graph to illustrate linear interpolation.

$$\text{difference} = 30° \begin{cases} \sin 30° = .500 \\ \sin 45° = ? \\ \sin 60° = .866 \end{cases} \text{difference} = .366$$

The angle 45° is halfway from 30° to 60°; then, in linear interpolation, sin 45° is taken halfway between sin 30° = .500 and sin 60° = .866. That is, $\sin 45° = .500 + \frac{1}{2}(.366) = .683$. From the tables the actual value of sin 45° is .707, which shows that linear interpolation is not always accurate.

In linear interpolation, the error is the vertical distance CD (Fig. 3-8) between the sine curve and the chord AB, at $\theta = 45°$. For very small differences in values of θ, points A and B become close together, forcing the error distance CD, to become very small.

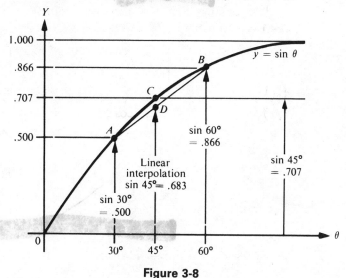

Figure 3-8

Example 2. Find sin 26° 34′.

Solution: In linear interpolation we make the assumption that sin 26° 34′ is $\frac{4}{10}$ of the way from sin 26° 30′ to sin 26° 40′

$$10′ \begin{cases} 4 \begin{cases} \sin 26° 30′ = .4462 \\ \sin 26° 34′ = \quad ? \end{cases} d \\ \sin 26° 40′ = .4488 \end{cases} .0026$$

$$\frac{d}{.0026} = \frac{4}{10}$$

$$d = \frac{4}{10}(.0026) = .0010 \text{ rounded off to four decimal places.}$$

Therefore,

$$\sin 26° 34' = .4462 + .0010 = .4472.$$

To find sin 26° 34′ on the calculator proceed as follows:

KEY OPERATION   DISPLAY

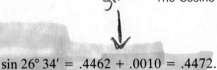

[C] 3 4 [÷] 6 0 [=] [+] 2 6 [=] [sin] | .4472388142 |

When rounded off to four decimal places we have .4472 which is the same number found by interpolation.

Example 3. Find cos 49° 18′.

Remember that as θ increases from 0° to 90°, the value of cos θ **decreases** from 1 to 0. (See Fig. 3-6.)

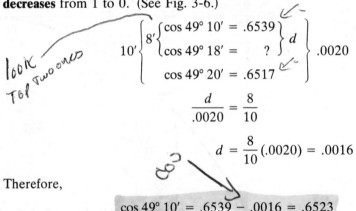

$$10' \left\{ 8' \left\{ \begin{array}{l} \cos 49° 10' = .6539 \\ \cos 49° 18' = \quad ? \end{array} \right\} d \right\} .0020$$
$$\cos 49° 20' = .6517$$

$$\frac{d}{.0020} = \frac{8}{10}$$

$$d = \frac{8}{10}(.0020) = .0016$$

Therefore,

$$\cos 49° 10' = .6539 - .0016 = .6523$$

(Note that for cosine we substract.)

Example 4. Find the acute angle θ for which

$$\sin \theta = .5750.$$

Solution.

$$10' \left\{ \begin{array}{l} \sin 35° 00' = .5736 \\ \sin \quad \theta \quad = .5750 \\ \sin 35° 10' = .5760 \end{array} \right.$$

14 in last two digits

24 in last two digits

Therefore,

$$\theta = 35° 00' + \frac{14}{24}(10')$$

$$= 35° 06' \text{ to the nearest minute.}$$

Example 5. Solve for $x = \sqrt{10.63}$.

Solution: Using Table 5, we find that $\sqrt{10.63}$ lies in the "$\sqrt{10n}$" column between $\sqrt{10.6}$ and $\sqrt{10.7}$.

$$10\left\{\begin{array}{l} 3\left\{\begin{array}{l} \sqrt{10.6} \ = 3.25576 \\ \sqrt{10.63} = ? \end{array}\right\} \\ \sqrt{10.7} \ = 3.27109 \end{array}\right\}.01533$$

$$\sqrt{10.63} = 3.25576 + \frac{3}{10}(.01533).$$

Therefore, $\sqrt{10.63} = 3.26$ (to three significant figures).

Example 6. Find a second-quadrant angle θ for which

$$\cos \theta = -.9258.$$

Solution: Let α be the related angle; then

$$\cos \alpha = +.9258.$$

$$10'\left\{\begin{array}{l} \left\{\begin{array}{l} \cos 22° 10' = .9261 \\ \cos \alpha \ \ \ \ \ = .9258 \end{array}\right\}3 \\ \cos 22° 20' = .9250 \end{array}\right\}11$$

$$\alpha = 22° 10' + \frac{3}{11}(10')$$

$$= 22° 10' + 3' = 22° 13'.$$

Then $\theta = 180° - 22° 13' = 179° 60' - 22° 13' = 157° 47'.$

Exercises 3-6

Use Table 1 and the given or related angle to find the value of each function in Exercises 1–12.

1. $\cos 22° 16'$.
2. $\cos 33° 19'$.
3. $\sin 49° 22'$.
4. $\sin 54° 04'$.
5. $\cos 75° 03'$.
6. $\cos 81° 57'$.
7. $\sin 186° 14'$.
8. $\sin 277° 12'$.
9. $\cos (-15° 13')$.
10. $\cos 3° 07'$.
11. $\sin 4° 09'$.
12. $\sin (-17° 17')$.

Use Table 1 to find the acute angle θ for each of the Exercises 13–18.

13. $\sin \theta = .3292$.
14. $\sin \theta = .5529$.
15. $\cos \theta = .9428$.
16. $\sin \theta = .9880$.
17. $\cos \theta = .5852$.
18. $\cos \theta = .6167$.

Use Table 1 to find a second-quadrant angle θ for each of the Exercises 19–24.

19. $\sin \theta = .6950$.   20. $\cos \theta = -.3621$.   21. $\cos \theta = -.9120$.

22. $\sin \theta = .9472$.   23. $\cos \theta = -.9882$.   24. $\cos \theta = -.6785$.

Solve each of the following equations for all of the values of θ for which $0° \leqq \theta < 360°$. Do not use tables.

25. $\cos \theta = 1$.   26. $\cos \theta = -1$.

27. $\cos \theta = 0$.   28. $\cos \theta = \frac{1}{2}\sqrt{2}$.

29. $\cos \theta = \frac{1}{2}\sqrt{3}$.   30. $\cos \theta = -\frac{1}{2}\sqrt{2}$.

31. $\cos \theta = -\frac{1}{2}\sqrt{3}$.   32. $2 \cos \theta = 1$.

33. $2 \cos \theta = -1$.   34. $\cos^2 \theta = 1$.

35. $\cos^2 \theta + \cos \theta = 0$.   36. $\cos^2 \theta - \cos \theta = 0$.

37. $4 \cos^2 \theta - 1 = 0$.   38. $4 \cos^2 \theta - 3 = 0$.

39. $2 \cos^2 \theta + \cos \theta - 1 = 0$.   40. $\cos^2 \theta + 2 \cos \theta + 1 = 0$.

41. $\cos^2 \theta - 2 \cos \theta + 1 = 0$.   42. $\cos^3 \theta - \cos \theta = 0$.

43. $2 \cos^3 \theta - \cos^2 \theta - \cos \theta = 0$.   44. $\cos^2 \theta + \cos \theta - 2 = 0$.

45. $\cos 2\theta = 0$.   46. $\cos 2\theta = 1$.

47. $\cos^2 2\theta + \cos 2\theta = 0$.   48. $\cos^2 2\theta - \cos 2\theta = 0$.

49. $2 \cos^2 2\theta + \cos 2\theta = 1$.   50. $\cos^2 \frac{1}{2}\theta - 2 \cos \frac{1}{2}\theta + 1 = 0$.

51. $\cos^3 \theta - 1 = 0$.   52. $(\cos \theta)(2 \cos \theta + 1)(\cos \theta - 1) = 0$.

53. $(2 \sin \theta - 1)(\cos \theta + 1) = 0$.   54. $\sin \theta \cos \theta + \cos \theta = \sin \theta + 1$.

3-7 APPLICATION OF THE SINE AND COSINE FUNCTIONS TO RIGHT TRIANGLES

In general, in any calculation the computed result should not show any more significant digits than the number warranted by the tables used nor by the least accurate of the measured data. (See Section 2-10 and 2-11.)

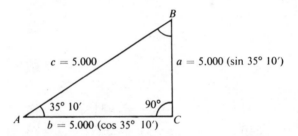

Figure 3-9

Example 1. Solve the right triangle ABC in which $c = 5.000$ and $A = 35° 10'$. (Fig. 3-9.)

Solution:

$$\cos 35° 10' = \frac{b}{5.000},$$

$$b = 5.000(\cos 35° 10') \approx 5.000(.8175) \approx 4.09.$$

$$\sin 35° 10' = \frac{a}{5.000},$$

$$a = 5.000(\sin 35° 10') \approx 5.000(.5760) \approx 2.88.$$

$$B = 90° - 35° 10' = 54° 50'.$$

Example 2. Solve the right triangle ABC in which $a = 3.621$ and $c = 10.00$. (Fig. 3-10.)

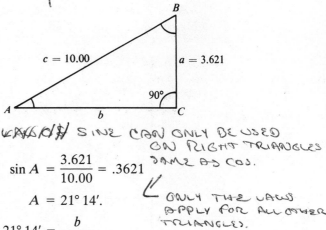

Figure 3-10

Solution:

$$\sin A = \frac{3.621}{10.00} = .3621$$

$$A = 21° 14'.$$

$$\cos 21° 14' = \frac{b}{10.00},$$

$$b = 10.00(\cos 21° 14')$$

$$= 10.00(.9321) \approx 9.321.$$

$$B = 90° - 21° 14'$$

$$= 89° 60' - 21° 14' = 68° 46'.$$

3-8 THE LAW OF COSINES

In any triangle, the square of any side is equal to the sum of the squares of the other two sides, minus twice the product of those sides times the cosine of their included angle.

As an equation, the Law of Cosines takes one of the following forms:

> THE LAW OF COSINES:
>
> $$a^2 = b^2 + c^2 - 2bc \cos A,$$
> $$b^2 = c^2 + a^2 - 2ca \cos B,$$
> $$c^2 = a^2 + b^2 - 2ab \cos C,$$

Proof.   In any given triangle let A be any angle. Place A in standard position with respect to a system of coordinates. Then A is at the origin, and one of the other vertices lies on the positive x-axis. As a matter of notation we shall call this vertex B, which then has coordinates $(c, 0)$. (If we denote this vertex by C, we merely interchange b and c in what follows.) It is immaterial whether A is acute, as in Figure 3-11a, or obtuse, as in Figure 3-11b, or is a right angle. The radius of the vertex C

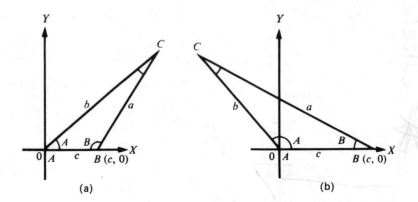

(a)   (b)

Figure 3-11

is b, and by the definition of the trigonometric functions, the coordinates of C are $(b \cos A, \, b \sin A)$. From the coordinates of B and C, we can find $|BC|^2 = a^2$, by the distance formula (Section 1-7). We get

$$a^2 = |BC|^2 = (b \cos A - c)^2 + (b \sin A - 0)^2$$
$$= b^2 \cos A - 2bc \cos A + c^2 + b^2 \sin^2 A$$
$$= b^2 (\sin^2 A + \cos^2 A) + c^2 - 2bc \cos A$$
$$= b^2 + c^2 - 2bc \cos A,$$

as we wished to show. Since A was *any* angle of the triangle, this proves all three forms of the equation of the Law of Cosines.*

The Law of Cosines is used (1) to solve a triangle when two sides and the included angle are known and (2) to solve for an angle when all three sides are known.

Example 1.   Given $b = 2.00$, $c = 5.00$, $A = 35° \, 10'$, find a.

Solution:

$$a^2 = (2)^2 + (5)^2 - 2(2)(5) \cos 35° \, 10'$$

$$\approx 4 + 25 - 20(.8175),$$

$$a = \sqrt{12.65}.$$

To find $\sqrt{12.65}$, use the square root table, Table 5, and interpolate as follows:

$$10\left\{5\left\{\begin{array}{l}\sqrt{12.60} = 3.54965 \\ \sqrt{12.65} = ? \\ \sqrt{12.70} = 3.56371\end{array}\right\}\right. .01406$$

Therefore,

$$a = \sqrt{12.65} = 3.54965 + .5(.01406)$$

$$= 3.54965 + .00703 = 3.55668 \approx 3.56.$$

Example 2.   Given $a = 3.0$, $b = 6.0$, and $c = 7.0$, solve the triangle ABC completely. Find the angles accurately to the nearest degree. (Fig. 3-12.)

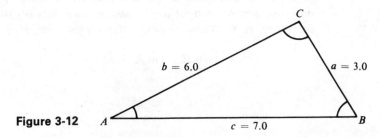

Figure 3-12

---

*This proof is taken, by permission, from R. W. Brink, *Plane Trigonometry*, Third Edition (New York: Appleton-Century-Crofts, 1959), p. 167.

Solution:

We first solve for A. Begin by writing the Law of Cosines for a, the side opposite the desired angle.

$$a^2 = b^2 + c^2 - 2bc \cos A,$$

$$3^2 = 6^2 + 7^2 - 2(6)(7) \cos A,$$

$$9 = 36 + 49 - 84 \cos A,$$

$$84 \cos A = 76,$$

$$\cos A = \frac{76}{84} \approx .9048;$$

$$A = 25° \qquad \text{(to the nearest degree).}$$

Having solved for A, we could use the Law of Sines to solve for another angle; however, it is better, when possible, to use the original data in solving for the remaining parts of a problem. Therefore, we use the Law of Cosines to find B and then check the results by using the Law of Sines.

$$b^2 = a^2 + c^2 - 2ac \cos B,$$

$$6^2 = 3^2 + 7^2 - 2(3)(7) \cos B,$$

$$36 = 9 + 49 - 42 \cos B,$$

$$42 \cos B = 22,$$

$$\cos B \approx .5238;$$

$$B = 58° \qquad \text{(to the nearest degree).}$$

The Law of Sines will not distinguish between an acute and an obtuse angle [$\sin \theta = \sin (180° - \theta)$]. Therefore, if we use the Law of Sines, we must choose an angle known to be acute. *A triangle can have only one obtuse or right angle, which must lie opposite the longest side*; therefore, B, not being opposite the longest side, is an acute angle.

$$\frac{\sin B}{b} = \frac{\sin A}{a},$$

$$\frac{\sin B}{6} = \frac{\sin 25°}{3},$$

$$\sin B = \frac{6 \sin 25°}{3} \approx 2(.4226) = .8452;$$

$$B = 58° \qquad \text{(to the nearest degree).}$$

Knowing two angles, we can find the third angle from the relationship

$$A + B + C = 180°,$$

$$25° + 58° + C = 180°,$$

$$C = 97°.$$

Exercises 3-8

In Exercises 1–4 solve each of the right triangles ABC having the given parts ($C = 90°$).

1. $c = 20.0, A = 24° 10'$.   2. $c = 30.0, A = 41° 20'$.

3. $c = 10.00, A = 67° 43'$.   4. $c = 10.00, A = 74° 17'$.

In Exercises 5–8 solve each of the oblique triangles ABC for the missing side.

5. $A = 27° 10', b = 6.00, c = 5.00$.

6. $A = 28° 40', b = 7.00, c = 6.00$.

7. $B = 110° 50', a = 3.00, c = 5.00$.

8. $B = 120° 10', a = 4.00, c = 5.00$.

In Exercises 9 and 10 solve completely each of the oblique triangles ABC.

9. $a = 5.0, b = 7.0, c = 10.0$.

10. $a = 4.0, b = 8.0, c = 10.0$.

11. Point A is 2000.0 feet from one end of a lake and 3000.0 feet from the other end. The lake subtends (that is, is opposite to) an angle of $114° 50'$ at point A. Find the length of the lake.

12. A 50.0-foot flagpole stands beside a road which is inclined $9° 10'$ from the horizontal. Point P is 110.0 feet down the road from the foot of the flagpole. Find the distance from the top of the flagpole to point P.

13. Prove that if R is the radius of the earth (regarded as a sphere), the radius of the parallel of latitude that passes through a place P of latitude L is given by $r = R \cos L$. (Fig. 3–13.)

14. Find the speed at which New York City is moving east due solely to the rotation of the earth about its axis. The latitude of New York City is approximately $40.5°$. Assume the radius of the earth to be 4000 miles. See Problem 13 and Figures 3-13 and 3-14.

15. Find the speed at which Los Angeles, California, is moving east due solely to the rotation of the earth about its axis. The latitude of Los Angeles is approximately $34°$. Assume the radius of the earth to be 4000 miles. See Problem 13 and Figures 3-13 and 3-14.

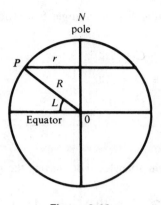

Figure 3-13

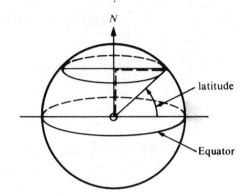

Figure 3-14

3-9 GRAPH OF THE COSINE FUNCTION

The unit circle can be used to show the increasing and decreasing values of the cosine function of an angle that changes from 0° to 360°. Let the variable angle θ be in standard position with respect to the coordinate system. Then, as was shown in Figure 3-4, the value of the cosine of θ is equal to the abscissa (x-value) of the point where the terminal side of θ crosses the unit circle. We shall show the relationship of θ and cos θ by means of a graph.

Example 1. Graph $y = \cos \theta$.

Top numbers

funcion escalides

de sagete

We take values of θ along the horizontal axis and the corresponding values of y along the vertical axis. On the θ-axis we choose a convenient length to represent the equal distances between 0°, 30°, 60°, etc. By plotting many points on the curve $y = \cos \theta$, we discover the shape of the curve. The coordinates of some of these points are given in the following table:

This being the normal cos values so if looking
for   y = 3cos Ø MULT. bottom nums by 3 and graph

θ	0°	30°	60°	90°	120°	150°	180°	210°	240°	270°	300°	330°	360°
y	1	$\frac{1}{2}\sqrt{3}$	$\frac{1}{2}$	0	$-\frac{1}{2}$	$-\frac{1}{2}\sqrt{3}$	-1	$-\frac{1}{2}\sqrt{3}$	$-\frac{1}{2}$	0	$\frac{1}{2}$	$\frac{1}{2}\sqrt{3}$	1

In Figure 3-15 the cosine function is graphed through one *complete cycle*. This curve repeats its cycle in an interval of 360°, and, therefore, the cosine has a period of 360°. The maximum deviation from the θ-axis is 1 unit. This deviation is called the *amplitude* of the curve, and thus the amplitude of this curve is 1.

Example 2. Graph $y = \sin \theta$ and $y = \cos \theta$, on the same coordinate axes, in the interval from −90° to 360°.

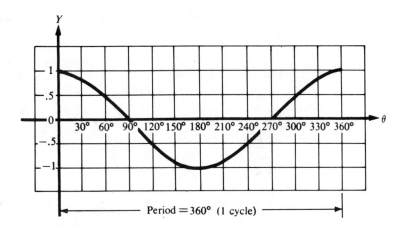

Figure 3-15

As an aid in determining the points on the respective curves, we keep in mind the unit circle, the 30°-60° right triangle, the 45° right triangle, and the principle of related angles. By plotting points of each equation, we sketch the curves as shown in Figure 3-16.

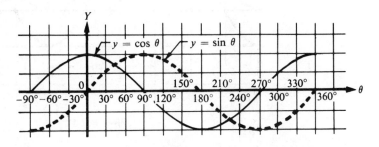

Figure 3-16

The student will observe that the basic sine curve and cosine curve have the same general shape (Fig. 3-16). The only difference between them is that the graph of $y = \cos \theta$ is shifted a distance of 90° to the left of the graph of $y = \sin \theta$.

The graph of $y = \cos \theta$ is therefore like a simple sine curve with amplitude 1 and **wave length** 360°, but we say that it **differs in phase** by 90° from the graph of $y = \sin \theta$.

Now that the general shape of the sine and cosine curves is known, the student should be able to sketch either curve quickly. To do this, locate the points where the required curve crosses the θ-axis, its high and low points, and a few general points; then draw a smooth curve through these points.

Example 3. Graph the two curves $y_1 = \cos \theta$ and $y = 2 \cos \theta$, on the same axes, in the interval from $-180°$ to $180°$.

Solution: We first graph $y_1 = \cos \theta$ as in Figure 3-16. For each point of this graph there is a corresponding point of the graph of $y = 2 \cos \theta$ having the same abscissa (or value of θ) and an ordinate (or value of y) that is twice the ordinate (or the value of y_1) on the first graph. After plotting a sufficient number of such points, we can draw the graph of $y = 2 \cos \theta$ through them. (Fig. 3-17.)

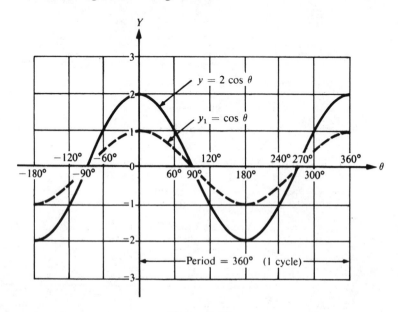

Figure 3-17

3-10 *GRAPHING BY THE ADDITION OF ORDINATES*

If a function is in the form of the sum of two or more simpler functions, its graph is usually best drawn by the method known as the *addition of ordinates,* or as the *composition of ordinates.* The method is illustrated in the following example.

Example. Graph the curve $y = \sin \theta + \cos \theta$ in the interval from $0°$ to $360°$. (See Fig. 3-18.)

Solution: We begin by drawing the graphs of the separate terms $y = \sin \theta$ and $y = \cos \theta$ on the same set of axes. We can do this quickly from our general knowledge of these simple functions. Then, for any value of θ, the height, or ordinate, of the graph of $y = \sin \theta + \cos \theta$ can be found

Cos = X

on (x y)

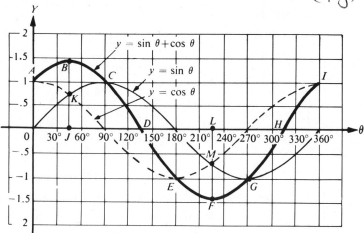

Figure 3-18

by adding the ordinates of the other two curves for that value of θ. In this way we can quickly find as many points as we wish on the desired curve and then draw it. We shall take a number of specific points on the curve to illustrate this method.

Let points A, B, C, D, E, F, G, H, and I be points on the composite curve $y = \sin \theta + \cos \theta$. These points can be located by adding the ordinate values of the component curves.

The value of y, for $y = \sin \theta$, is zero when $\theta = 0°$, 180°, and 360°. Therefore, points A, E, and I of the composite curve are on the cosine curve for these values of θ.

The value of y, for $y = \cos \theta$, is zero when $\theta = 90°$ and 270°. Therefore, points C and G of the composite curve are on the sine curve for these values of θ.

The ordinates of the component curves are equal and positive when $\theta = 45°$. Therefore, the ordinate value of point B on the composite curve is twice the length of JK.

The ordinates of the component curves are equal and negative when $\theta = 225°$. Therefore, the ordinate value of point F on the composite curve is negative and twice the length of LM.

At points D and H the ordinates of the component curves are numerically equal but opposite in sign; thus, the sum of these component ordinates is zero.

For most purposes the composite curve can be determined by locating a sufficient number of points, as we have done, and then drawing a smooth freehand curve through these points. For more accurate work the student would find a pair of draftsman's dividers helpful.

Exercises 3-10

In Exercises 1–8 find the maximum and minimum values of each of the functions of θ.

1. $2 + \sin \theta$.

2. $1 + \cos \theta$.

3. $1 - \cos \theta$.

4. $1 - \sin \theta$.

5. $3 + \sin 2\theta$.

6. $2 + \cos 2\theta$.

7. $2 - \cos^2 \theta$.

8. $2 - \sin^2 \theta$.

In Exercises 9–18 graph each of the following functions through one complete cycle.

9. $y = 3 \cos \theta$.

10. $y = \frac{1}{2} \cos \theta$.

11. $y = 1 + \cos \theta$.

12. $y = -1 + \cos \theta$.

13. $y = \sin \theta + 2 \cos \theta$.

14. $y = 2 \sin \theta + \cos \theta$.

15. $y = \sin \theta - \cos \theta$.

16. $y = \cos \theta - \sin \theta$.

17. $y = \cos \theta + \frac{1}{2} \sin \theta$.

18. $y = \sin \theta + \frac{1}{2} \cos \theta$.

REVIEW EXERCISES

In Exercises 1–6 find the value of each term and then combine their values. Do not use tables.

1. $\cos 60° + \cos 120° + \cos 90° + \cos 135°$.

2. $\cos 180° + \cos 240° + \cos 300° + \cos 0°$.

3. $\cos 45° + \cos 225° + \cos (-60°) + \cos 360°$.

4. $\cos (-180°) + \cos (-300°) + \cos (-90°) + \cos 420°$.

5. $\sin^2 45° + \cos^2 45°$.

6. $4(\cos^2 10° + \sin^2 10°)$.

7. Let θ be an angle in standard position with its terminal side passing through the point $(-4, 3)$.
   (a) Find $\sin \theta$.   (b) Find $\cos \theta$.

In Exercises 8–11 use Table 1 as an aid in finding the value of each of the following functions:

8. $\cos 55° 14'$.

9. $\cos 140° 25'$.

10. $\cos (-95° 47')$.

11. $\cos 730° 38'$.

In Exercises 12 and 13 find the angle that satisfies the indicated conditions.

12. $\cos \theta = .9957$. $0° < \theta < 90°$.

13. $\cos \theta = -.8843$. $90° < \theta < 180°$.

14. Find the value of $\cos 63° 16'$.

In Exercises 15–18 solve each equation for all the values of θ for which $0° \leqq \theta < 360°$. Do not use tables.

15. $2 \cos^2 \theta - \cos \theta = 0$.

16. $2 \cos^2 \theta = 1$.

17. $(\cos \theta)(\cos \theta + 1)(\cos \theta - 1) = 0$.

18. $\cos 2\theta + 1 = 0$.

19. A 50.0-foot flagpole stands beside a road which is inclined $12° 40'$ from the horizontal. Point P is 100.0 feet up the road from the foot of the flagpole. Find the distance from the top of the flagpole to point P.

20. Find the maximum value of the expression $(2 - \cos \theta)$.

21. Find the minimum value of the expression $(1 + \cos 2\theta)$.

22. Graph $y = 1 + 2 \cos \theta$ through one complete cycle.

CHAPTER 3: DIAGNOSTIC TEST

The purpose of this test is to see how well you understand the work covered in Chapter 3. We recommend that you work this test before your instructor tests you on this chapter. Allow yourself about 50 minutes to do the test.

Solutions to the problems, together with section references, are given in the Answer Section at the end of this book. We suggest that you study the sections referred to for the problems you do incorrectly.

---

1. Find the value of each of the following functions. Do not use tables.
(a) $\cos 60°$. (b) $\cos 150°$. (c) $\cos 180°$.
(d) $\cos(-90°)$. (e) $2 \cos(-120°)$. (f) $\cos 450°$.
(g) $\sin^2 16° + \cos^2 16°$. (h) $3 \sin^2 20° + 3 \cos^2 20°$.

2. Let θ be an angle in standard position with its terminal side passing through the point $(\sqrt{3}, -1)$.
(a) Find $\sin \theta$. (b) Find $\cos \theta$.

3. Use Table 1 as an aid in finding the values of each of the following functions.
(a) $\cos 5° 11'$. (b) $\cos(-78° 16')$. (c) $\cos 23° 14'$.

4. Use Table 1 to find the angle that satisfies each of the following conditions:
(a) $\cos \theta = .4669$ $0° < \theta < 90°$.
(b) $\cos \theta = -.5831$ $90° < \theta < 180°$.

In Problems 5–7 solve each equation for all the values of θ for which $0° \leqq \theta < 360°$. Do not use tables.

5. $2 \cos^2 \theta + \cos \theta = 0$.

6. $\cos 2\theta = 1$.

7. $4 \cos^2 \theta = 3$.

8. Find the maximum value of the expression $(3 - \cos 2\theta)$.

9. Given triangle ABC with $A = 42° \, 11'$, $b = 40.00$, and $c = 30.00$, solve for side a.

10. Graph $y = -1 + \cos \theta$ through one complete cycle.

4

The Tangent Function

Let θ be an angle placed in standard position (Fig. 4-1) and P be a point on the terminal side of θ. Associated with P are three numbers: the x- and y-coordinates, and r, the radius.

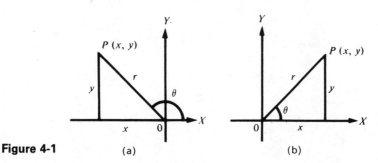

Figure 4-1   (a)   (b)

As was explained in Chapters 2 and 3, six ratios called trigonometric functions can be made with the numbers x, y, and r. In this chapter we shall learn about the ratio of the ordinate to the abscissa of P, which is called the **tangent function**. The radius, r, does not come into the definition of the tangent function.

4-1 DEFINITION OF THE TANGENT OF AN ANGLE

1. Place an angle θ in standard position.
2. Let point P, whose coordinates are (x, y) where $x \neq 0$, be a point on the terminal side of angle θ.

Then the tangent of θ (written $\tan \theta$) is defined as the ratio of y to x,

$$\tan \theta = \frac{y}{x} = \frac{\text{ordinate}}{\text{abscissa}}, \qquad x \neq 0.$$

Special case Let θ be an acute angle of a right triangle. Let this triangle be placed in the first quadrant with θ in standard position. Let the point (x, y) be the end of the hypotenuse in the first quadrant. With the triangle in this position, x becomes the side of the triangle adjacent to θ and y the side opposite θ. Then, from the definition of the tangent of an angle,

$$\tan \theta = \frac{y}{x} = \frac{\text{side opposite}}{\text{side adjacent}}.$$

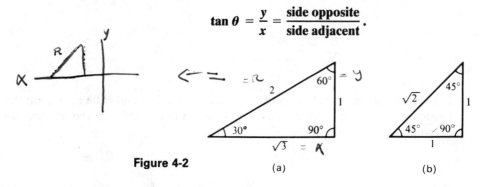

Figure 4-2

(a)   (b)

Examples.   (See Fig. 4-2.)

1. $\tan 30° = \dfrac{1}{\sqrt{3}} = \tfrac{1}{3}\sqrt{3},$

2. $\tan 60° = \dfrac{\sqrt{3}}{1} = \sqrt{3},$

3. $\tan 45° = \dfrac{1}{1} = 1.$

The student will recall that $\sin \theta$ and $\cos \theta$ exist for all values of θ. The tangent function of an odd multiple of 90° is not defined; for such an angle, the point (x, y) of our definition would lie on the y-axis, and hence x would equal 0 and $\tan \theta$ would involve $y/0$, which is not defined. The sine and cosine functions vary from minus 1 to plus 1; the tangent function varies from "minus infinity" to "plus infinity." In essence, this means that we can make $\tan \theta$ as large in absolute value as we please by taking θ sufficiently close to plus or minus 90°. When $\theta = 90°$, $\tan \theta$ is undefined. These are some of the characteristics of $\tan \theta$ that distinguish it from $\sin \theta$ and $\cos \theta$.

4-2 THE SIGN OF THE TANGENT OF AN ANGLE

As defined, $\tan \theta = y/x$ and is therefore positive when the values of y and x have the same sign and negative when the values of y and x have opposite signs. Therefore, $\tan \theta$ is positive when θ is in the first or third quadrant and negative when θ is in the second or fourth quadrant (Fig. 4-3).

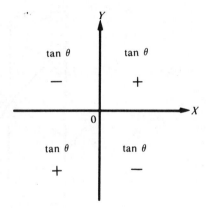

Figure 4-3

4-3 THE UNIT CIRCLE AND THE TANGENT FUNCTION

We construct a circle with center at the origin and radius equal to one unit. AB is the horizontal diameter and lines l_1 and l_2 are drawn tangent to the circle at points A and B. Let θ represent any angle placed in standard position, and let point $P(x, y)$ be the intersection of the terminal side of θ and one of the lines l_1 or l_2. In these special cases (Fig. 4-4a, b, c) the value of x in $\tan \theta = \dfrac{y}{x}$ becomes ±1. Thus,

$$\tan \boldsymbol{\theta} = \frac{\mathbf{y}}{\pm\mathbf{1}} = \pm\mathbf{y}$$

(only when $x = \pm 1$).

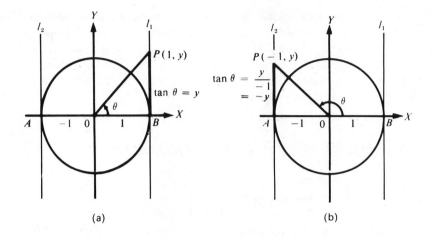

(a)   (b)

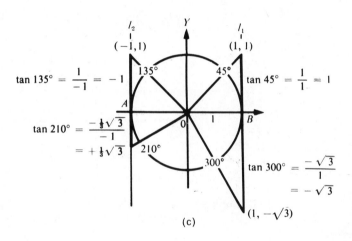

(c)

Figure 4-4

If an angle is in standard position, its tangent is numerically equal to the ordinate of the point of intersection of its terminal side and a line parallel to and at the distance of one unit to the right or to the left of the y-axis; the tangent is positive if the angle terminates in the first or third quadrant and negative if the angle terminates in the second or fourth quadrant.

Examples.

1. $\tan 0° = 0$.

2. $\tan 30° = \dfrac{1}{\sqrt{3}} \approx .577$.

3. $\tan 45° = 1.000$.

4. $\tan 60° = \sqrt{3} \approx 1.732$.

5. $\tan 80° \approx 5.671$.

6. $\tan 89° \, 50' \approx 343.8$.

7. tan 90° undefined.

8. tan 100° ≈ −5.671.

9. tan 180° = 0.

10. tan 270° undefined.

11. tan 300° = − $\sqrt{3}$ ≈ −1.732.

12. tan 330° = − $\dfrac{1}{\sqrt{3}}$ ≈ −.577.

To find the tangent of an angle place the angle in standard position; then determine its quadrant and related angle.

Example 13. Find tan 120°.
Tan 120° is negative because 120° is in the second quadrant. The related angle is 60°. Therefore, tan 120° = −tan 60° = − $\sqrt{3}$.

Example 14. Find tan (−120°).
Tan (−120°) is positive because −120° is in the third quadrant. The related angle is 60°. Therefore, tan (−120°) = tan 60° = $\sqrt{3}$.

Example 15. Find tan 339° 10′.
Tan 339° 10′ is negative because 339° 10′ is in the fourth quadrant. The related angle = (360° − 339° 10′) = 20° 50′. Locate 20° 50′ in the degree column of Table 1, and read the value of tan 20° 50′ = .3805 to the right of 20° 50′ in the tangent column. The tangents of angles in the fourth quadrant are negative; therefore, tan 339° 10′ = −.3805.

In Example 16 we see how to find θ when tan θ is known.

Example 16. Find θ where 90° < θ < 180° and tan θ = −.6494.
We first find the related angle of θ from tan θ = +.6494. Locate .6494 in the tangent column of Table 1, and then read 33° 00′ in the degree column to the left of .6494. The desired angle is (180° − 33° 00′) = 147° 00′.

Exercises 4-3

In Exercises 1–14 find the value of each term, and then combine the values.

1. tan 0° + tan² 30° + tan 45° + tan 135° + tan 180°.

2. tan 315° + tan 225° + tan 60° + tan (−60°).

3. tan² 210° + tan 420° + tan (−135°).

4. tan 750° + tan (−30°) − tan 150°.

5. tan 180° − tan 240° + tan 225°.

6. sin 30° + cos 30° + tan 30°.

7. $\sin 45° + \tan 45° + \cos 45°$.

8. $\sin 0° + \tan 0° + \cos 0°$.

9. $\sin 315° + \cos 315° + \tan 315°$.

10. $\sin^2(-150°) + \cos^2(-150°) + \tan(-135°)$.

11. $\sin^2 60° + \cos^2 60° + \tan^2 60°$.

12. $\sin^3 45° + \cos^3 45° + \tan^3 60°$.

13. $\sin^2(-135°) - \cos^3(-45°) - \tan^3(-180°)$.

14. $\tan(-150°) + \tan(-210°) - \cos(-120°)$.

In Exercises 15–30 use Table 1 to find the value of each of the functions.

15. $\tan 35°$.	16. $\sin 35°$.	17. $\cos 35°$.
18. $\tan 70°$.	19. $\tan 85°$.	20. $\tan 100°$.
21. $\tan 125°$.	22. $\tan 200°$.	23. $\tan 220°$.
24. $\tan 280°$.	25. $\tan 310° 20'$.	26. $\tan 190° 40'$.
27. $\tan(-50°)$.	28. $\tan(-80°)$.	29. $\tan(-150° 50')$.

30. $\tan(-285° 20')$.

In each of the Exercises 31–36 determine the quadrant in which θ lies if the functions are positive or negative as indicated.

31. $\sin \theta$ and $\cos \theta$ are both negative.

32. $\sin \theta$ and $\tan \theta$ are both negative.

33. $\cos \theta$ and $\tan \theta$ are both negative.

34. $\sin \theta$ is negative and $\tan \theta$ is positive.

35. $\cos \theta > 0$ and $\tan \theta < 0$.

36. $\sin \theta < 0$ and $\cos \theta < 0$.

In each of the Exercises 37–42 use the given information to find the value of the indicated expressions.

37. Given $\sin \theta = \frac{3}{5}$ and $0° < \theta < 90°$, find the value of $(\cos \theta + \tan \theta)$.

38. Given $\tan \theta = -\frac{3}{4}$ and $90° < \theta < 180°$, find the value of $(5 \sin \theta - 10 \cos \theta)$.

39. Given $\cos \theta = \frac{3}{5}$ and $270° < \theta < 360°$, find the value of $9 \tan^2 \theta - 5 \sin \theta$.

40. Given $\sin \theta = -\frac{5}{13}$ and $180° < \theta < 270°$, find the value of $(13 \cos \theta + 12 \tan \theta)^2$.

41. Given $\tan \theta = \frac{1}{7}$ and θ not in Q I, find the value of $(\sin \theta + \cos \theta)^2$.

42. Given $\tan \theta = -\frac{3}{5}$ and θ not in Q IV, find the value of $(\sin \theta + \cos \theta)^2$.

4-4 ANGLES OF ELEVATION AND DEPRESSION

Suppose that an observer is at a point O and a certain object is at a point P. The line OP is then called the **line of sight** from O to P. Let OH be drawn horizontally in the vertical plane through OP. Then the acute angle HOP, which the line of sight makes with the horizontal, is called the **angle of elevation** of P from O if P is higher than O, and it is called the **angle of depression** of P from O if P is lower than O. (See Fig. 4-5.)

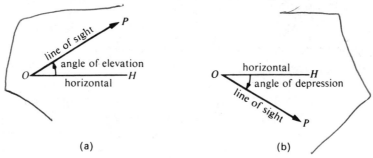

(a)　　　　　　　　　　　(b)

Figure 4-5

Example. From an airplane the angle of depression of a boat is 40°, and at the same time the angle of elevation of the airplane, as observed from the boat, is 40°.

4-5 THE BEARING OF A LINE

The direction of a line on the earth's surface (or the direction of one point with respect to another) is generally given by means of the angle which the line makes with the true north–south line. This angle is called the **bearing of the line**. The **bearing angle** of a line is the smallest angle which that line makes

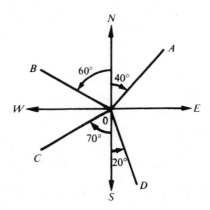

Figure 4-6

with the north–south line. It can never be more than 90 degrees. The bearing of a line is written as the bearing angle with the notation of the directional quadrant in which the line lies. Thus, in Figure 4-6 the bearing of *OA* is *N* 40° *E*, that of *OB* is *N* 60° *W*, that of *OC* is *S* 70° *W*, and that of *OD* is *S* 20° *E*.

In some other systems, such as astronomy and navigation, the bearing is measured clockwise from the north up to 360°. Thus, the direction *N* 60° *W* could also be expressed as 300°, a due west direction could be expressed as 270°, and so forth. This method is used by most branches of the United States Armed Forces.

4-6 APPLICATION OF THE TANGENT FUNCTION TO RIGHT TRIANGLES

Example 1. Solve for side *a* of the right triangle *ABC* in which *b* = 10.00 and *A* = 35° 10′. (Fig. 4-7.)

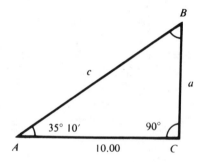

Figure 4-7

Solution:

$$\tan 35° 10′ = \frac{a}{10.00},$$

$$a = 10.00(\tan 35° 10′)$$

$$\approx 10.00(.7046) = 7.046$$

$$\approx 7.05 \text{ (to two decimal places)}.$$

Example 2. Solve for angle *A* of the right triangle *ABC* in which *a* = 3.00 and *b* = 4.00. (Fig. 4-8.)

Solution:

$$\tan A = \frac{3.00}{4.00} = .7500.$$

From Table 1, we have tan 36° 50′ = .7490 and tan 37° 00′ = .7536.

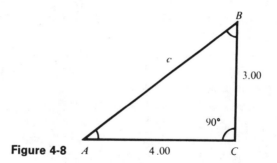

Figure 4-8

Since tan A is between these numbers and is nearer to .7490 than to .7536, we take $A = 36° 50'$ (accurate to 10′).

The sides were expressed accurately to three significant digits which limits the accuracy of the angle to the nearest 10 minutes. See Section 2-11.

Exercises 4-6

In Exercises 1–4 solve the right triangles ABC for side a. Angle $C = 90°$.

1. $A = 14° 50'$, $b = 10.00$.   2. $A = 28° 10'$, $b = 100.0$.

3. $A = 77° 10'$, $b = 25.00$.   4. $A = 81° 30'$, $b = 20.00$.

In Exercises 5–8 use Table 1 to solve the right triangles ABC for angle A. Angle $C = 90°$.

5. $a = 7.465$, $b = 10.00$.   6. $a = 3.814$, $b = 10.00$.

7. $a = 1.043$, $b = 10.00$.   8. $a = 2.079$, $b = 10.00$.

9. A road changes altitude 67.0 feet for every 1000.0 feet of horizontal change. Find the angle of inclination of the road.

10. A tree, standing on level ground, casts a 91-foot shadow when the angle of elevation of the sun is 55°. Find the height of the tree.

11. A flagpole, standing on level ground, casts a 65-foot shadow when the angle of elevation of the sun is 61°. Find the height of the flagpole.

12. A surveyor wishes to find the distance from point A to point B which is across a pond and due south of A. From point C, which is due east of B, he measures the distance BC which is 344 feet. With an instrument at C he finds that angle $BCA = 61° 10'$. Find the distance AB.

13. The top of an observation tower is 95.0 feet above the shore of a lake. Find the distance of a buoy from the foot of the tower if the angle of depression of the buoy, as seen from the top of the tower, is 21° 30′.

14. A ship is 25 miles to the east and 35 miles to the south of a certain port. Find its distance and bearing from the port.

15. A ship is 15 miles to the west and 25 miles to the north of a certain port. Find its distance and bearing from the port.

16. A road of uniform inclination along a mountainside rises 52.3 feet in a distance of 1000.0 feet measured along the road. Find the angle of inclination of the road.

17. From the top of a mountain, the angles of depression of two successive milestones in the horizontal plane which is below and in the same vertical plane as the observer are $24° \, 10'$ and $16° \, 50'$. Find the height of the mountain above the horizontal plane.

18. A tower stands at point P, which is 845 feet due north of point B. Point C is due west of B. From C the bearing of P is $N \, 32° \, 20' \, E$ and the angle of elevation of the top of the tower is $20° \, 20'$. Find the height of the tower.

19. A tower stands at point P, which is 900.0 feet due north of point B. Point C is due east of B. From C the bearing of P is $N \, 36° \, 00' \, W$ and the angle of elevation of the top of the tower is $18° \, 10'$. Find the height of the tower.

20. A ship, 15.2 miles west of a shore that runs due north and south, heads for a port on the shore on a course bearing $156° \, 20'$ (using the second system of giving bearings). What is its distance from the port?

Solve the following equations for all the values of θ for which $0° \leq \theta < 360°$; use Table 1 when necessary.

21. $\tan \theta = 1$.

22. $\tan \theta = \sqrt{3}$.

23. $\tan \theta = -1$.

24. $\tan \theta = -\sqrt{3}$.

25. $\tan \theta = \dfrac{1}{\sqrt{3}}$.

26. $\tan \theta = -\dfrac{1}{\sqrt{3}}$.

27. $\tan \theta = 0$.

28. $\tan^2 \theta = 1$.

29. $\tan^2 \theta = 3$.

30. $\tan^2 \theta = \frac{1}{3}$.

31. $\tan^2 \theta + \tan \theta = 0$.

32. $\tan 2\theta = 1$.

33. $\tan \frac{1}{2}\theta = 1$.

34. $\tan 2\theta = 0$.

35. $\tan \frac{1}{2}\theta = 0$.

36. $\tan^2 \theta - \tan \theta = 0$.

37. $\tan^2 \theta + 2 \tan \theta + 1 = 0$.

38. $\tan^3 \theta - \tan \theta = 0$.

39. $\tan^2 2\theta + \tan 2\theta = 0$.

40. $\tan \theta = .3640$.

41. $\tan \theta = .7002$

42. $\tan \theta = 2.7475$.

43. $\tan \theta = 5.6713$.

44. $\tan \theta = 57.290$.

45. $\tan \theta = -.3476$.

46. $\tan \theta = -.1435$.

Some physical quantities are completely determined when their **magnitudes**, in terms of specific units, are given. Examples of physical quantities of this kind are feet, inches, gallons, dollars, etc. Other quantities, such as forces and velocities, in which the direction as well as magnitude is important are called **vector quantities.**

> **Definition 1.** *A **vector** is a directed line segment; it is characterized by its length and direction, its actual position being immaterial. A vector shows both direction and magnitude.*

It is customary to represent a vector by an arrow whose direction represents the direction of the vector and whose length (in terms of some chosen unit of length) represents the magnitude.

A vector is determined by two points, the **initial point** and the **terminal point** of the vector (Fig. 4-9). Boldface type is usually used to denote vectors. For example, we have labeled the vector in Figure 4-9a as **A**, and we may also call it **PQ** to emphasize that it is the vector from P to Q. It is customary to denote the magnitude of the vector **A** by the symbol $|A|$. We often refer to this number as the **absolute value** of **A**. If **A** is the vector **PQ**, then the vector **QP** is denoted by $-A$ (see Fig. 4-9b).

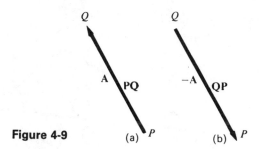

Figure 4-9   (a)   (b)

> **Definition 2.** *Two vectors are said to be equal if, and only if, they have the same magnitude and the same direction.*

Vectors **A** and **B** in Figure 4-10 illustrate a condition in which **A** = **B**. Notice that we do not require that equal vectors coincide, but they must be parallel, have the same length, and point in the same direction.

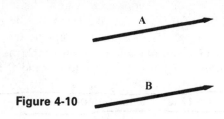

Figure 4-10

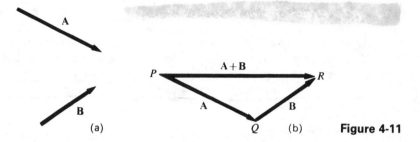

Figure 4-11

We obtain the vector sum **A** + **B** by placing the initial point of **B** on the terminal point of **A** and joining the initial point of **A** to the terminal point of **B**, as shown in Figure 4-11b.

Thus, if P, Q, and R are points such that **A** = **PQ** and **B** = **QR**, then **A** + **B** = **PR**; that is,

$$\mathbf{PQ} + \mathbf{QR} = \mathbf{PR}.$$

In a similar way we get the sum **B** + **A** by placing the initial point of **A** on the terminal point of **B** and joining the initial point of **B** to the terminal point of **A**. Figure 4-12 shows how we construct **A** + **B** and **B** + **A**, and from this figure it is apparent that the **commutative law** of addition,

$$\mathbf{A} + \mathbf{B} = \mathbf{B} + \mathbf{A},$$

holds for vector addition.

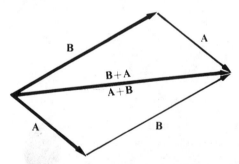

Figure 4-12

An alternate method for finding the sum of two vectors is known as the **parallelogram law for the composition of forces**. Let vectors **P** and **Q** represent two forces. We move these vectors until the initial point of each falls on some point such as O (Fig. 4-13). We let A be the terminal point of **P** and B the terminal point of **Q**. We then complete the parallelogram with the sides adjacent to OA and OB and draw **OC** = **R**, the diagonal of the parallelogram. It is shown in mechanics that if vectors **P** and **Q** represent two forces, they may be replaced in their effect by an equivalent force **R**, called the **resultant** of **P** and **Q**. **P** and **Q** are called components of **R**.

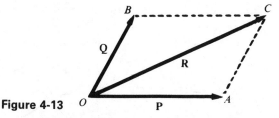

Figure 4-13

It is often required to **resolve** a given vector quantity into components in two given directions. For example, in Figure 4-14 let it be required to find the components of **R** in the directions OA' and OB'. We complete the parallelogram by drawing lines through C parallel to $A'O$ and $B'O$. Let B and C be the respective points where these lines intersect $B'O$ and $A'O$. Then the line segments OB and OA represent the required vector components of **R**.

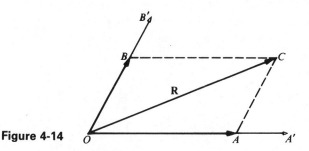

Figure 4-14

Example 1. An airplane has a heading (bearing) of 37° and an air speed of 550 miles per hour. Find the east and north components of the velocity of the airplane (assuming zero wind velocity).

Solution: In Figure 4-15, **OC** represents the velocity of the airplane. *CA* and *CB* are drawn parallel to the *N*- and *E*-axes, respectively. Then

$\quad$ **OA** = **BC** = $550 \times \sin 37° \approx 330$ miles per hour (the speed that the airplane is traveling east), and

$\quad$ **OB** = $550 \times \cos 37° \approx 440$ miles per hour (the speed that the airplane is traveling north).

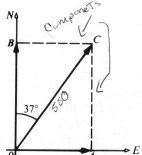

Figure 4-15

Example 2.   Find the tension on each arm of a 150-pound gymnast as he hangs by his hands from a high horizontal bar. His hands are separated such that his arms form 43° angles with the bar.

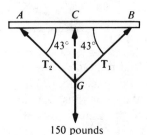

150 pounds   **Figure 4-16**

Solution: In Figure 4-16, **GC** represents the vertical upward force required to support the weight of the gymnast. Since $\mathbf{T}_1$ and $\mathbf{T}_2$ each support half the load, in computing $|\mathbf{T}_1|$ we take $\frac{1}{2}|\mathbf{GC}| = 75$ pounds. Then

$$\sin 43° = \frac{75}{|\mathbf{T}_1|},$$

$|\mathbf{T}_1| = \dfrac{75}{\sin 43°} = \dfrac{75}{.6820} = 110$ pounds (to the nearest pound). Thus, the pull on each arm is 110 pounds.

Example 3.   Find the force (independent of friction) required to pull a 1650-pound boat up a ramp that is inclined 17° 00′ to the horizontal.

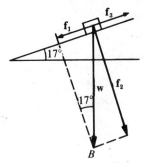

Figure 4-17

Solution: In Figure 4-17 the vertical vector **w** (representing the 1650-pound gravitational force) is drawn; then, from point B, lines perpendicular and parallel to the direction of the ramp are drawn. Vectors $\mathbf{f}_1$ and $\mathbf{f}_2$ are components of **w**. Then $|\mathbf{f}_3|$, the force required to pull the boat up the ramp, is equal to $|\mathbf{f}_1|$, the force causing the boat to roll down the ramp.

$$\sin 17° 00' = \frac{|\mathbf{f_1}|}{|\mathbf{w}|},$$

$$|\mathbf{f_1}| = |\mathbf{w}| \cdot \sin 17° 00' \approx 1650(.2924) \approx 482 \text{ pounds}$$

(to the nearest pound); this is the force required to pull the boat up the ramp.

Example 4. Two forces, $\mathbf{f_1} = 250$ pounds and $\mathbf{f_2} = 150$ pounds, act on a point P. Find the magnitude of the equilibrant when the angle between $\mathbf{f_1}$ and $\mathbf{f_2}$ is 64°. Find the angle between the equilibrant and $\mathbf{f_1}$. (The equilibrant is the force that could be applied to prevent the motion that $\mathbf{f_1}$ and $\mathbf{f_2}$ jointly tend to produce. See Fig. 4-18.)

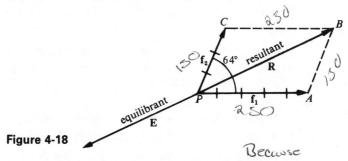

Figure 4-18

Solution: In Figure 4-18 we construct $\mathbf{f_1}$ and $\mathbf{f_2}$ with their initial point at P making an angle of 64° with each other. Then $\mathbf{R}$, the diagonal of the parallelogram constructed on $\mathbf{f_1}$ and $\mathbf{f_2}$, is the resultant of the two given forces. $\mathbf{E}$ is the equilibrant which has the same magnitude as $\mathbf{R}$ but acts in the opposite direction. In triangle PAB, angle $PAB = 180° - 64° = 116°$ and $AB = 150$; therefore, using the Law of Cosines (Section 3-9), we have

$$|\mathbf{R}|^2 = (250)^2 + (150)^2 - 2(250)(150)\cos 116°$$

$$\approx 62,500 + 22,500 - (75,000)(-.4384)$$

$$\approx 117,800.$$

Thus, $|\mathbf{R}| = \sqrt{117,880} \approx 340$ pounds (to the nearest pound). Therefore, the magnitude of the equilibrant is 340 pounds.

The angle formed by $\mathbf{E}$ and $\mathbf{f_1}$ is the supplement of $\angle APB$. We apply the Law of Sines (Section 2-14) to triangle APB to find $\angle APB$.

$$\frac{\sin \angle APB}{150} = \frac{\sin 116°}{340},$$

$$\sin \angle APB = \frac{(150)(\sin 116°)}{340} \approx .3965.$$

Thus, $\angle APB = 23°$ (to the nearest degree). Therefore, the angle between **E** and $\mathbf{f}_1$ is $(180° - 23°) = 157°$.

Exercises 4-7

Unless otherwise indicated, consider data accurate to the nearest unit.

1. Given two forces, 12 pounds and 16 pounds, each acting upon the point P, find their resultant (a) when the angle between them in 180°; (b) when the included angle is 120°; (c) when it is 90°; (d) when it is 51°; (e) when it is zero.

2. A force of 30 pounds acts easterly upon the point P. A second force of 50 pounds acts $S\ 40°\ W$. Represent graphically and show the length and direction (or bearing) of the equilibrant. What is the magnitude of the resultant?

3. Forces of 30 and 40 pounds act on an object at an angle of 70° with respect to each other. Find the magnitude of the resultant force and the angles the resultant force makes with respect to the given forces.

4. A force of 75 pounds makes an angle of 35° with the vertical. Find the horizontal and vertical components of the force.

5. A force of 55 pounds makes an angle of 55° with the horizontal. Find the horizontal and vertical components of the force.

6. Forces of 125 pounds and 150 pounds act on an object at an angle of 140° with respect to each other. Find the magnitude of the resultant force and the angles the resultant force makes with respect to the given forces.

7. An inclined plane is 12 feet long, and one end is 4 feet higher than the other. A weight of 300 pounds rests on the plane. Find the value of the force perpendicular to the plane and the value of the force needed to keep the weight from sliding down the plane (disregard the effect of friction).

8. The resultant of two equal forces acting at right angles upon an object is 120 pounds. Find the magnitude of each force.

9. A man pushes with a force of 70 pounds against the handle of a lawn mower. If the handle makes an angle of 35° with the level of the ground, find both the horizontal and the vertical components of the 70-pound force.

10. A 30.0-pound picture is hung from a hook on the wall by means of a wire. Find the tension in the wire if the two divisions of the wire make an angle of 110° at the hook.

11. A person who weighs 180.0 pounds sits in the center of a hammock. The ropes supporting the hammock make angles of 58° 40′ with the posts to which they are attached. Find the tension on each rope.

12. A child weighing 100.0 pounds sits on a swing. Find the tension on the ropes when a man pushes against the swing with a horizontal force of 35.0 pounds.

13. An airplane heads due north at a rate which in still air would carry it 250 miles per hour. A wind, of constant velocity 50 miles per hour, is blowing from the direction of 250°. Find the actual speed of the plane over the ground and its course (direction or bearing of flight).

14. An airplane heads due south at a rate which in still air would carry it 240 miles per hour. A wind, of constant velocity 40 miles per hour, is blowing from the direction 245°. Find the actual speed of the plane over the ground and its course (direction or bearing of flight).

15. An airplane is climbing with an air speed of 450 miles per hour at an angle of 12° from the horizontal. Find its rate of vertical rise.

16. Find the force (independent of friction) required to pull a 2500-pound boat up a ramp that is inclined 15° to the horizontal.

4-8 GRAPH OF THE TANGENT FUNCTION

The graph of the tangent function always slopes upward toward the right, except at points where there is a sudden break in the graph.

The student will recall that the tangent function is undefined for all odd numbered multiples of 90° (±90°, ±270°, etc.); therefore, there is no point on the curve for any of these values. We use U as a symbol for *undefined* in the following table and draw vertical lines at the places that separate the branches of the tangent curve. (See Fig. 4-19.)

Lines such as the dotted vertical lines shown in Fig. 4-19 are called **asymptotes**. Notice that the horizontal distance between the graphs and the vertical lines becomes less and less as we take values of θ closer and closer to the value of θ for the respective vertical lines.

By making use of our previous knowledge of the values of the tangent function of certain angles, we trace out the graph of the curve $y = \tan \theta$.

A brief table of points on the curve $y = \tan \theta$ follows:

θ	−270°	−240°	−225°	−210°	−180°	−150°	−135°	−120°	−90°	−45°	−30°	0°
y	U	$-\sqrt{3}$	-1	$-\frac{1}{3}\sqrt{3}$	0	$\frac{1}{3}\sqrt{3}$	1	$\sqrt{3}$	U	-1	$-\frac{1}{3}\sqrt{3}$	0
$\tan \theta$	U	-1.732	-1	$-.577$	0	$.577$	1	1.732	U	-1	$-.577$	0

θ	30°	45°	60°	90°	120°	135°	150°	180°	210°	225°	240°	270°
y	$\frac{1}{3}\sqrt{3}$	1	$\sqrt{3}$	U	$-\sqrt{3}$	-1	$-\frac{1}{3}\sqrt{3}$	0	$\frac{1}{3}\sqrt{3}$	1	$\sqrt{3}$	U
$\tan \theta$	$.577$	1	1.732	U	-1.732	-1	$-.577$	0	$.577$	1	1.732	U

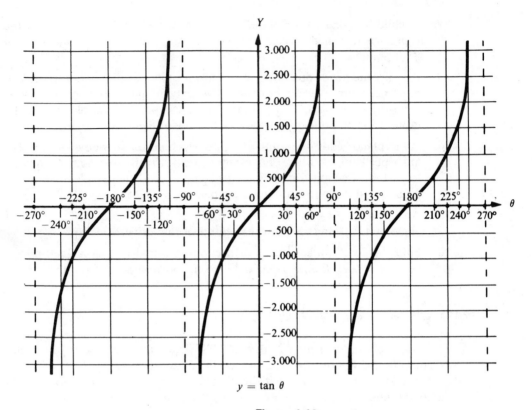

$$y = \tan \theta$$

Figure 4-19

Exercises 4-8

1. Graph $y = \frac{1}{2} \tan \theta$ from $\theta = 90°$ to $270°$.
2. Graph $y = 1 + \tan \theta$ from $\theta = -90°$ to $270°$.
3. Graph $y = \tan \theta - 1$ from $\theta = -90°$ to $270°$.
4. Graph $y = \frac{1}{4} \tan \theta$ from $\theta = 90°$ to $270°$.
5. Graph $y = 3 - \tan \theta$ from $\theta = -90°$ to $270°$.
6. Graph $y = 1 - \tan \theta$ from $\theta = -90°$ to $270°$.

REVIEW EXERCISES

In Exercises 1–4 find the value of each term and then combine their values.

1. $\tan 180° + \tan 135° + \tan 225° + \tan 315°$.
2. $\tan 0° - \tan 60° + \tan (-120°) + \tan 45°$.

3. $\tan 30° + \tan 150° + \tan (-45°) + \tan^2 120°$.

4. $\tan 240° + \tan (-60°) + \tan^2 30° + \tan^2 60°$.

In Exercises 5–7 use Table 1 to find the value of each function.

5. $\tan 38° \; 11'$.   6. $\tan (-134° \; 16')$.

7. $\tan 215° \; 48'$.

In Exercises 8 and 9 use Table 1 to find the angle that satisfies the given conditions.

8. $\tan \theta = .6289$. $180° < \theta < 270°$.

9. $\tan \theta = -1.2915$. $90° < \theta < 180°$.

In Exercises 10–12 determine the quadrant in which θ lies if the functions are positive or negative as indicated.

10. $\cos \theta$ and $\tan \theta$ are both negative.

11. $\sin \theta < 0$ and $\tan \theta > 0$.

12. $\cos \theta > 0$ and $\tan \theta < 0$.

In Exercises 13 and 14 use the given information to find the value of the indicated expressions.

13. Given $\tan \theta = \frac{4}{3}$ and $180° < \theta < 270°$, find the value of $(\cos \theta - \sin \theta)$.

14. Given $\sin \theta = \frac{3}{5}$ and $90° < \theta < 180°$, find the value of $\cos^2 \theta(1 + \tan^2 \theta)$.

15. Solve for side a of the right triangle ABC in which $b = 10.00$, $A = 36° \; 10'$, and $C = 90°$.

16. A ship is 25 miles east and 15 miles north of a port. Find the bearing of the ship from the port. Express your answer accurate to the nearest degree.

In Exercises 17–19 solve each equation for all values of θ where $0° \leqq \theta < 360°$. Do not use tables.

17. $\tan^2 \theta - \tan \theta = 0$.   18. $\tan \frac{1}{2}\theta = 0$.

19. $3 \tan^2 \theta = 1$.

20. A man pushes with a force of 25 pounds against the handle of a lawn mower. If the handle makes an angle of 40° with the level of the ground, find both the horizontal and the vertical components of the 25-pound force. Consider data to be of two-figure accuracy.

21. Graph $y = \frac{1}{3} \tan \theta$ from $\theta = -90°$ to $270°$.

22. Graph $y = 2 + \tan \theta$ from $\theta = -90°$ to $270°$.

CHAPTER 4: DIAGNOSTIC TEST

The purpose of this test is to see how well you understand the work covered in Chapter 4. We recommend that you work this test before your instructor tests you on this chapter. Allow yourself approximately 50 minutes to do the test.

Solutions to the problems, together with section references, are given in the Answer Section at the end of this book. We suggest that you study the sections referred to for the problems you do incorrectly.

---

1. Find the value of $\tan \theta$ where θ is in standard position and its terminal side passes through the point $(-3, 4)$.

2. Find the value of each of the following functions. Do not use tables.
   (a) $\tan (-135°)$.　　　(b) $\tan 120°$.　　　(c) $\tan 360°$.

3. Determine the quadrant in which θ lies to satisfy the following conditions.
   (a) $\sin \theta < 0$ and $\tan \theta < 0$.
   (b) $\cos \theta < 0$ and $\tan \theta > 0$.

4. Given $\sin \theta = -\frac{4}{5}$ and $270° < \theta < 360°$, find the value of $(5 \cos \theta - 3 \tan \theta)$.

5. Solve each of the following equations for all values of θ where $0° \leqq \theta < 360°$.
   (a) $\tan^2 \theta - \sqrt{3} \tan \theta = 0$.
   (b) $\tan^2 \theta - 2 \tan \theta + 1 = 0$.

6. Graph $y = \tan \theta - 2$ from $\theta = -90°$ to $270°$.

7. Use Table 1 to find the value of $\tan 143° 18'$.

8. Use Table 1 to find θ when $\tan \theta = .7199$ and $0° < \theta < 90°$. Express your answer accurate to the nearest minute.

9. Given triangle ABC with $a = 138.9$, $b = 100.0$, and $C = 90°$, solve for angle A. (Use Table 1.)

10. Two forces, $\mathbf{f}_1 = 20.0$ pounds and $\mathbf{f}_2 = 10.0$ pounds, act on a point P. Find the magnitude of the equilibrant when the angle between $\mathbf{f}_1$ and $\mathbf{f}_2$ is $60° 00'$. (The equilibrant is the force that could be applied to prevent the motion that $\mathbf{f}_1$ and $\mathbf{f}_2$ jointly tend to produce.) Express your answer accurate to three significant digits.

5

The Reciprocal Functions

Let θ be an angle placed in standard position (Fig. 5-1) and P be a point on the terminal side of θ. Associated with P are three numbers: the x- and y-coordinates, and r, the radius.

As was explained previously, six ratios called trigonometric functions can be made with the numbers x, y, and r. In this chapter we shall learn about the

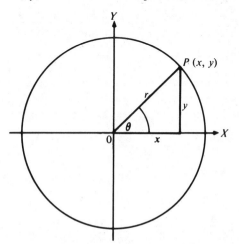

Figure 5-1

reciprocals of the sine, cosine, and tangent functions. These functions are often called the **reciprocal functions.** The **reciprocal of a number** is 1 divided by the number. For example, the reciprocal of x is $\dfrac{1}{x}$. The **reciprocal of a fraction** is formed by inverting the given fraction. For example, the reciprocal of $\dfrac{a}{b}$ is $\dfrac{b}{a}$.

5-1 *DEFINITIONS OF THE COTANGENT, SECANT, AND COSECANT FUNCTIONS*

We shall now define the three reciprocal functions of an angle θ, which are the **cosecant of θ,** the **secant of θ,** and the **cotangent of θ,** written csc θ, sec θ, and cot θ, respectively.

1. Place an angle θ in standard position.
2. Let point P, whose coordinates are (x, y), be a point on the terminal side of angle θ, not at the origin.
3. Let r be the distance of P from the origin.

Definition. *For any angle, the sine and the cosecant, the cosine and the secant, and the tangent and the cotangent are, respectively, reciprocals of each other:*

$$\csc \theta = \frac{1}{\sin \theta} = \frac{\textbf{radius}}{\textbf{ordinate}} = \frac{r}{y}, \qquad (y \neq 0),$$

$$\sec \theta = \frac{1}{\cos \theta} = \frac{\textbf{radius}}{\textbf{abscissa}} = \frac{r}{x}, \qquad (x \neq 0),$$

$$\cot \theta = \frac{1}{\tan \theta} = \frac{\textbf{abscissa}}{\textbf{ordinate}} = \frac{x}{y}. \qquad (y \neq 0).$$

Now that you know the values of the sin, cos, and tan of many special angles you can quickly find the csc, sec, and cot of these same angles by writing their respective reciprocals. For example, because $\sin 30° = \frac{1}{2}$, $\csc 30° = \frac{2}{1}$, etc.

Examples. (See Fig. 5-2.)

1. $$\csc 30° = \frac{1}{\sin 30°} = \frac{1}{1/2} = \frac{2}{1},$$

2. $$\sec 30° = \frac{1}{\cos 30°} = \frac{1}{\frac{1}{2}\sqrt{3}} = \frac{2}{\sqrt{3}},$$

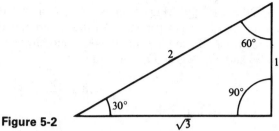

Figure 5-2

3.
$$\cot 30° = \frac{1}{\tan 30°} = \frac{1}{1/\sqrt{3}} = \frac{\sqrt{3}}{1}.$$

More Examples. (See Fig. 5-3.)

4. $\cos \alpha = \frac{4}{5}$; $\sec \alpha = \frac{5}{4}$.
5. $\tan \alpha = \frac{3}{4}$; $\cot \alpha = \frac{4}{3}$.
6. $\sin \alpha = \frac{3}{5}$; $\csc \alpha = \frac{5}{3}$.

Note that the absolute value of sec α and of csc α is each greater than or equal to 1.

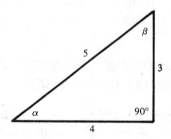

Figure 5-3

Special cases. Let θ be an acute angle of a right triangle (Fig. 5-1). Let this triangle be placed in the first quadrant with θ in standard position. Let the point (x, y) be at the end of the hypotenuse in the first quadrant. With the triangle in this position, x becomes the side of the triangle adjacent to θ, and y the side opposite θ. Then, from the definitions of cosecant, secant, and cotangent of an angle,

$$\csc \theta = \frac{r}{y} = \frac{\text{hypotenuse}}{\text{side opposite } \theta},$$

$$\sec \theta = \frac{r}{x} = \frac{\text{hypotenuse}}{\text{side adjacent to } \theta},$$

$$\cot \theta = \frac{x}{y} = \frac{\text{side adjacent to } \theta}{\text{side opposite } \theta}.$$

When $P(x, y)$ (Fig. 5-1) is not on a coordinate axis, r is always the hypotenuse of a right triangle and thus is greater than $|x|$ or $|y|$. When $P(x, y)$ is on any one of the coordinate axes, then either $r = |x|$ or $r = |y|$. Therefore, $r \geqq |x|$ and $r \geqq |y|$, and as a result,

$$|\csc \theta| = \left|\frac{r}{y}\right| \geqq 1,$$

and

$$|\sec \theta| = \left|\frac{r}{x}\right| \geqq 1.$$

As $|y| \to 0$, $|\csc \theta| \to \infty$ (read "as the absolute value of y approaches zero, $|\csc \theta|$ increases without limit").

Sin θ and cos θ exist for all values of θ, but the other four trigonometric functions of θ have limitations.

A fraction has no meaning and is undefined if its denominator is zero. For a quadrantal angle the terminal side lies on one of the coordinate axes, and either x or y is zero. When $x = 0$, tan θ and sec θ are therefore undefined, and when $y = 0$, cot θ and csc θ are undefined. Consequently,

tan θ *and* sec θ *are undefined for* 90°, 270°, *and their coterminal angles*;
cot θ *and* csc θ *are undefined for* 0°, 180°, *and their coterminal angles*.

With these exceptions, the domain of definition of the trigonometric functions includes all angles θ.

5-2 THE SIGNS OF THE COTANGENT, SECANT, AND COSECANT OF AN ANGLE

Inverting a fraction does not change its sign; therefore, a trigonometric function of any angle and the reciprocal function of the same angle have the same sign.

In Figure 5-4 we show the quadrants in which the respective functions are positive or negative. Thus, *sin θ and csc θ are positive in the quadrants above*

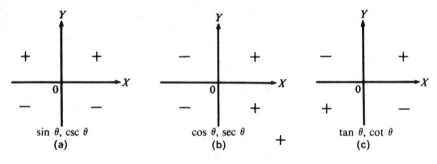

Figure 5-4

the x-axis; cos θ and sec θ are positive in the quadrants to the right of the y-axis; and tan θ and cot θ are positive in the first and third quadrants.

5-3 TRIGONOMETRIC FUNCTIONS OF COMPLEMENTARY ANGLES

Let ABC be a right triangle with a and b the lengths of the sides opposite the acute angles A and B, and let c be the length of the side opposite the right angle C.

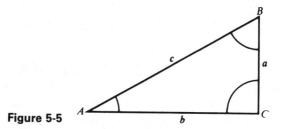

Figure 5-5

Two angles are said to be **complementary** if their sum is equal to 90°. Thus, in the right triangle ABC (Fig. 5-5), A and B are complementary angles, since $A + B = 90°$.

The following are cofunctions each to the other:

sine and **co**sine
tangent and **cot**angent
secant and **co**secant

Theorem. *Any trigonometric function of an acute angle is equal to the corresponding cofunction of its complementary angle.* Thus, sin 80° = cos 10°, tan 30° = cot 60°, sec 20° = csc 70°, etc.

Proof. If A and B are any two complementary acute angles, they can be taken as the acute angles of the triangle ABC (Fig. 5-5). We use the definitions of the trigonometric functions of an acute angle in a right triangle, namely,

$$\sin \theta = \frac{\text{side opposite } \theta}{\text{hypotenuse}}, \qquad \cos \theta = \frac{\text{side adjacent to } \theta}{\text{hypotenuse}}, \text{ etc.,}$$

to show that

$$\sin A = \frac{a}{c} = \cos B, \qquad \cos A = \frac{b}{c} = \sin B,$$

$$\tan A = \frac{a}{b} = \cot B, \qquad \cot A = \frac{b}{a} = \tan B,$$

$$\sec A = \frac{c}{b} = \csc B, \qquad \csc A = \frac{c}{b} = \sec B.$$

Example 1.   Find the acute angle x for which

$$\sec x = \csc 23° \; 15'.$$

Solution: Since secant and cosecant are cofunctions, the statement $\sec x = \csc 23° \; 15'$ implies

$$x + 23° \; 15' = 90° \qquad \text{or} \qquad x = 89° \; 60' - 23° \; 15'.$$

Therefore,

$$x = 66° \; 45'.$$

Example 2.   Find the acute angle x for which

$$\cos \left(\frac{3}{2}x + 20° \right) = \sin (50° - x).$$

Solution: For acute angles, the statement $\cos (\frac{3}{2}x + 20°) = \sin (50° - x)$ implies,

$$(\tfrac{3}{2}x + 20°) + (50° - x) = 90°.$$

Thus, $$\tfrac{1}{2}x + 70° = 90°,$$

$$\tfrac{1}{2}x = 20°.$$

Therefore,

$$x = 40°.$$

The cofunction relations hold true for any two angles whose sum is 90°. For example,

$$\sin (-30°) = \cos 120° \qquad (-30° + 120° = 90°),$$

$$-\frac{1}{2} = -\frac{1}{2}.$$

Example 3.   Solve $\sec^3 \theta + \sec^2 \theta - 2 \sec \theta = 0$ for θ where $0° \leq \theta < 360°$.

Solution: Factor $\sec^3 \theta + \sec^2 \theta - 2 \sec \theta = 0$,

$$(\sec \theta)(\sec^2\theta + \sec \theta - 2) = 0,$$

$$(\sec \theta)(\sec \theta + 2)(\sec \theta - 1) = 0.$$

Then equate each factor to zero and solve for θ.

$\sec \theta = 0$	$\sec \theta + 2 = 0$	$\sec \theta - 1 = 0$,
not possible	$\sec \theta = -2$	$\sec \theta = 1$.
$(\sec \theta \geqq 1)$		

Thus, $\theta = 120°, 240°$ $\theta = 0°$.

Therefore, $\theta = 0°, 120°,$ and $240°$.

Exercises 5-3

In Exercises 1–20 find the value of each term and then combine the values.

1. $\csc 30° + \sec 60° + \cot 45°$.

2. $\sec 30° + \csc 60° + \cot 90°$.

3. $\csc 90° + \sec 0° + \cot 270°$.

4. $\cot(-45°) - \sec(-60°) - \csc(-30°)$.

5. $\sec(-180°) - \csc(-90°) - \cot(-270°)$.

6. $\sec^2 30° - \tan^2 30°$.

7. $\sec^2 60° - \tan^2 60°$.

8. $\csc^2 45° - \cot^2 45°$.

9. $\csc^2 135° - \cot^2 135°$.

10. $\sin 150° \csc 150° + \sec 60° \cos 60°$.

11. $\tan 135° \cot 135° + \csc 210° \sin 210°$.

12. $(\sec 150°)(\tan 150° + \cot 150°)$.

13. $(\csc 60°)(\cot 60° + \tan 60°)$.

14. $\sin 720° + \cot 225° - \sec^2 225°$.

15. $\cot(-30°) - \sec(-60°) - \csc(-30°)$.

16. $\csc(-150)° - \cot(-135°) - \sec(-300°)$.

17. $2 \csc^2 60° + 4 \sec^2 45° + 6 \cot 135°$.

18. $4 \cot^2 120° + 2 \csc^2 150° - 6 \sec^2 210°$.

19. $3 \tan 135° + 2 \sec^2 135° - 2 \cot 225°$.

20. $2 \sec^2 60° \csc^3 45° - 2 \sec^3 45° \sin^2 30°$.

Each of the following points is on the terminal side of an angle θ, in standard position. Find the six trigonometric functions of θ when they exist.

21. (4, 3).   22. (12, 5).   23. (−12, 5).
24. (−4, 3).   25. (4, −3).   26. (5, −12).
27. (−12, −5).   28. (−4, −3).   29. (−3, 2).
30. (−2, 3).   31. (−2, 0).   32. (0, −2).

In Exercises 33–38 name the quadrant in which θ must terminate in order to satisfy the indicated conditions.

33. $\cos \theta$ is negative and $\tan \theta$ is positive.

34. $\csc \theta$ is positive and $\cos \theta$ is negative.

35. $\cot \theta$ is negative and $\cos \theta$ is positive.

36. $\sec \theta$ is positive and $\cot \theta$ is negative.

37. $\cot \theta$ is negative and $\sin \theta$ is positive.

38. $\tan \theta$ is positive and $\csc \theta$ is negative.

In Exercises 39–48 identify each as being possible or impossible.

39. $\sin \theta = 2$.   40. $\tan \theta = 0$.
41. $\cot \theta = 500$.   42. $\sec \theta = .500$.
43. $\csc \theta = .313$.   44. $\cos \theta = 2$.
45. $\tan \theta = .001$.   46. $\csc \theta = .614$.
47. $\cot \theta < 0$ and $\sin \theta < 0$.   48. $\tan \theta < 0$ and $\csc \theta < 0$.

In Exercises 49–56 state whether θ is close to $0°$ or close to $90°$. ($0° \leq \theta \leq 90°$.)

49. $\cos \theta = .99$.   50. $\sin \theta = .99$.   51. $\csc \theta = 1.01$.
52. $\sec \theta = 1.01$.   53. $\sin \theta = .01$.   54. $\cos \theta = .01$.
55. $\tan \theta = .01$.   56. $\tan \theta = 45$.

In Exercises 57–62 find the acute angle θ for which each of the following statements is true.

57. $\sin \theta = \cos 15° \, 10'$.   58. $\cos \theta = \sin 89°$.
59. $\cot \theta = \tan (2\theta + 30°)$.   60. $\tan \theta = \cot (50° - \frac{1}{2}\theta)$.
61. $\sec (\frac{1}{4}\theta + 25°) = \csc (5° + \frac{1}{2}\theta)$.   62. $\csc (\theta - 40°) = \sec (\frac{1}{4}\theta + 55°)$.

In Exercises 63–66 prove each statement. Do not use tables.

63. $\tan 89° = \dfrac{1}{\tan 1°}$.   64. $\sec 87° = \dfrac{1}{\sin 3°}$.

65. $\cos 87° = \dfrac{1}{\text{) } \csc 3°}.$ **66.** $\sin 88° = \dfrac{1}{\sec 2°}.$

5-4 FINDING THE VALUES OF COSECANT, SECANT, AND COTANGENT OF ANGLES ON THE CALCULATOR

If you have the sin, cos, and tan keys on your calculator, you can use your $1/x$ key to convert them to the respective csc, sec, and cot values.

Example 1. Find csc 35°.

KEY OPERATION   DISPLAY

$\boxed{\text{C}}$ 3 5 $\boxed{\sin}$ $\boxed{1/x}$ $\boxed{1.743446796}$

Example 2. Find sec 156°.

$\boxed{\text{C}}$ 1 5 6 $\boxed{\cos}$ $\boxed{1/x}$ $\boxed{-1.094636279}$

Example 3. Find cot 284°.

$\boxed{\text{C}}$ 2 8 4 $\boxed{\tan}$ $\boxed{1/x}$ $\boxed{-2.493280028}$

Example 4. Find the acute angle θ for which $\csc \theta = 1.74344796$. (This is the inverse of Example 1.)

$\boxed{\text{C}}$ 1 $\boxed{\cdot}$ 7 4 3 4 4 6 7 9 6 $\boxed{1/x}$ $\boxed{\text{INV}}$ $\boxed{\sin}$ $\boxed{34.99999999}$

Use your calculator to verify the following answers. The answers have been rounded off to five significant digits.

1. $\csc 74° = 1.0403.$
2. $\sec 39° = 1.2868.$
3. $\cot 52° = .78129.$
4. $\csc 137° = 1.4663.$
5. $\sec 156° = -1.0946.$
6. $\cot 175° = -11.430.$
7. $\csc 1234° = 2.2812.$
8. $\cot 256.4° = .24193.$
9. Given $\sec \theta = 2.5674$ where $0° < \theta < 90°$, find θ.
   Answer: $67.077°$.
10. Given $\csc \theta = -1.6403$ where $-90° < \theta < 0°$, find θ.
    Answer: $-37.564°$.

5-5 GRAPH OF THE COTANGENT FUNCTION

The graph of the cotangent function always slopes downward toward the right except at points where the function is not defined. The cotangent function is undefined for 0° and 180° and for angles coterminal with those angles.

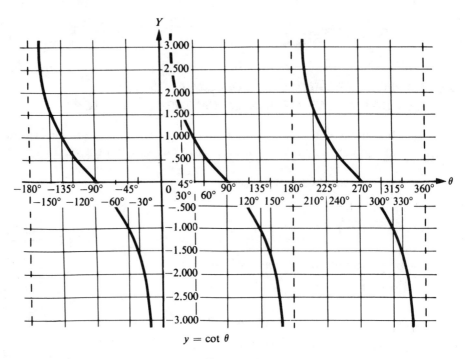

$$y = \cot \theta$$

Figure 5-6

The student is asked to form a table showing the values, if they exist, of cot θ for θ = 0°, ±30°, ±45°, ±60°, ±90°, ±120°, ±135°, ±159°, ±180°, ±210°, ±225°, ±240°, ±270°, ±300°, ±315°, ±330°, ±360°, and with these values to verify the graph of $y = \cot \theta$ as shown in Figure 5-6.

5-6 GRAPH OF THE COSECANT FUNCTION

The graph of $y = \csc \theta$ can be obtained from that of $y = \sin \theta$, of which it is the reciprocal: $y = \csc \theta = 1/\sin \theta$. When $\sin \theta = \pm 1$, csc θ has the same value. When $\sin \theta = \pm\frac{1}{2}$, $\csc \theta = \pm 2$. As θ tends toward 0°, or any whole multiple of 180°, |csc θ| becomes infinite, that is, increases without limit.

Draw the graph of $y = \sin \theta$, and construct vertical lines at the points where it crosses the θ-axis. These vertical lines separate the branches of the graph of $y = \csc \theta$. (See Fig. 5-7.)

A brief table of points on the $y = \csc \theta$ curve is given below.

θ	−180°	−150°	−90°	−30°	0°	30°	90°	150°	180°	210°	270°	330°	360°
y	U	−2	−1	−2	U	2	1	2	U	−2	−1	−2	U

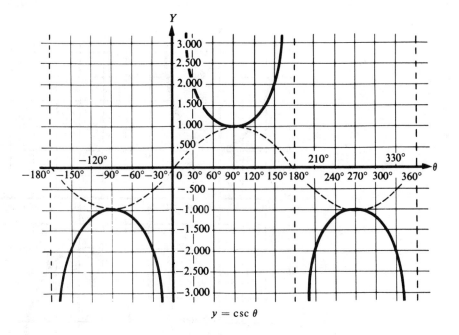

$$y = \csc \theta$$

Figure 5-7

5-7 GRAPH OF THE SECANT FUNCTION

The graph of $y = \sec \theta$ is obtained from that of its reciprocal, $y = \cos \theta$, in a manner similar to that used to obtain the graph of $y = \csc \theta$: $y = \sec \theta = 1/\cos \theta$.

Draw the graph of $y = \cos \theta$, and construct vertical lines at the points where it crosses the θ-axis. These vertical lines separate the branches of the $y = \sec \theta$ curve. (See Fig. 5-8.)

A brief table of points on the $y = \sec \theta$ curve is given below.

θ	$-90°$	$-60°$	$0°$	$60°$	$90°$	$120°$	$180°$	$240°$	$270°$
y	U	2	1	2	U	-2	-1	-2	U

You should practice drawing the six basic trigonometric curves until you can make a quick rough sketch of each from memory.

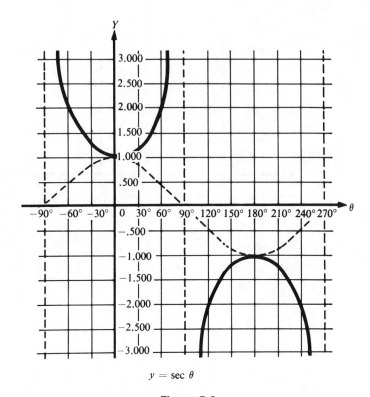

$$y = \sec \theta$$

Figure 5-8

Exercises 5-7

Graph each of the following functions in the indicated interval:

1. $y = \frac{1}{3}\cot \theta$ from $\theta = -180°$ to $360°$.

2. $y = 1 + \frac{1}{2}\cot \theta$ from $\theta = -180°$ to $360°$.

3. $y = 1 + \sec \theta$ from $\theta = -90°$ to $270°$.

4. $y = 1 + \csc \theta$ from $\theta = -180°$ to $360°$.

5. $y = -\sec \theta$ from $\theta = -90°$ to $270°$.

6. $y = -\csc \theta$ from $\theta = -180°$ to $360°$.

7. $y = 2 \csc \theta$ from $\theta = -180°$ to $360°$.

8. $y = 2 \sec \theta$ from $\theta = -90°$ to $270°$.

9. $y = \csc 2\theta$ from $0°$ to $180°$.

10. $y = \sec 2\theta$ from $-45°$ to $135°$.

11. $y = \cot 2\theta$ from $0°$ to $90°$.

12. $y = \cot \frac{1}{2}\theta$ from $-360°$ to $360°$.

REVIEW EXERCISES

In Exercises 1–4 find the value of each expression. Do not use tables.

1. (a) $\cot 135°$. (b) $\sec(-60°)$. (c) $\csc 150°$. (d) $\sec 135°$.

2. (a) $\sec 180°$. (b) $\csc 90°$. (c) $\cot 90°$. (d) $\sec(-45°)$.

3. (a) $\csc^2 135° - \cot^2 135°$. (b) $\sec^2 240° - \tan^2 240°$.

4. (a) $\dfrac{1}{\csc^2 30°} + \dfrac{1}{\sec^2 30°}$. (b) $\sec^2 360° + \cot^2 45°$.

5. Let θ be an angle in standard position with its termial side passing through the point $(3, -4)$.
   (a) Find $\sec \theta$. (b) Find $\csc \theta$. (c) Find $\cot \theta$.

In Exercises 6–9 name the quadrant in which θ must terminate to satisfy the following indicated conditions.

6. $\csc \theta > 0$, $\cot \theta < 0$.

7. $\csc \theta < 0$, $\sec \theta > 0$.

8. $\cot \theta > 0$, $\csc \theta < 0$.

9. $\sec \theta > 0$, $\cot \theta > 0$.

In Exercises 10–13 identify each as being possible or impossible.

10. $\sec \theta = \frac{1}{2}$.   11. $\csc \theta = 2$.

12. $\cot \theta = .2867$.   13. $\csc \theta = .4167$.

14. Solve $\tan(\theta + 55°) = \cot(\frac{1}{2}\theta + 5°)$ for θ where $0° < \theta < 90°$.

15. Solve $\csc(\frac{1}{2}\theta + 71°) = \sin(\theta + 10°)$ for θ where $0° < \theta < 90°$.

16. Solve $\sec \theta = -1$ for θ where $0° \leqq \theta < 360°$.

17. Solve $\cot \theta = 1$ for θ where $0° \leqq \theta < 360°$.

18. Solve $\sec^2 \theta - 3 \sec \theta + 2 = 0$ for θ were $0° \leqq \theta < 360°$.

19. Prove $\cos 86° = \dfrac{1}{\csc 4°}$.

20. Graph $y = \csc \theta$ from $\theta = 0°$ to $360°$.

21. Graph $y = 1 + \cot \theta$ from $\theta = 0°$ to $360°$.

CHAPTER 5: DIAGNOSTIC TEST

The purpose of this test is to see how well you understand reciprocal functions. We recommend that you work this test before your instructor tests you on this chapter. Allow yourself approximately 45 minutes to do the test.

Solutions to the problems, together with section references, are given in the Answer Section at the end of this book. We suggest that you study the sections referred to for the problems you do incorrectly.

In Problems 1 and 2 find the value of each expression. Do not use tables.

1. (a) $\csc 210°$.   (b) $\cot 225°$.   (c) $\sec 300°$.   (d) $\csc (-30°)$.

2. (a) $\cot^2 60°$.   (b) $\sec^2 30°$.   (c) $\cot^2 30°$.   (d) $\csc^2 60°$.

3. Let θ be an angle in standard position with its terminal side passing through the point $(-4, -3)$.
    (a) Find $\cot \theta$.   (b) Find $\sec \theta$.   (c) Find $\csc \theta$.

4. Name the quadrant in which $\sec \theta > 0$ and $\csc \theta < 0$.

5. Identify the following as being possible or impossible.
    (a) $\csc \theta = .6896$.   (b) $\cot \theta = .5364$.
    (c) $\sec \theta = 10$.   (d) $\sin \theta = 5$.

6. Solve $\sin (\tfrac{3}{4}\theta + 12°) = \cos (\theta + 8°)$ for θ where $0° < \theta < 90°$.

7. Solve $\csc^2 \theta - \csc \theta - 2 = 0$ for θ where $0° \leq \theta < 360°$.

8. Graph $y = \sec \theta - 1$ from $\theta = -90°$ to $270°$.

6

Radian Measure of Angles

6-1 THE RADIAN

The two most common ways of measuring angles are degree measure and **radian** measure. You are already familiar with the system of degree measure. Many of the formulas in calculus and science are simplified by the use of radian measure. For that reason radian measure is introduced in trigonometry.

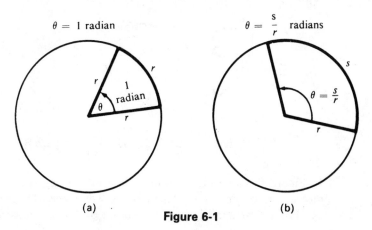

(a)   (b)

Figure 6-1

113

If an angle is placed with its vertex at the center of a circle, it subtends on the circle an arc whose length is proportional to the size of the angle (Fig. 6-1). This gives us a way of defining units for measuring angles.

The unit of measure in the degree system and in the radian system can both be defined by relating an arc of a circle to its central angle. An arc $\frac{1}{360}$ of the circumference of a circle subtends a central angle in the circle of one degree.

(1)

> **Radian defined.** *A **radian** is an angle which, if its vertex is placed at the center of a circle, subtends on the circle an arc equal to the radius of the circle (Fig. 6-1a). Thus, the radian measure θ of a central angle of a circle is the ratio of the intercepted arc s to the radius r,*
>
> $$\theta = \frac{s}{r}.$$

As a consequence of the definition of a radian, the number of radians in 360° is equal to the number of times the radius of a circle can be laid off along the circumference of that circle. The length of the circumference $= 2\pi r$. Therefore, the central angle (θ) which intercepts 360° of arc contains

$$\frac{c}{r} = \frac{2\pi r}{r} = 2\pi \text{ radians} \approx 6.28 \text{ radians.} \qquad \text{(Fig. 6-2.)}$$

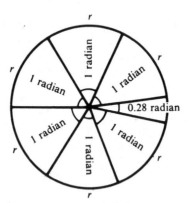

A complete circle has
2π (approximately 6.28) radians.   **Figure 6-2**

In Figure 6-3a we placed a unit circle on the xy-coordinate system with its center at the origin. A stretched string was placed tangent to the circle at point A. Point A was used as the zero point on the string, and using the radius of the circle as a unit, we marked off positive and negative lengths along the string. Then holding the center of the string fixed at A, we wrapped the positive part of the string counterclockwise and the negative part clockwise around the unit circle (Fig. 6-3b). This gave us arc lengths on the unit circle of 1 radian, $\frac{\pi}{2}$ radians, 2 radians, etc. Because the circumference of the unit circle is 2π, a positive length of 2π, on the string, completely wraps the circle one time, laying off an arc length of 2π radians which is equivalent to 360°.

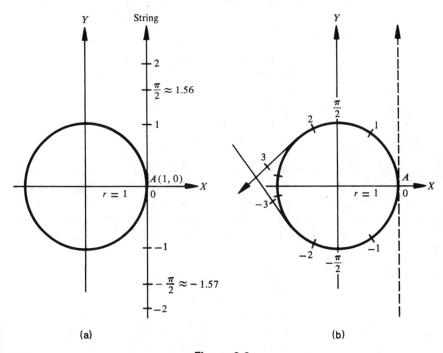

(a)   (b)

Figure 6-3

Thus, 2π radians $= 360°$;

therefore,

(2)

$$\pi \text{ radians} = 180°.$$

and

(3)

$$1 \text{ radian} = \frac{180°}{\pi} \approx 57.2958° \approx 57° \; 17.75'.$$

Since $180° = \pi$ radians,

(4)
$$1° = \frac{\pi}{180} \approx .017453 \text{ radian.}$$

The relation $(180° = \pi$ radians) is an important relation to remember because it can be used for deriving equivalent radian and degree measures.

Examples.

1. $180° = \pi$ radians.
2. $90° = \frac{1}{2}(180°) = \pi/2$ radians.
3. $60° = \frac{1}{3}(180°) = \pi/3$ radians.
4. $45° = \frac{1}{4}(180°) = \pi/4$ radians.
5. $30° = \frac{1}{6}(180°) = \pi/6$ radians.
6. $270° = 3(90°) = 3\pi/2$ radians.

Since they will be used often in this and other related subjects, you should memorize these radian-degree equivalences; to assist in memorizing these equivalences, it is helpful to place them on a circle, as shown in Figure 6-4. The radian measures of some of these angles are usually read "π over 6, π over 4, π over 2, 2π over 3," and the like.

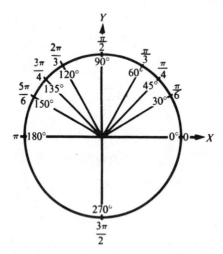

Figure 6-4

When no other unit of angular measure is indicated, it is assumed that the angle is expressed in radian measure. For example, if we write sin 1.5, the 1.5 is understood to mean 1.5 radians.

Example 7. Find tan 1.60.

Solution: The 1.60 is understood to be 1.60 radians. From Table 3 we find that tan 1.60 = −34.233.

Example 8. Express 37.2° in radians.

First Solution: By Formula (4), 1° ≈ .017453 radian. Therefore,

$$37.2° = 37.2 × .017453 \text{ radians}$$

$$= .649 \text{ radians (to three decimal places).}$$

Second Solution: From Table 4,

$$37° = .64577 \text{ radian,}$$

$$(.2)° = (.2 × 60)' = 12' = \underline{.00349} \text{ radian.}$$

Therefore,   37.2° = .64926, or 0.649 radian

(to three decimal places).

Example 9. Express 2.45 radians in degree measure accurate to the nearest tenth of a minute.

First Solution: By Formula (3), 1 radian ≈ 57.2958°. Therefore,

2.45 radians = 2.45 × 57.2958° = 140.37471° = 140° + (.37471 × 60)′

$$= 140° \, 22.5'.$$

Second Solution: From Table 3,

$$1.45 \text{ radians} = \quad 83° \, 04.7', \text{ and}$$

$$\underline{1.00 \text{ radian} \quad = \quad 57° \, 17.7'.}$$

Therefore,   2.45 radians = 140° 22.4′.

The slight difference in the results of the first and second solutions is due to the rounded off conversion numbers used. More accurately, 1 radian = 57.2957795°.

When Table 3 is used, it is assumed that the numbers in the radian column are exact, and the numbers in the degree column are rounded to the nearest tenth of a minute.

Example 10. Find the value of sin $\frac{1}{6}\pi$.

Solution: $\frac{1}{6}\pi$ is assumed to be in radian measure and is equivalent to 30°. Therefore, sin $\frac{1}{6}\pi$ = .5000.

Example 11. Find the value of cos 1.243.

Solution: 1.243 is assumed to be in radian measure. We use Table 3 and find that the value of cos 1.243 lies between the table entries for cos 1.24 and cos. 1.25. We interpolate as follows:

$$\left.\begin{array}{l} \cos 1.24 \; = .32480 \\ \cos 1.243 = \\ \cos 1.25 \; = .31532 \end{array}\right\} .00948.$$

Therefore, cos 1.243 = .32480 − .3(.00948) ≈ .32196.

Example 12.   Find sin 4.14.

Solution: Since 4.14 radians is not given in the table, we must determine the quadrant and solve for the related angle. $180° = \pi \approx 3.14$ radians. Therefore, 4.14 radians is a third quadrant angle. To find the related angle, subtract $\pi \approx 3.14$ from 4.14.

$$4.14 - 3.14 = 1.00 \text{ (related angle).}$$

Therefore, sin 4.14 = −sin 1.00 = −.84147.

Because the given number (4.14) of Example 12 is a three significant digit number and we used a three significant digit number for π, we cannot expect the answer to be accurate to more than three significant digits. Using a hand calculator we read sin 4.14 = −.840609436 ≈ −.841. As you can see, the answer shown in Example 12 would round off to this same number.

Exercises 6-1

The following angles are given in radian measure. Express them in degree measure either exactly or to the nearest tenth of a minute. Draw each angle, indicating its approximate magnitude by a curved arrow. Use Table 3 when necessary.

1. $\pi/3$.

2. $\pi/4$.

3. $\pi/2$.

4. $\frac{2}{3}\pi$.

5. $\frac{5}{6}\pi$.

6. π.

7. 2π.

8. $\frac{3}{2}\pi$.

9. $\frac{7}{6}\pi$.

10. $\frac{4}{5}\pi$.

11. $\frac{5}{3}\pi$.

12. $\frac{11}{6}\pi$.

13. $\frac{5}{2}\pi$.

14. $\frac{13}{6}\pi$.

15. 1.15.

16. 1.03.

17. 2.03.

18. 2.14.

19. .758.

20. .109.

Assume the following angles to be exact and express each in radian measure either exactly in terms of π or to four decimal places. Use Table 4 when necessary.

21. 30°.	22. 60°.	23. 45°.	24. 90°.
25. 150°.	26. 135°.	27. 225°.	28. 300°.
29. 270°.	30. 240°.	31. 330°.	32. 315°.
33. 18°.	34. 10°.	35. 75° 15'.	36. 47° 35'.
37. 14° 57'.	38. 21° 53'.	39. 125° 14.7'.	40. 135° 33.6'.

In Exercises 41–50 find the value of each term, and then combine the values.

41. $\sin \pi/6 + \tan^2 \pi/3 + \cos \pi/2 + \sec \pi/3$.

42. $\cot^2 \pi/6 + \sec \pi/2 + \sin 2\pi + \cos \pi$.

43. $\csc \pi + \sin 2\pi + \cos \pi/2 - \tan \pi/4$.

44. $\cos^2 \frac{5}{6}\pi + \sin^2 \frac{5}{6}\pi + \tan \frac{3}{4}\pi - \cot \frac{5}{4}\pi$.

45. $\sin (-\pi/2) + \cos (-\pi/3) - \sec^2 \frac{5}{4}\pi + \csc \frac{11}{6}\pi$.

46. $\tan (-\pi/4) + \cot (-\frac{3}{4}\pi) + \sin^2 \frac{13}{6}\pi - \cos (-\pi)$.

47. $\sqrt{3} \tan \frac{5}{6}\pi + \sqrt{2} \sin \frac{3}{4}\pi + 4 \sec^2 \frac{5}{4}\pi$.

48. $\sqrt{2} \cos (-\pi/4) - \sqrt{3} \tan \frac{4}{3}\pi - \csc^2 (-\frac{5}{4}\pi)$.

49. $\sin^2 \frac{7}{3}\pi + \cos^2 \frac{7}{3}\pi - \tan (-\frac{3}{4}\pi)$.

50. $\sec^2 \pi/4 - \tan^2 \pi/4 - \sin^2 \frac{15}{4}\pi$.

In Exercises 51–58 use Table 3 to find the value of each function.

51. $\sin 1.47$.	52. $\cos 1.19$.
53. $\tan .43$.	54. $\sin .67$.
55. $\cos 4.54$.	56. $\tan 2.86$.
57. $\sin .272$.	58. $\sin 1.102$.

6-2 ARCS AND ANGLES

Let s be the length of a circular arc that is intercepted by a central angle θ on a circle whose radius is r. Then, using Formula (1), Section 6-1,

$$\theta = \frac{s}{r},$$

we obtain

---

arc length

(1) $s = r\theta.$

That is,

       arc length = radius $\times$ central angle in radians.

(Note, s and r may be measured in any unit of length, but they must be
expressed in the same unit.)

Example 1.   To the nearest tenth of an inch, find the length of an arc of
a circle of radius 10 inches, which is intercepted by a central angle of
125°.

Solution: We must convert 125° to radian measure, which can be done
by using either Formula (4) Section 6-1, or Table 4. By Formula (4),
$1° \approx .017453$ radians. Therefore,

$$125° = 125 \times .017453 = 2.181625 \approx 2.18 \text{ radians.}$$

Or from Table 4, we find

$$90° = 1.57080 \text{ radians,}$$

$$35° = \quad .61087 \text{ radian,}$$

Thus, $\overline{125° = 2.18167} \approx 2.18$ radians.

Therefore, the arc length s is

$$s = r\theta = 10 \times 2.18 \text{ inches} = 21.8 \text{ inches (to the nearest}$$
$$\text{tenth of an inch).}$$

Example 2.   As a wheel of 20-inch diameter rotates, a point on the rim
travels a linear distance of 200 inches each second.

(a) Find the angular velocity in radians and in degrees per second.
(b) How many revolutions does the wheel make each second?

Solution:
(a) *The* **angular velocity** *is the* **angular displacement** *of any radius of the
wheel, relative to some fixed direction, during a particular unit of time.* It
can, for example, be expressed in terms of degrees per second, revolutions per minute, or similar units.

    Using Formula (1), Section 6-1, we have

$$\theta = \frac{s}{r} = \frac{200}{10} = 20 \text{ radians per second.}$$

CONTINUE

Now,

$$1 \text{ radian} \approx 57.2958°;$$

therefore, 20 radians = $(20 \times 57.2958)° = 1145.9°$ (accurate to the nearest tenth of a degree).

(b) *One revolution is made when the wheel rotates through 2π radians.* Therefore, the number of revolutions (R) can be found by the formula

(2) $$R = \frac{\theta}{2\pi}.$$

(Note, R and θ may be any unit of time, but they must be expressed in the same unit of time. θ must be expressed in radians.)

$$R = \frac{\theta}{2\pi} = \frac{20}{2\pi} = 3.18 \text{ revolutions per second.}$$

Example 3.   Find the height of a tower that stands 1 mile away, on a horizontal plane, if the angle of elevation of the top of the tower is 1° 24′.

Solution: *When the central angle is relatively small, the length of the intercepted arc may be taken as a close approximation of the length of its chord* (Fig. 6-5).

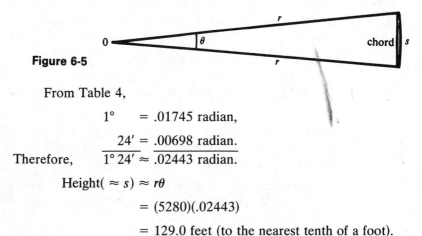

Figure 6-5

From Table 4,

$$1° \quad = .01745 \text{ radian,}$$

$$\overline{24' = .00698 \text{ radian.}}$$

Therefore, $\overline{1° 24' \approx .02443 \text{ radian.}}$

$$\text{Height}(\approx s) \approx r\theta$$

$$= (5280)(.02443)$$

$$= 129.0 \text{ feet (to the nearest tenth of a foot).}$$

Exercises 6-2

In the following examples assume the data to be exact and give answers to the nearest unit.

1. Find the number of radians in the central angle that subtends an arc of 8 inches on a circle of diameter 2 feet. Express the angle in radians, accurate to the nearest hundredth of a radian.

2. A pendulum 25 inches long oscillates 3° 30′ on each side of its vertical position. Find the length of arc through which the end of the pendulum swings.

3. The hour hand of a clock is 5 inches long. Through what distance does the tip of the hour hand travel in 55 minutes?

4. The latitude of Los Angeles, California, is 34° 03′. Find the surface (arc) distance from Los Angeles to the equator. Assume the earth to be a sphere of diameter 7920 statute miles.

5. Find the number of feet in a nautical mile. A nautical mile is the length of arc on a great circle of the earth which is subtended by an angle of 1′ at the center of the earth. Assume the earth to be a sphere of diameter 7920 statute miles (1 statute mile = 5280 feet).

6. From a distance 1000 feet a building subtends an angle of 5°. Find the approximate height of the building.

7. As viewed from the earth, the sun subtends an angle of approximately 32′. If the sun is 93,000,000 miles from the earth, find the diameter of the sun.

8. An automobile is traveling 60 miles per hour. The effective radius of the wheels is 13 inches. Find (a) the rate at which the wheels turn (angular speed) in radians per second; (b) the revolutions per minute of the wheels.

9. A diesel locomotive is traveling 80 miles per hour. The diameter of its drive wheels is 36 inches. Find (a) the rate at which the wheels turn (angular speed) in radians per second; (b) the revolutions per minute of the wheels.

10. Determine the speed of a point on the equator of the earth (in miles per hour) due to the rotation of the earth. Assume the diameter of the earth at the equator to be 7927 statute miles.

11. The latitude of Los Angeles, California, is 34° 03′. Find the surface (arc) distance from Los Angeles to the north pole. Assume the earth to be a sphere of diameter 7920 statute miles.

12. As viewed from the earth, the moon subtends an angle of approximately 31′. If the moon is 240,000 miles from the earth, find the diameter of the moon.

13. The latitude of Los Angeles, California, is 34° 03′ and that of Portland, Oregon, is 45° 30′. Find, in both statute and nautical miles, the surface (arc) distance that Portland is due north of Los Angeles. Assume the earth to be a sphere of diameter 7920 statute miles. (See Exercise 5 for the definition of a nautical mile.)

14. The latitude of Chicago, Illinois, is 41° 50′ and that of Mexico City, Mexico, is 19° 26′. Find, in both statute and nautical miles, the surface (arc) distance that Chicago is due north of Mexico City. Assume the earth to be a sphere of diameter 7920 statute miles. (See Exercise 5 for definition of a nautical mile.)

15. A belt traveling at the rate of 30 feet per second drives a pulley at a speed of 600 revolutions per minute. Find the radius of the pulley.

16. The diameters of the front and rear wheels of a tractor are 30 inches and 50 inches, respectively. Find the revolutions per minute of the wheels when the tractor is traveling 10 miles per hour.

17. If an automobile wheel, with an effective radius of 13 inches, rotates at 900 revolutions per minute, what is the speed of the car in miles per hour?

18. A tower at a distance of 2500 feet subtends an angle of 5°. How high is the tower?

19. Two pulleys of diameters 10 inches and 20 inches, respectively, are connected by an open belt (not crossing). If the centers of the pulleys are 20 inches apart, find the length of the belt.

20. If the belt of Exercise 19 crossed between the pulleys, what would be its length?

6-3 AREAS OF SECTORS AND SEGMENTS OF CIRCLES

*A **sector** of a circle is a region bounded by two radii and an arc of the circle.* (See Fig. 6-6.)

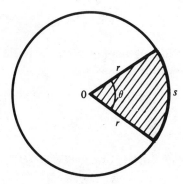

Figure 6-6

It was proved in plane geometry that the areas of two sectors of the same circle are to each other as their corresponding angles. Let K be the area of a sector in a circle of radius r whose corresponding angle is θ, expressed in radian measure. If we take the entire circle as a sector whose angle is 2π (radians), we have

$$\frac{\text{area of sector } K}{\text{area of circle}} = \frac{\text{central angle of sector}}{\text{central angle of circle}},$$

$$\frac{K}{\pi r^2} = \frac{\theta}{2\pi}.$$

Therefore,

(1)

> *area of a sector*:
>
> $$K = \frac{1}{2}r^2\theta.$$

A **segment** of a circle is a region bounded by an arc and a chord of the circle. (See Fig. 6-7.)

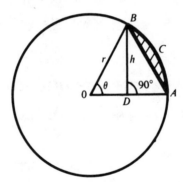

Figure 6-7

We denote the radius of the circle by r and the central angle of the arc by θ, expressed in radians.

Area of segment ACB = area of sector $OACB$ − area of triangle OAB

$$= \frac{1}{2}r^2\theta - \frac{1}{2}rh.$$

But h (the altitude of triangle OAB) = $r \sin \theta$. Therefore,

$$\text{area of segment } ACB = \frac{1}{2}r^2\theta - \frac{1}{2}r(r \sin \theta)$$

$$= \frac{1}{2}r^2\theta - \frac{1}{2}r^2 \sin \theta.$$

Hence,

(2)

$$\text{area of segment} = \frac{1}{2}r^2(\theta - \sin \theta).$$

Example. In a circle of radius 10.0 inches find the area of the sector and the segment whose central angle is 125° 00′.

Solution: Using Table 4, we find

$$90° = 1.57080 \text{ radians,}$$

$$\frac{35° = .61087 \text{ radian.}}{125° = 2.18167 \text{ radians.}}$$

Therefore,

$$\text{Area of sector} = \tfrac{1}{2}r^2\theta$$

$$= \tfrac{1}{2}(10^2)2.18167$$

$$\approx 109 \text{ square inches}$$

(to the nearest tenth of a square inch).

$$\text{Area of segment} = \tfrac{1}{2}r^2(\theta - \sin \theta)$$

$$= \tfrac{1}{2}(10^2)(2.18167 - \sin 125°)$$

$$= \tfrac{1}{2}(10^2)(2.18167 - .8192)$$

$$= 68.1 \text{ square inches}$$

(to the nearest square inch).

Exercises 6-3

In the following exercises assume the data to be exact and give answers to the nearest unit.

1. Find the areas of the sector and segment subtended by a central angle of 100° in a circle of radius 10 inches.

2. Find the areas of the sector and segment subtended by a central angle of $\pi/3$ radians in a circle of radius 12 inches.

3. Find the areas of the sector and segment subtended by a central angle of 4 radians in a circle of radius 10 inches.

4. Find the areas of the sector and segment subtended by a central angle of 280° in a circle of radius 10 inches.

5. Find the area of a circular segment whose chord is 5 inches from the center of the circle if the radius is 10 inches.

6. A cylindrical tank placed with its axis horizontal is 10 feet in diameter and 20 feet long. If this tank is partly filled with water to a depth of 4 feet, how many gallons of water are in the tank? (Take $7\frac{1}{2}$ gallons to 1 cubic foot.)

7. Find the number of gallons of water in the tank of Exercise 6 if the tank is filled to a depth of 7 feet.

8. Find the area common to two intersecting circles of radii 8 feet and 10 feet if their common chord is 10 feet long.

REVIEW EXERCISES

Assume the given numbers to be exact and round answers to table accuracy.

In Exercises 1–8 the angles are given in radian measure. Express them in degree measure either exactly or to the nearest tenth of a minute.

1. $\dfrac{3\pi}{4}$.   2. $\dfrac{2\pi}{3}$.   3. $\dfrac{\pi}{3}$.   4. $\dfrac{\pi}{6}$.

5. $2\frac{1}{2}\pi$.   6. $1\frac{1}{3}\pi$.   7. 1.25.   8. 2.34.

In Exercises 9–12 assume the given angles to be exact and express each in radian measure either exactly, in terms of π, or to five significant digits.

9. $225°$.   10. $315°$.   11. $50°\,43.5'$.   12. $84°\,15.8'$.

In Exercises 13 and 14 find the value of each term and then combine their values.

13. $\cos\dfrac{\pi}{3} + \tan\pi + \csc\dfrac{\pi}{6} + \cot\dfrac{\pi}{2}$. $\csc = \dfrac{1}{\sin\theta}$ $\sin\dfrac{\pi}{6}$ or $\sin 30° = 0.5$ or $\frac{1}{2}$ $\csc = \dfrac{2}{1}$

14. $\sin^2\dfrac{\pi}{3} + \cos^2\dfrac{\pi}{3} + \sec^2\dfrac{\pi}{4} - \tan^2\dfrac{\pi}{4}$.

In Exercises 15–18 use tables to find the value of each function.

15. $\sin\dfrac{\pi}{10}$.   16. $\cos\dfrac{\pi}{20}$.   17. $\tan 1.75$.   18. $\sin 2.55$.

19. Find the number of radians in the central angle that subtends a 7.5-inch arc on a circle that has a 10-inch diameter.

20. Find the area of a sector bounded by a 12-inch arc and 6-inch radii.

21. The southernmost city in the world, Punta Arenas, Chile, has a latitude of S $53°\,12'$. Find the surface (arc) distance from Punta Arenas to the south pole. Express this distance in both nautical and statute miles. (The latitude of the south pole is S $90°$. A minute of latitude is one nautical mile, and a nautical mile is approximately 1.152 statute miles.)

22. Find the area of a segment subtended by a central angle of 120° in a circle of radius 10 inches.

23. Find the revolutions per minute of the wheels of an automobile when it is traveling 55 miles per hour. The radius of its wheels is 12 inches.

CHAPTER 6: DIAGNOSTIC TEST

The purpose of this test is to see how well you understand the work covered in Chapter 6. We recommend that you work this test before your instructor tests you on this chapter. Allow yourself approximately 50 minutes to do the test.

Solutions to the problems, together with section references, are given in the Answer Section at the end of this book. We suggest that you study the sections referred to for the problems you do incorrectly.

Assume the given numbers to be exact and round answers to table accuracy.

---

1. The following angles are given in radian measure. Express them in degree measure either exactly or to the nearest tenth of a minute.

    (a) $1\frac{1}{2}\pi$.　　　(b) $\dfrac{\pi}{10}$.　　　(c) 4π.　　　(d) 1.5.

2. Assume the following angles to be exact and express each in radian measure in terms of π.
    (a) 135°.　　　(b) 180°.　　　(c) 300°.

3. Express 125° 41′ in radians. Use Table 4.

4. Find the value of each of the following functions. Do not use tables.

    (a) $\sin \dfrac{\pi}{6}$.　　　(b) $\cos \dfrac{2\pi}{3}$.　　　(c) $\tan \dfrac{3\pi}{4}$.

5. Use tables to find the value of each of the following functions.
    (a) sin 1.19.
    (b) cos 1.80. (Hint: Use $\pi = 3.14$.)
    (c) tan 3.28. (Hint: Use $\pi = 3.14$.)

6. Find the number of radians in the central angle that subtends an arc of 13 inches on a circle of diameter 20 inches.

7. A pendulum 30 inches long oscillates 4° 30′ on each side of its vertical position. Find the length of the arc through which the end of the pendulum swings.

8. Find the area of a sector having a central angle of 71° 15′ and radii of 10 inches.

9. Find the area of a sector having a central angle of 2π radians and radii of 10 inches.

10. Find the area of a circular segment subtended by a central angle of 1.40 radians in a circle of radius 10 inches.

7

The Fundamental Relations Between the Trigonometric Functions

7-1 THE FUNDAMENTAL RELATIONS

We recall from previous chapters that if θ is any angle in standard position (Fig. 7-1a), then

(a) $\sin \theta = \dfrac{y}{r}$,   (b) $\cos \theta = \dfrac{x}{r}$,   (c) $\tan \theta = \dfrac{y}{x}$,

(d) $\cot \theta = \dfrac{x}{y}$,   (e) $\sec \theta = \dfrac{r}{x}$,   (f) $\csc \theta = \dfrac{r}{y}$.

$P(x, y)$ is the point where the termial side of θ meets the circle with radius r and center at $(0, 0)$ (Fig. 2-4). Exceptions occur for certain quadrantal angles (see Sections 4-1 and 5-1).

Since x, y, and r are numerically equal to the sides of a right triangle,

(g) $$x^2 + y^2 = r^2,$$

and in the unit circle,

(h) $$x^2 + y^2 = 1.$$

From these eight equations we obtain eight **fundamental relations**, or **fundamental identities**, that connect the functions of any angle for which both members of the identities are defined.

Reciprocal identities:

(1) $\csc \theta = \dfrac{1}{\sin \theta}$,   or $\sin \theta = \dfrac{1}{\csc \theta}$,   or $\sin \theta \cdot \csc \theta = 1$.

(2) $\sec \theta = \dfrac{1}{\cos \theta}$,   or $\cos \theta = \dfrac{1}{\sec \theta}$,   or $\cos \theta \cdot \sec \theta = 1$.

(3) $\cot \theta = \dfrac{1}{\tan \theta}$,   or $\tan \theta = \dfrac{1}{\cot \theta}$,   or $\tan \theta \cdot \cot \theta = 1$.

Quotient identities:

(4) $\tan \theta = \dfrac{\sin \theta}{\cos \theta}$.

(5) $\cot \theta = \dfrac{\cos \theta}{\sin \theta}$.

Square identities:

(6) $\sin^2 \theta + \cos^2 \theta = 1$.

(7) $1 + \tan^2 \theta = \sec^2 \theta$.

(8) $1 + \cot^2 \theta = \csc^2 \theta$.

Proof. Because corresponding sides of similar triangles have equal ratios, we can simplify the task of proving these identities by referring to the unit circle (Fig. 7-1b).

Identities (1), (2), (3), the "reciprocal relations," are already familiar to us. To prove these, refer to Figures 7-1a and 7-1b.

(1) $\csc \theta = \dfrac{r}{y} = \dfrac{1}{y/r} = \dfrac{1}{\sin \theta}$.

(2) $\sec \theta = \dfrac{r}{x} = \dfrac{1}{x/r} = \dfrac{1}{\cos \theta}$.

(3) $\cot \theta = \dfrac{x}{y} = \dfrac{1}{y/x} = \dfrac{1}{\tan \theta}$.

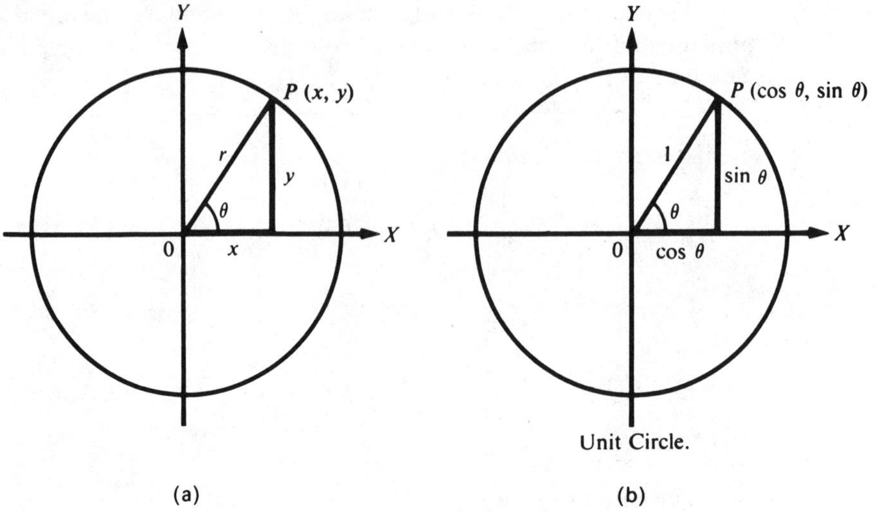

(a)   (b)

Figure 7-1

We prove Identities (4) and (5) by using the unit circle (Fig. 7-1b) and by applying the definitions of $\tan \theta$ and $\cot \theta$:

(4) $$\tan \theta = \frac{y}{x} = \frac{\text{ordinate}}{\text{abscissa}} = \frac{\sin \theta}{\cos \theta}.$$

(5) $$\cot \theta = \frac{x}{y} = \frac{\text{abscissa}}{\text{ordinate}} = \frac{\cos \theta}{\sin \theta}.$$

The proof for Identity (6) is given in Section 3-5.
To prove Identity (7), divide each term of Identity (6) by $\cos^2 \theta$.

$$\frac{\sin^2 \theta}{\cos^2 \theta} + \frac{\cos^2 \theta}{\cos^2 \theta} = \frac{1}{\cos^2 \theta},$$

which gives

$$\left(\frac{\sin \theta}{\cos \theta}\right)^2 + \left(\frac{\cos \theta}{\cos \theta}\right)^2 = \left(\frac{1}{\cos \theta}\right)^2.$$

But $\sin \theta / \cos \theta = \tan \theta$ (Identity 4) and $1/\cos \theta = \sec \theta$ (Identity 2); therefore,

$$\tan^2 \theta + 1 = \sec^2 \theta,$$

or $$1 + \tan^2 \theta = \sec^2 \theta.$$

The proof for Identity (8) can be obtained by dividing each term of Identity (6) by $\sin^2 \theta$; then simplify as was done in the proof of Identity (7).

These basic fundamental identities should be memorized at once. The student should be able to recognize them in other forms. For example,

$$\cos \theta = \frac{1}{\sec \theta},$$

$$\sin \theta = \pm\sqrt{1 - \cos^2 \theta},$$

$$\sec^2 \theta - \tan^2 \theta = 1,$$

$$\sin^2 \theta = \frac{1}{\csc^2 \theta}.$$

Example 1. Express each of the other functions in terms of $\sin \theta$.

Solution: From Identity (6) $\sin^2 \theta + \cos^2 \theta = 1$; thus,

$$\cos \theta = \pm\sqrt{1 - \sin^2 \theta}.$$

From Identity (4)

$$\tan \theta = \frac{\sin \theta}{\cos \theta} = \pm\frac{\sin \theta}{\sqrt{1 - \sin^2 \theta}}.$$

From Identity (3)

$$\cot \theta = \frac{\cos \theta}{\sin \theta} = \pm\frac{\sqrt{1 - \sin^2 \theta}}{\sin \theta}.$$

From Identity (2)

$$\sec \theta = \frac{1}{\cos \theta} = \pm\frac{1}{\sqrt{1 - \sin^2 \theta}}.$$

From Identity (1)

$$\csc \theta = \frac{1}{\sin \theta}.$$

Example 2. Use the fundamental identities to find the remaining functions when $\tan \theta = 3/4$ and $180° < \theta < 270°$.

Solution: From Identity (7)

$$\sec^2 \theta = 1 + \tan^2 \theta$$

$$= 1 + \left(\frac{3}{4}\right)^2 = 1 + \frac{9}{16} = \frac{25}{16},$$

$$\sec \theta = \pm\sqrt{\frac{25}{16}} = \pm\frac{5}{4}.$$

Therefore, $\sec \theta = \pm\frac{5}{4}$. But when $180° < \theta < 270°$, $\sec \theta$ is negative. Hence,

$$\sec \theta = -\frac{5}{4}.$$

From Identity (3)

$$\cot \theta = \frac{1}{\tan \theta} = \frac{4}{3}.$$

From Identity (2)

$$\cos \theta = \frac{1}{\sec \theta} = -\frac{4}{5}.$$

From Identity (6)

$$\sin^2 \theta + \cos^2 \theta = 1,$$

$$\sin^2 \theta + \left(-\frac{4}{5}\right)^2 = 1,$$

$$\sin^2 \theta = 1 - \frac{16}{25} = \frac{9}{25},$$

$$\sin \theta = \pm\sqrt{\frac{9}{25}} = \pm\frac{3}{5}.$$

But in the third quadrant, $\sin \theta$ is negative. Therefore,

$$\sin \theta = -\frac{3}{5}.$$

From Identity (1)

$$\csc \theta = \frac{1}{\sin \theta} = \frac{1}{-3/5} = -\frac{5}{3}.$$

Example 3. Use the fundamental identities to simplify the following expressions:

(a) $\tan \theta \cot \theta + \dfrac{\sin^2 \theta}{\cos^2 \theta}.$

Solution: From Identities (3), (4), and (7)

$$\tan \theta \cot \theta + \frac{\sin^2 \theta}{\cos^2 \theta}$$

$$= 1 \quad + \tan^2 \theta = \sec^2 \theta.$$

(b) $\dfrac{\sin \theta}{1 + \cos \theta} + \dfrac{1 + \cos \theta}{\sin \theta}$.

Solution:

$$\frac{\sin \theta}{1 + \cos \theta} + \frac{1 + \cos \theta}{\sin \theta} = \frac{\sin^2 \theta + (1 + \cos \theta)^2}{\sin \theta (1 + \cos \theta)}$$

$$= \frac{\sin^2 \theta + 1 + 2 \cos \theta + \cos^2 \theta}{\sin \theta (1 + \cos \theta)}$$

$$= \frac{\sin^2 \theta + \cos^2 \theta + 1 + 2 \cos \theta}{\sin \theta (1 + \cos \theta)}$$

$$= \frac{2 + 2 \cos \theta}{\sin \theta (1 + \cos \theta)} = \frac{2(1 + \cos \theta)}{\sin \theta (1 + \cos \theta)}$$

$$= \frac{2}{\sin \theta} = 2 \left(\frac{1}{\sin \theta} \right) = 2 \csc \theta.$$

Exercises 7-1

In each of the Exercises 1–6 use the fundamental identities to find the remaining functions of the angle satisfying the given conditions.

1. $\cos \theta = -\frac{3}{5}$, θ in second quadrant.
2. $\sin \theta = -\frac{5}{13}$, θ in fourth quadrant.
3. $\tan \theta = \frac{4}{3}$, θ in first quadrant.
4. $\sec \theta = 2$, θ in fourth quadrant.
5. $\cot \theta = \frac{5}{12}$, $\sin \theta < 0$.
6. $\csc \beta = \frac{5}{4}$, $\cos \beta > 0$.

In each of the Exercises 7–12 reduce the given expression to an equivalent expression involving only $\sin x$ and $\cos x$, and then simplify it.

7. $\dfrac{\tan x \csc x}{\sec x}$.

8. $\dfrac{\cot x \sec x}{\csc x}$.

9. $\tan x + \cot x$.

10. $\dfrac{1}{\csc^2 x} + \dfrac{1}{\sec^2 x}$.

11. $\dfrac{\csc x}{\tan x + \cot x}$.

12. $\dfrac{\sec x \csc x}{\sec^2 x + \csc^2 x}$.

In each of the Exercises 13–24 perform the indicated operations and use the fundamental identities to simplify the expressions.

13. $(1 + \tan \theta)^2 - 2 \sin \theta \sec \theta$.
14. $\sin \theta \csc \theta + \cos^2 \theta + \sec \theta \cos \theta + \sin^2 \theta$.
15. $\csc^2 \theta (\sec \theta - \cos \theta)$.

16. $\sin \theta \sec \theta \cot \theta$.

17. $(\sin \theta + \cos \theta)^2 - (\sin \theta - \cos \theta)^2$.

18. $\dfrac{1 + \tan^2 \theta}{\csc^2 \theta}$.

19. $\dfrac{1}{\sin \theta - 1} - \dfrac{1}{\sin \theta + 1}$.

20. $\dfrac{\sin \theta}{\csc \theta} + \dfrac{\cos \theta}{\sec \theta}$.

21. $\cot \theta + \dfrac{\sin \theta}{1 + \cos \theta}$.

22. $\dfrac{3 - \cot \theta}{3 \sec \theta - \csc \theta}$.

23. $\dfrac{\cot^3 x - \tan^3 x}{\cot x - \tan x} - \sec^2 x$.

24. $\dfrac{\sin x + \tan x \sin x + \cos x}{\sec x} - \sin x \cos x$.

7-2   IDENTITIES AND EQUATIONS

An equation that is true for all numbers for which the two sides are defined is called an identical equation, or an **identity.** Thus, the statement

$$(x + 1)^2 = x^2 + 2x + 1$$

is an identity, since it is true for all values of x. The trigonometric statement

$$\sin^2 x + \cos^2 x = 1$$

is an identity because it is true for every value of x.

An equation that is false for at least one number at which both sides are defined is called a *conditional equation*, or just an **equation.** Thus, the statement

$$x^2 - x - 6 = 0$$

is a conditional equation, because it is true only for $x = 3$ and for $x = -2$, and we say that both 3 and -2 *satisfy* the equation $x^2 - x - 6 = 0$. To say that a certain value of a variable "satisfies" an equation means that the equation becomes true when that value is substituted for the variable in the equation. If the members of a conditional equation can be made identical by replacing the unknown by a number, that number is a **root** of the equation. The process of finding the roots of an equation is known as **solving** the equation. The trigonometric equation

$$\sin x = \cos x$$

is conditional, because it is true only for the angles $x = 45°$ and $x = 225°$ and for values of x coterminal with those values.

7-3 *PROVING TRIGONOMETRIC IDENTITIES*

The process of transforming one side of an identity into the form of the other side is called *proving the identity.*

As the student advances through his study of mathematics and science, he will find that the changing of a trigonometric expression from one form to another is a necessary step in the solution of many problems. Proving identities is, therefore, an important part of this course.

Basic to proving trigonometric identities is the ability of the student to recognize the fundamental trigonometric identities in all of their various forms. When a student is asked to prove an identity, he should strive by one or more changes to transform one side of the identity into the form of the other side *which should be left unchanged. This method should be followed even though other logically correct methods may be available.*

In solving equations we generally follow a definite procedure. In proving trigonometric identities no definite procedure can be given; nevertheless, the following suggestions should prove helpful:

1. It is usually best to start with the more complicated member and reduce it to the simpler member.
2. The student should keep constantly in mind the expression he is trying to arrive at and look for fundamental relations that will lead him to this final expression.
3. Factoring, adding fractions, multiplying two or more expressions, etc., will often bring obvious simplifications to light.
4. If possible, avoid introducing radicals.
5. Sometimes it is useful to express all of the functions on one side in terms of sines and cosines.

Example 1.   Prove the identity

$$\frac{\csc x}{\cot x} = \sec x.$$

Solution: Since the left-hand side is the more complicated, we will start with it and, by successive applications of the fundamental identities, reduce it to the form of the right-hand side.

$$\frac{\csc x}{\cot x} = \frac{1/\sin x}{\cos x/\sin x} = \frac{1}{\sin x} \cdot \frac{\sin x}{\cos x} = \frac{1}{\cos x} = \sec x.$$

Example 2.   Prove the identity

$$\csc^2 x = \frac{\sin^4 x - \cos^4 x}{\sin^2 x - \cos^2 x} + \cot^2 x.$$

Solution: Since the right-hand side is the more complicated, we will start with it and, by factoring and applying the fundamental identities, reduce it to the form of the left-hand side.

$$\csc^2 x = \frac{\sin^4 x - \cos^4 x}{\sin^2 x - \cos^2 x} + \cot^2 x$$

$$= \frac{(\sin^2 x + \cos^2 x)(\cancel{\sin^2 x - \cos^2 x})}{(\cancel{\sin^2 x - \cos^2 x})} + \cot^2 x$$

$$= \sin^2 x + \cos^2 x + \cot^2 x$$

$$= 1 + \cot^2 x = \csc^2 x.$$

Example 3.   Prove the identity

$$\frac{\tan x + \sin x}{\tan x - \sin x} = \frac{\sec x + 1}{\sec x - 1}.$$

Both sides of this identity appear to be as complicated as each other. We show two solutions.

First Solution: Reduce the right-hand side to the form of the left-hand side. We observe that if the numerator and denominator of $(\sec x + 1)/(\sec x - 1)$ are multiplied by $\sin x$, the last term of both becomes $\sin x$, which is part of the expression we want. We make this multiplication and find that the desired $\tan x$ expression appears.

$$\frac{\sin x(\sec x + 1)}{\sin x(\sec x - 1)} = \frac{\sin x \sec x + \sin x}{\sin x \sec x - \sin x}.$$

But

$$\sin x \sec x = \sin x\left(\frac{1}{\cos x}\right) = \frac{\sin x}{\cos x} = \tan x.$$

Thus,

$$\frac{\sin x \sec x + \sin x}{\sin x \sec x - \sin x} = \frac{\tan x + \sin x}{\tan x - \sin x}.$$

Second Solution: We observe that if the numerator and denominator of $(\tan x + \sin x)/(\tan x - \sin x)$ are divided by $\sin x$, the last term of both becomes 1, which is part of the right-hand expression we want. We make this division and find that the desired expression, $\sec x$, results.

$$\frac{(\tan x/\sin x) + (\sin x/\sin x)}{(\tan x/\sin x) - (\sin x/\sin x)} = \frac{(\tan x/\sin x) + 1}{(\tan x/\sin x) - 1}.$$

But $\qquad \dfrac{\tan x}{\sin x} = \dfrac{(\sin x/\cos x)}{\sin x} = \dfrac{\sin x}{\cos x} \cdot \dfrac{1}{\sin x} = \dfrac{1}{\cos x} = \sec x.$

Thus, $\qquad \dfrac{(\tan x/\sin x) + 1}{(\tan x/\sin x) - 1} = \dfrac{\sec x + 1}{\sec x - 1}.$

Exercises 7-3

Prove the following identities by operating on one side of the identity and leaving the other side unchanged.

1. $\dfrac{\sin x}{\tan x} = \cos x.$

2. $\dfrac{\cos \alpha}{\cot \alpha} = \sin \alpha.$

3. $\dfrac{\sin^2 \beta + \cos^2 \beta}{1 + \tan^2 \beta} = \cos^2 \beta.$

4. $\dfrac{\sec^2 \theta - \tan^2 \theta}{1 + \cot^2 \theta} = \sin^2 \theta.$

5. $\dfrac{(\sin x + \cos x)^2}{\sin x} = \csc x + 2 \cos x.$

6. $\cos x (1 + \tan x)^2 = \sec x + 2 \sin x.$

7. $\dfrac{1 - \cos^2 x}{\sin x (\csc x + \cot x)} = 1 - \cos x.$

8. $\dfrac{1 - \sin^2 A}{\cos A (\sec A + \tan A)} = 1 - \sin A.$

9. $\sec B \csc B = \tan B + \cot B.$

10. $\dfrac{1}{\tan A} + \tan A = \sec A \csc A.$

11. $\dfrac{\sin^3 x - \cos^3 x}{\sin x - \cos x} = 1 + \sin x \cos x.$

12. $\dfrac{\sec^4 x - \tan^4 x}{\sec^2 x + \tan^2 x} + \tan^2 x = \sec^2 x.$

13. $\dfrac{\csc^4 x - \cot^4 x}{\csc^2 x + \cot^2 x} + \cot^2 x = \csc^2 x.$

14. $\dfrac{1}{1 + \cos A} + \dfrac{1}{1 - \cos A} = 2 \csc^2 A.$

15. $\dfrac{1}{\sin x - 1} - \dfrac{1}{\sin x + 1} = -2 \sec^2 x.$

16. $\dfrac{1 + \cos x}{1 - \cos x} = \dfrac{\sec x + 1}{\sec x - 1}.$

17. $\dfrac{\tan B + 1}{\tan B - 1} = \dfrac{\sec B + \csc B}{\sec B - \csc B}.$

18. $\dfrac{\tan^2 A - 1}{\sec^2 A} = \dfrac{\tan A - \cot A}{\tan A + \cot A}.$

19. $\dfrac{\cos A}{1 + \sin A} + \dfrac{1 + \sin A}{\cos A} = 2 \sec A.$

20. $\dfrac{\sin x}{1 + \cos x} + \dfrac{1 + \cos x}{\sin x} = 2 \csc x.$

21. $\tan \theta + \dfrac{\cos \theta}{1 + \sin \theta} = \sec \theta.$

22. $\csc \alpha - \dfrac{\sin \alpha}{1 + \cos \alpha} = \cot \alpha.$

23. $\dfrac{\tan^2 \theta - 1}{1 + \tan^2 \theta} = 1 - 2 \cos^2 \theta.$

24. $\dfrac{\cot^2 \beta - 1}{1 + \cot^2 \beta} = 1 - 2 \sin^2 \beta.$

25. $\dfrac{1 + \sin \beta}{\cos \beta} = \dfrac{\cos \beta}{1 - \sin \beta}.$

26. $\dfrac{\sec \alpha - 1}{\tan \alpha} = \dfrac{\tan \alpha}{\sec \alpha + 1}.$

27. $\dfrac{\cot \theta - \cos \theta}{\cos^3 \theta} = \dfrac{\csc \theta}{1 + \sin \theta}.$

28. $\dfrac{\tan \theta - \sin \theta}{\sin^3 \theta} = \dfrac{\sec \theta}{1 + \cos \theta}.$

29. $\cot^2 x - \tan^2 x = \dfrac{\cot x - \tan x}{\sin x \cos x}.$

30. $\sin x - \sin x \cos^2 x = \sin^3 x.$

31. $1 - 2 \sin^2 x + \sin^4 x = \cos^4 x.$

32. $\dfrac{\sec^2 x - 1}{\cot x} = \tan^3 x.$

33. $\dfrac{\csc^2 x - 1}{\tan x} = \cot^3 x.$

34. $\dfrac{1 - 2 \cos^2 \theta}{\sin \theta \cos \theta} = \tan \theta - \cot \theta.$

[Handwritten annotations:]
Since you are solving for cos and sin. They do not change

$\dfrac{COT - COS}{COS^3} = \dfrac{CSC}{1 + Sin}$

$\dfrac{COS}{SIN}$

35. $(x \sin \theta + y \cos \theta)^2 + (x \cos \theta - y \sin \theta)^2 = x^2 + y^2.$

36. $(1 + \sin \beta + \cos \beta)^2 = 2(1 + \sin \beta)(1 + \cos \beta).$

37. $\dfrac{\tan x + \sec x}{\sec x - \cos x + \tan x} = \csc x.$

38. $\sec^4 x - \tan^4 x = 1 + 2 \tan^2 x.$

39. $\dfrac{\sin A \cos B + \cos A \sin B}{\cos A \cos B - \sin A \sin B} = \dfrac{\tan A + \tan B}{1 - \tan A \tan B}.$

40. $\dfrac{\cos A \cos B - \sin A \sin B}{\cos A \cos B + \sin A \sin B} = \dfrac{1 - \tan A \tan B}{1 + \tan A \tan B}.$

7-4 TRIGONOMETRIC EQUATIONS

The student will recall that we solved trigonometric equations in Section 2-16 and again in Chapters 3, 4, and 5. Now, with our added knowledge of the eight fundamental identities, we can solve more complicated trigonometric equations.

Example 1. Solve the following equation for values of x in the interval $0° \leqq x < 360°$:

$$2 \cos^2 x = 1 + \sin x.$$

Solution: In problems like this it is generally best to express all terms in the same function. Here we replace $\cos^2 x$ by $1 - \sin^2 x$.

$$2 \cos^2 x = 1 + \sin x,$$
$$2(1 - \sin^2 x) = 1 + \sin x,$$
$$2(1 + \sin x)(1 - \sin x) = 1 + \sin x.$$

We rearrange all terms to the left-hand member (in order to obtain 0 on the right) and factor. We get

$$(1 + \sin x)[2(1 - \sin x) - 1] = 0,$$
$$(1 + \sin x)(1 - 2 \sin x) = 0.$$

We set each factor equal to 0, for, provided that all of the factors are defined, a product is equal to 0 if and only if at least one of its factors is equal to 0.

$$\sin x + 1 = 0, \quad \text{or} \quad 2 \sin x - 1 = 0,$$

$$\sin x = -1; \qquad \sin x = \frac{1}{2};$$

$$x = 270°. \qquad x = 30° \text{ and } 150°.$$

The complete solution is therefore $x = 30°, 150°,$ and $270°$.

Example 2. Solve the equation

$$3 \tan^2 \theta = 7 \sec \theta - 5,$$

for $0 \leqq \theta < 2\pi$.

Solution: From Identity (7) we replace $\tan^2 \theta$ with $\sec^2 \theta - 1$, and then we simplify and factor.

$$3 \tan^2 \theta = 7 \sec \theta - 5,$$
$$3(\sec^2 \theta - 1) = 7 \sec \theta - 5,$$
$$3 \sec^2 \theta - 3 = 7 \sec \theta - 5,$$
$$3 \sec^2 \theta - 7 \sec \theta + 2 = 0,$$
$$(3 \sec \theta - 1)(\sec \theta - 2) = 0.$$

Then, if the equation has a solution, either

$$3 \sec \theta - 1 = 0 \quad \text{or} \quad \sec \theta - 2 = 0.$$

But $3 \sec \theta - 1 \neq 0$, since $\sec \theta$ would then be $\frac{1}{3}$, which is impossible; the absolute value of the secant function cannot be less than one (see Fig. 5-8, Section 5-6). When the other factor, $\sec \theta - 2$, equals 0,

$$\sec \theta = 2.$$

Hence,

$$\theta = \frac{\pi}{3} \quad \text{and} \quad \frac{5\pi}{3}.$$

Example 3. Solve the equation

$$2 \sin x \cos x = \sin x,$$

for $0° \leqq x < 360°$.

A word of caution. Reducing an equation by dividing both sides of the equation by an expression containing the unknown may give a *defective equation*. That is, the reduced equation may not have all the roots of the original equation. In this example if we proceed to simplify the equation $2 \sin x \cos x = \sin x$ by dividing both sides by $\sin x$, the resulting equation

$$2 \cos x = 1$$

will not give **all** the roots of the original equation. One way to avoid this loss of roots is to solve by factoring.

$$2 \sin x \cos x = \sin x,$$
$$2 \sin x \cos x - \sin x = 0,$$
$$\sin x (2 \cos x - 1) = 0.$$

$$\sin x = 0; \qquad\qquad \text{or} \qquad 2\cos x - 1 = 0;$$

$$x = 0°, 180°. \qquad\qquad\qquad \cos x = \tfrac{1}{2}; x = 60°, 300°.$$

The complete solution is therefore $x = 0°, 60°, 180°, 300°$.

Example 4.   Solve the equation $\sin^2 x = 1$ for $0° \leqq x < 360°$.

Solution: If we solve by taking the square root of both sides of the equation, then we must use both $\sin x = +1$ and $\sin x = -1$. The solution by factoring will give the same results.

$$\sin^2 x = 1.$$

$$(\sin x + 1)(\sin x - 1) = 0,$$

$$\sin x + 1 = 0, \qquad\qquad \text{or} \qquad\qquad \sin x - 1 = 0,$$

$$\sin x = -1; \qquad\qquad\qquad\qquad \sin x = 1;$$

$$x = 270°. \qquad\qquad\qquad\qquad\qquad x = 90°.$$

The complete solution is therefore $x = 90°$ and $270°$.

As in any algebraic solution of an equation, if we divide both members of an equation by a common factor involving the unknown, or if we take the square root of both members, we must be careful not to throw away any roots.

Example 5.   Solve the equation $\tan x = \sec x$ for $0° \leqq x < 360°$.

Solution: Write $\tan x = \sin x / \cos x$ and $\sec x = 1/\cos x$, and then multiply both members of the resulting equation by $\cos x$. We obtain

$$\frac{\sin x}{\cos x} = \frac{1}{\cos x};$$

hence, $\qquad\qquad\qquad\qquad \sin x = 1, x = 90°.$

However, $x = 90°$ does not satisfy the original equation since neither $\tan 90°$ nor $\sec 90°$ is defined. We introduced the extraneous root $x = 90°$ when we multiplied both members of

$$\frac{\sin x}{\cos x} = \frac{1}{\cos x}$$

by $\cos x$, since $\cos 90° = 0$. The equation has no roots.

If both members of an equation are multiplied by an expression involving the unknown (in order, for example, to get rid of fractions), or if we square both members (to free the equation of radicals, perhaps), the resulting equation may have roots that do not satisfy the original equation. The roots that do not satisfy the original equation are called *extraneous roots.* The best test is to substitute each of the supposed roots in the original equation and then reject any that fail to satisfy the equation.

Example 6.   Solve the equation

$$\sin^2 \theta - 2 \sin \theta - 1 = 0,$$

for $0° \leq \theta < 360°$.

Solution: This trigonometric equation is a quadratic equation that does not have simple factors. We shall use the quadratic formula to solve this equation.

The roots of the quadratic equation

$$ax^2 + bx + c = 0$$

where $a \neq 0$, are

$$x = \frac{-b \pm \sqrt{b^2 - 4ac}}{2a}.$$

Let $\sin \theta$ represent x in the quadratic formula; then, for

$$(\sin \theta)^2 - 2(\sin \theta) - 1 = 0,$$

substitute $\begin{cases} a = 1 \\ b = -2 \\ c = -1 \\ x = \sin \theta \end{cases}$ into $x = \dfrac{-b \pm \sqrt{b^2 - 4ac}}{2a}.$

We have

$$\sin \theta = \frac{2 \pm \sqrt{4 + 4}}{2} = 1 \pm \sqrt{2}.$$

Hence, $\sin \theta = 1 + \sqrt{2}$ or $\sin \theta = 1 - \sqrt{2}$

$$\approx 2.41421 \qquad\qquad \approx -.41421.$$

(From Table 5, $\sqrt{2} \approx 1.41421$.)

But $\sin \theta$ cannot be numerically greater than 1; therefore, we take $\sin \theta = -.4142$. From Table 1 we find $\sin 24° 28' = .4142$. But $\sin \theta$ is negative; therefore, θ is in the third or fourth quadrant. The desired roots are

$$\theta = 180° + 24° 28' = 204° 28',$$

$$\theta = 360° - 24° 28' = 335° 32'.$$

Exercises 7-4

Solve the following equations for x in the interval $0° \leq x < 360°$.

1. $(\tan x - \sqrt{3})(\sin x + 1) = 0.$
2. $(2 \cos x - \sqrt{3})(2 \sin x + \sqrt{2}) = 0.$
3. $(\sec x - 2)(\csc x + 1) = 0.$
4. $\sin x(\cot x - 1)(\cos x - 1) = 0.$
5. $2 \sin^2 x = 1 + \cos x.$
6. $2 \cos^2 x = 1 - \sin x.$
7. $4 \sin^2 x = 3.$
8. $2 \cos^2 x = 1.$
9. $2 \tan^2 x = \sec x - 1.$
10. $2 \tan^2 x + 3 \sec x = 0.$
11. $\sin x \cos x = \cos x.$
12. $\sin x = \sin x \cos x.$
13. $\sin x + \csc x + 2 = 0.$
14. $\cos x + \sec x + 2 = 0.$
15. $\sin x - \sqrt{1 - \sin^2 x} = 0.$
16. $tan x + \sqrt{1 - 2 \tan^2 x} = 0.$
17. $\sin^2 x = 2(\sin x + 1).$
18. $\cos^2 x = 1 + \cos x.$
19. $\sqrt{3}(\sec x + 1) = \tan x.$
20. $\sin x \tan x - 1 = \tan x - \sin x.$

REVIEW EXERCISES

In Exercises 1 and 2 use the fundamental identities to find the indicated functions of the angle satisfying the given conditions.

1. Given $\sin \theta = \frac{3}{5}$ and θ in the second quadrant, find the following:
   (a) $\cos \theta.$ (b) $\sec \theta.$ (c) $\tan \theta.$ (d) $\cot \theta.$ (e) $\csc \theta.$
2. Given $\cos \theta = -\frac{3}{5}$ and θ in the third quadrant, find the following:
   (a) $\sin \theta.$ (b) $\tan \theta.$ (c) $\cot \theta.$ (d) $\sec \theta.$ (e) $\csc \theta.$

In Exercises 3–12 perform the indicated operations and use the fundamental identities to simplify the expressions.

3. $\sec \theta \tan \theta \cos^2 \theta \csc \theta.$
4. $\sin^2 \theta \csc \theta \cot \theta \sec \theta.$
5. $(\sin \theta + \cos \theta)^2 - 2 \sin \theta \cos \theta.$
6. $(1 + \cot \theta)^2 - \csc^2 \theta.$
7. $\dfrac{1}{1 - \cos \theta} + \dfrac{1}{1 + \cos \theta}.$
8. $\dfrac{1}{1 - \sin \theta} + \dfrac{1}{1 + \sin \theta}.$
9. $\dfrac{1 + \tan^2 \theta}{\sin^2 \theta + \cos^2 \theta}.$
10. $\cos x(\tan x + \cot x).$
11. $\dfrac{(\sin x + \cos x)^2}{\sin x} - \csc x.$
12. $\dfrac{\sin^3 x - \cos^3 x}{\sin x - \cos x} - \sin x \cos x.$

In Exercises 13–20 solve the equations for x in the interval $0° \leq x < 360°$.

13. $(\sec x + 2)(\tan x - 1) = 0.$

14. $\cos x(2 \sin x - 1)(\cot x + 1) = 0.$

15. $2 \sin^2 x = 1.$

16. $\csc^2 x = 4.$

17. $\tan^2 x - \cos^2 x = \sin^2 x.$

18. $\sin x + \cot^2 x = \csc^2 x.$

19. $\cos x - \sqrt{1 - \cos^2 x} = 0.$

20. $\cot x + \sqrt{1 - 2 \cot^2 x} = 0.$

Prove the following identities by operating on one side of the identity and leaving the other side unchanged.

21. $\dfrac{1 - 2 \cos^2 \theta}{\sin \theta \cos \theta} + \cot \theta = \tan \theta.$

22. $\dfrac{1}{\sin x + 1} - \dfrac{1}{\sin x - 1} = 2 \sec^2 x.$

CHAPTER 7: DIAGNOSTIC TEST

The purpose of this test is to see how well you understand the work covered in Chapter 7. We recommend that you work this test before your instructor tests you on this chapter. Allow yourself approximately 50 minutes to do the test.

Solutions to the problems, together with section references, are given in the Answer Section at the end of this book. We suggest that you study the sections referred to for the problems you do incorrectly.

1. Given $\tan \theta = -\frac{3}{4}$ and θ in the fourth quadrant, use the fundamental identities to find the following:
   (a) $\cot \theta.$ (b) $\csc \theta.$ (c) $\sin \theta.$

In Problems 2–6 use the fundamental identities to simplify the expressions.

2. $\csc \theta \tan \theta \sec \theta \cot^2 \theta.$

3. $(1 + \cot^2 \alpha) \sin^2 \alpha + \tan^2 \alpha.$

4. $(\sin^2 x + \cos^2 x)(\sin^2 x - \cos^2 x) + \cos^2 x.$

5. $\dfrac{\sec \theta}{\tan \theta + \cot \theta}.$

6. $\dfrac{\cos \theta - \sin \theta}{\cos \theta} + \tan \theta.$

In Problems 7–9 solve for x in the interval $0° \leqq x < 360°.$

7. $\sec^2 x = 4.$

8. $(\sin x + \cos x)^2 = 1.$
9. $\cos^2 x = 3 \sin^2 x.$
10. Prove the following identity by reducing the expressions on the left side to that on the right side.

$$\frac{\sin \theta}{1 + \cos \theta} + \cot \theta = \csc \theta.$$

8

Combination of Angles

In Chapter 7 we developed eight fundamental trigonometric identities. In this chapter we shall add many other important trigonometric identities to our collection.

8-1 FUNCTIONS OF (−θ)

The identities that we wish to develop in this section are:

$$(9) \qquad \sin(-\theta) = -\sin\theta,$$

$$(10) \qquad \cos(-\theta) = \cos\theta,$$

$$(11) \qquad \tan(-\theta) = -\tan\theta,$$

$$(12) \qquad \cot(-\theta) = -\cot\theta,$$

$$(13) \qquad \sec(-\theta) = \sec\theta,$$

$$(14) \qquad \csc(-\theta) = -\csc\theta.$$

In the above identities observe that changing the sign of the angle changes the sign of all of the functions except the cosine and secant.

Examples.

1. $\sin(-30°) = -\sin 30°;$
   $$-\tfrac{1}{2} = -(+\tfrac{1}{2}).$$

2. $\cos(-60°) = \cos 60°;$
   $$\tfrac{1}{2} = \tfrac{1}{2}.$$

3. $\tan(-135°) = -\tan 135°;$
   $$1 = -(-1).$$

4. $\cot(-45°) = -\cot 45°;$
   $$-1 = -(+1).$$

5. $\sec(-240°) = \sec 240°;$
   $$-2 = -2.$$

6. $\csc(-90°) = -\csc 90°;$
   $$-1 = -(+1).$$

7. $\sin[-(-568°)] = -\sin 28°$ and $\sin(-568°) = \sin 28°.$ Therefore, $\sin[-(-568°)] = -\sin(-568°).$

8. $\cos[-(-568°)] = -\cos 28°$ and $\cos(-568°) = -\cos 28°.$ Therefore, $\cos[-(-568°)] = \cos(-568°).$

Proof. Let $θ$ and $(-θ)$ be angles in standard position. Let the terminal sides of these angles meet the unit circle at the points P and Q, whose coordinates are (x, y) and (x', y') respectively. Then, as defined in Sections 2-5 and 3-4,

$$x = \cos θ, \qquad y = \sin θ,$$
$$x' = \cos(-θ), \qquad y' = \sin(-θ).$$

In each figure of Figure 8-1 the x-axis is the perpendicular bisector of line segment PQ at M. Then

$$x' = x, \quad \text{and} \quad y' = -y;$$

thus, $$\cos(-θ) = \cos θ, \qquad \sin(-θ) = -\sin θ.$$

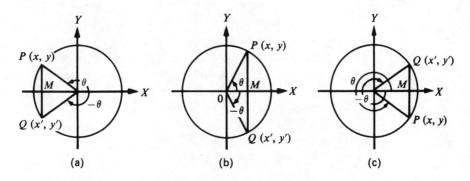

(a)   (b)   (c)

Figure 8-1

We use these results and Identities (1), (2), (3), and (4) of Section 7-1 to find the other trigonometric functions of the negative of an angle.

$$\tan(-\theta) = \frac{\sin(-\theta)}{\cos(-\theta)} = \frac{-\sin\theta}{\cos\theta} = -\tan\theta.$$

$$\cot(-\theta) = \frac{1}{\tan(-\theta)} = \frac{1}{-\tan\theta} = -\cot\theta.$$

$$\sec(-\theta) = \frac{1}{\cos(-\theta)} = \frac{1}{\cos\theta} = \sec\theta.$$

$$\csc(-\theta) = \frac{1}{\sin(-\theta)} = \frac{1}{-\sin\theta} = -\csc\theta.$$

8-2   cos (A + B) AND cos (A − B)

If A and B are any two angles, then

(15) $\cos(A + B) = \cos A \cos B - \sin A \sin B,$

and

(16) $\cos(A - B) = \cos A \cos B + \sin A \sin B.$

Stated in words:

The cosine of the sum of two angles is equal to the product of the cosines of the two angles, minus the product of their sines.

The cosine of the difference of two angles is equal to the product of the cosines of the two angles, plus the product of their sines.

Proof for Identity (16)

$$\cos(A - B) = \cos A \cos B + \sin A \sin B.$$

Let α and θ be any two angles in standard position in a unit circle. Let P_1, P_2, P_3, and P_4 be points on the unit circle. Point P_1, on the initial side of α and θ, has coordinates $(1, 0)$; point P_2, on the terminal side of α, has coordinates $(\cos\alpha, \sin\alpha)$; point P_3, on the terminal side of θ, has coordinates $(\cos\theta, \sin\theta)$; and point P_4, on the terminal side of $(\alpha + \theta)$, has coordinates $[\cos(\alpha + \theta), \sin(\alpha + \theta)]$. (See Fig. 8-2.)

Chord P_1P_2 equals chord P_3P_4, and by using the formula for the distance between two points (Section 1-5), we find

$$\sqrt{(\cos\alpha - 1)^2 + (\sin\alpha - 0)^2}$$

$$= \sqrt{[\cos(\alpha + \theta) - \cos\theta]^2 + [\sin(\alpha + \theta) - \sin\theta]^2}.$$

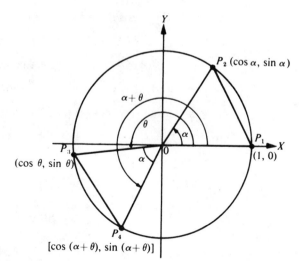

Figure 8-2

$[\cos (\alpha + \theta), \sin (\alpha + \theta)]$

By squaring both members, we obtain

$$\cos^2 \alpha - 2 \cos \alpha + 1 + \sin^2 \alpha$$
$$= \cos^2 (\alpha + \theta) - 2 \cos (\alpha + \theta) \cos \theta + \cos^2 \theta$$
$$+ \sin^2 (\alpha + \theta) - 2 \sin (\alpha + \theta) \sin \theta + \sin^2 \theta,$$

$$2 - 2 \cos \alpha = 2 - 2[\cos (\alpha + \theta) \cos \theta + \sin (\alpha + \theta) \sin \theta],$$

which reduces to

(1) $$\cos \alpha = \cos (\alpha + \theta) \cos \theta + \sin (\alpha + \theta) \sin \theta.$$

Equation (1) is true for all angles α and θ. Therefore, if A and B are any angles and we set $\alpha = A - B$ and $\theta = B$ in Equation (1), we obtain

$$\cos (A - B) = \cos A \cos B + \sin A \sin B.$$

We use Identity (16) and Identities (9) and (10) of Section 8-1 to prove

$$\cos (A + B) = \cos A \cos B - \sin A \sin B.$$

Proof. We substitute $(-B)$ for $(+B)$ in Identity (16) and proceed as follows:

$$\cos (A - B) = \cos A \cos B + \sin A \sin B,$$
$$\cos [A - (-B)] = \cos A \cos (-B) + \sin A \sin (-B).$$

But $\cos(-B) = \cos B$ and $\sin(-B) = -\sin B$. Therefore,

$$\cos(A + B) = \cos A \cos B + \sin A(-\sin B),$$

or $\qquad \cos(A + B) = \cos A \cos B - \sin A \sin B.$

The student is warned that, in general,

$$\cos(A - B) \neq \cos A - \cos B,$$

and $\qquad \cos(A + B) \neq \cos A + \cos B.$

For example,

$$\cos(60° + 30°) \neq \cos 60° + \cos 30°,$$

$$\cos 90° \neq \frac{1}{2} + \frac{1}{2}\sqrt{3},$$

$$0 \neq \frac{1 + \sqrt{3}}{2}.$$

Example 1. Show that Identity (15) is true for $A = 30°$ and $B = 60°$.

Solution:

$$\cos(A + B) = \cos A \cos B - \sin A \sin B,$$

$$\cos(30° + 60°) = \cos 30° \cos 60° - \sin 30° \sin 60°,$$

$$\cos 90° = \frac{1}{2}\sqrt{3} \cdot \frac{1}{2} - \frac{1}{2} \cdot \frac{1}{2}\sqrt{3},$$

$$0 = 0.$$

Example 2. Compute the value of $\cos 15°$ from the functions of $60°$ and $45°$.

Solution: We use Identity (16).

$$\cos(A - B) = \cos A \cos B + \sin A \sin B,$$

$$\cos(60° - 45°) = \cos 60° \cos 45° + \sin 60° \sin 45°,$$

$$\cos 15° = \frac{1}{2} \cdot \frac{1}{2}\sqrt{2} + \frac{1}{2}\sqrt{3} \cdot \frac{1}{2}\sqrt{2},$$

$$= \frac{\sqrt{2}(1 + \sqrt{3})}{4}.$$

Example 3. Given $\sin A = \frac{3}{5}$, with A in Q I, and $\cos B = -\frac{5}{13}$, with B in Q III, find (a) $\cos(A + B)$ and (b) $\cos(A - B)$.

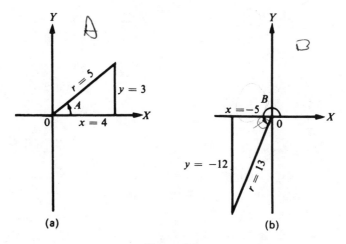

Figure 8-3

Solution: Cos A and sin B must be determined before we can apply Identities (15) and (16). We find these from the figures shown in Figure 8-3. Thus,

$$\cos A = \frac{4}{5} \quad \text{and} \quad \sin B = -\frac{12}{13}.$$

(a)

$$\cos (A + B) = \cos A \cos B - \sin A \sin B \text{ one of each.}$$

$$= \left(\frac{4}{5}\right)\left(-\frac{5}{13}\right) - \left(\frac{3}{5}\right)\left(-\frac{12}{13}\right)$$

$$= -\frac{20}{65} + \frac{36}{65} = \frac{16}{65}.$$

(b)

$$\cos (A - B) = \cos A \cos B + \sin A \sin B$$

$$= \left(\frac{4}{5}\right)\left(-\frac{5}{13}\right) + \left(\frac{3}{5}\right)\left(-\frac{12}{13}\right)$$

$$= -\frac{20}{65} - \frac{36}{65} = -\frac{56}{65}.$$

Example 4. Use Identity (16) to prove that

$$\cos (90° − \theta) = \sin \theta.$$

Proof. This relation was shown to be true for θ an acute angle (Section 5-3), but now, by applying Identity (16), we prove the relation to be true for any value of θ. Substitute 90° for A and θ for B in Identity (16).

$$\cos (A − B) = \cos A \cos B + \sin A \sin B;$$

therefore,

$$\cos(90° - \theta) = \cos 90° \cos \theta + \sin 90° \sin \theta.$$

But $\qquad\qquad\qquad \cos 90° = 0 \qquad$ and $\qquad \sin 90° = 1.$

Hence, $\qquad\qquad\quad \cos(90° - \theta) = 0 \cdot \cos \theta + 1 \cdot \sin \theta$

$$= \sin \theta.$$

Exercises 8-2

1. By substitution, verify that Identity (15) is true for $A = 45°$ and $B = 45°$.
2. By substitution, verify that Identity (16) is true for $A = 60°$ and $B = 30°$.
3. Compute the value of $\cos 105°$ from the functions of $60°$ and $45°$.
4. Compute the value of $\cos 75°$ from the functions of $30°$ and $45°$.
5. Given $\sin A = \frac{4}{5}$, with A in Q II, and $\cos B = \frac{12}{13}$, with B in Q I. Find (a) $\cos(A + B)$ and (b) $\cos(A - B)$.
6. Given $\sin A = -\frac{1}{3}$, with A in Q III, and $\cos B = -\frac{1}{5}$, with B in Q II. Find (a) $\cos(A + B)$ and (b) $\cos(A - B)$.

Prove the identities in Exercises 7–10.

7. $\cos(180° + \theta) = -\cos \theta.$ $\qquad$ 8. $\cos(\frac{1}{2}\pi + \theta) = -\sin \theta.$
9. $\cos(\pi - \theta) = -\cos \theta.$ $\qquad\quad$ 10. $\cos(\frac{3}{2}\pi - \theta) = -\sin \theta.$

Simplify the following by reducing each to a single term.

11. $\cos \frac{1}{3}\pi \cos \frac{2}{3}\pi - \sin \frac{1}{3}\pi \sin \frac{2}{3}\pi.$
12. $\cos \frac{4}{3}\pi \cos \frac{1}{3}\pi + \sin \frac{4}{3}\pi \sin \frac{1}{3}\pi.$
13. $\cos 3\theta \cos 2\theta + \sin 3\theta \sin 2\theta.$
14. $\cos 100° \cos 40° + \sin 100° \sin 40°.$
15. $\cos 50° \cos 10° - \sin 50° \sin 10°.$
16. $\cos \frac{3}{2}\theta \cos \frac{1}{2}\theta - \sin \frac{3}{2}\theta \sin \frac{1}{2}\theta.$

8-3   sin (A + B) AND sin (A − B)

If A and B are any two angles, then

(17) $\qquad\qquad$ **$\sin(A + B) = \sin A \cos B + \cos A \sin B,$**

and

(18) $\qquad\qquad$ **$\sin(A - B) = \sin A \cos B - \cos A \sin B.$**

Stated in words:

> *The sine of the sum of two angles is equal to the sine of the first angle times the cosine of the second, plus the cosine of the first angle times the sine of the second.*
>
> *The sine of the difference of two angles is equal to the sine of the first angle times the cosine of the second, minus the cosine of the first angle times the sine of the second.*

In Example 4, Section 8-2, it was shown that

(1) $$\cos (90° - \theta) = \sin \theta \qquad \text{for all values of } \theta.$$

By substituting $(90° - \theta)$ for θ in (1), we obtain

(2) $$\sin (90° - \theta) = \cos \theta \qquad \text{for all values of } \theta.$$

We substitute $(A + B)$ for θ in (1) to prove Identity (17).

$$\sin \theta = \cos (90° - \theta).$$

Then $$\sin (A + B) = \cos [90° - (A + B)]$$

$$= \cos [(90° - A) - B]$$

$$= \cos (90° - A) \cos B + \sin (90° - A) \sin B.$$

But $\cos (90° - A) = \sin A$ and $\sin (90° - A) = \cos A$. Therefore,

$$\sin (A + B) = \sin A \cos B + \cos A \sin B.$$

In Identity (17) we replace B by $(-B)$ to obtain

$$\sin [A + (-B)] = \sin A \cos (-B) + \cos A \sin (-B).$$

But $\cos (-B) = \cos B$, and $\sin (-B) = -\sin B$; therefore,

$$\sin (A - B) = \sin A \cos B - \cos A \sin B.$$

Identities (15), (16), (17), and (18) are basic identities and should be memorized before proceeding any further.

8-4   tan $(A + B)$ AND tan $(A - B)$

The following identities are true for all values of the angles A and B for which both members are defined.

$$(19) \qquad \tan (A + B) = \frac{\tan A + \tan B}{1 - \tan A \tan B},$$

$$(20) \qquad \tan (A - B) = \frac{\tan A - \tan B}{1 + \tan A \tan B}.$$

Stated in words:

> The tangent of the sum of two angles is equal to the tangent of the first angle plus the tangent of the second, divided by 1 minus the product of the tangents of the two angles (whenever all of these tangents are defined).
>
> The tangent of the difference of two angles is equal to the tangent of the first angle minus the tangent of the second, divided by 1 plus the product of the tangents of the two angles (whenever all of these tangents are defined).

Proof.   Our proof depends on the identity $\tan \theta = \sin \theta / \cos \theta$; Identities (15), (16), (17), and (18); and the formulas for $\sin (A \pm B)$ and cos $(A \pm B)$. Thus, for (19), we set

$$\tan (A + B) = \frac{\sin (A + B)}{\cos (A + B)} = \frac{\sin A \cos B + \cos A \sin B}{\cos A \cos B - \sin A \sin B}.$$

We divide the numerator and denominator of this fraction by $\cos A \cos B$ and obtain

$$\tan (A + B) = \frac{\left(\dfrac{\sin A \cos B}{\cos A \cos B}\right) + \left(\dfrac{\cos A \sin B}{\cos A \cos B}\right)}{\left(\dfrac{\cos A \cos B}{\cos A \cos B}\right) - \left(\dfrac{\sin A \sin B}{\cos A \cos B}\right)}$$

$$= \frac{\left(\dfrac{\sin A}{\cos A}\right) + \left(\dfrac{\sin B}{\cos B}\right)}{1 - \left(\dfrac{\sin A}{\cos A}\right) \cdot \left(\dfrac{\sin B}{\cos B}\right)} = \frac{\tan A + \tan B}{1 - \tan A \tan B}.$$

For Identity (20), we make a similar reduction of

$$\tan (A - B) = \frac{\sin (A - B)}{\cos (A - B)}$$

$$= \frac{\sin A \cos B - \cos A \sin B}{\cos A \cos B + \sin A \sin B} = \frac{\tan A - \tan B}{1 + \tan A \tan B}.$$

Example. Find tan 15° from the known values of the functions of 60° and 45°.

Solution:

$$\tan 15° = \tan (60° - 45°)$$

$$= \frac{\tan 60° - \tan 45°}{1 + \tan 60° \tan 45°}$$

$$= \frac{\sqrt{3} - 1}{1 + \sqrt{3}} = \frac{(\sqrt{3} - 1)}{(\sqrt{3} + 1)} \cdot \frac{(\sqrt{3} - 1)}{(\sqrt{3} - 1)}$$

$$= \frac{3 - 2\sqrt{3} + 1}{3 - 1} = \frac{4 - 2\sqrt{3}}{2}$$

$$= 2 - \sqrt{3} \approx .2679.$$

Handwritten annotation: To find which one to use TAN(A+B) or TAN(A-B), to find 15° you need to find the difference between 60 and 45 so then is TAN(A-B)

Exercises 8-4

1. Given $\sin A = -\frac{3}{5}$, with A in Q III, and $\tan B = \frac{4}{3}$, with B in Q I, find the sine, cosine, and tangent of $(A + B)$ and of $(A - B)$, and indicate the quadrant for each of the angles $(A + B)$ and $(A - B)$.

2. Find the sine, cosine, and tangent of $\frac{2}{3}\pi$ from functions of π and $\frac{1}{3}\pi$.

3. Find the sine, cosine, and tangent of $\frac{4}{3}\pi$ from functions of π and $\frac{1}{3}\pi$.

4. Find the sine, cosine, and tangent of 105° from functions of 60° and 45°.

Prove the following identities.

5. $\tan (\frac{3}{4}\pi - \theta) = \dfrac{\tan \theta + 1}{\tan \theta - 1}.$

6. $\sin (\pi + \theta) = -\sin \theta.$

7. $\tan (\pi + \theta) = \tan \theta.$

8. $\sin (\frac{3}{2}\pi - \theta) = -\cos \theta.$

9. $\tan (\frac{1}{4}\pi + \theta) = \dfrac{1 + \tan \theta}{1 - \tan \theta}.$

10. $\cos (\frac{3}{2}\pi + \theta) = \sin \theta.$

11. $\sin (\theta - \frac{1}{2}\pi) = -\cos \theta.$

12. $\sin (x + y) + \sin (x - y) = 2 \sin x \cos y.$

13. $\sin (x + y) - \sin (x - y) = 2 \cos x \sin y.$

14. $\cos (x + y) - \cos (x - y) = -2 \sin x \sin y.$

15. $\cos (x + y) + \cos (x - y) = 2 \cos x \cos y.$

16. $\sin (x - y) \cos y + \cos (x - y) \sin y = \sin x.$

17. $\sin (x + y) \sin y + \cos (x + y) \cos y = \cos x.$

In each of the Exercises 18–27 simplify the given expression and reduce it to a single term.

18. $\sin 7x \cos 6x - \cos 7x \sin 6x$.   19. $\cos 5x \cos 4x - \sin 5x \sin 4x$.

20. $\dfrac{\tan x + \tan 3x}{1 - \tan x \tan 3x}$.   21. $\dfrac{\tan x - \tan 2x}{1 + \tan x \tan 2x}$.

22. $\cos^2 x - \sin^2 x$. (Write as $\cos x \cos x - \sin x \sin x$.)
23. $2 \sin x \cos x$. (Write as $\sin x \cos x + \cos x \sin x$.)
24. $2 \sin 3x \cos 3x$. (Write as $\sin 3x \cos 3x + \cos 3x \sin 3x$.)
25. $\cos^2 3x - \sin^2 3x$. (Write as $\cos 3x \cos 3x - \sin 3x \sin 3x$.)
26. $\sin x \sin (x + y) + \cos x \cos (x + y)$.
27. $\sin (\theta + 60°) - \cos (\theta + 30°)$.
28. Replace B by A in Identity (17) and thus show that $\sin 2A = 2 \sin A \cos A$.
29. Replace B by A in Identity (15) and thus show that $\cos 2A = \cos^2 A - \sin^2 A$.
30. Replace B by A in Identity (19) and thus show that

$$\tan 2A = \frac{2 \tan A}{1 - \tan^2 A}.$$

8-5 TRIGONOMETRIC FUNCTIONS OF TWICE AN ANGLE

If we replace B by A in Identities (15), (17), and (19), $(A + B)$ becomes equal to $2A$, and we obtain the following important identities.

(21)	$\sin 2A = 2 \sin A \cos A$.
(22a)	$\cos 2A = \cos^2 A - \sin^2 A$
(22b)	$= 1 - 2 \sin^2 A$
(22c)	$= 2 \cos^2 A - 1$.
(23)	$\tan 2A = \dfrac{2 \tan A}{1 - \tan^2 A}$.

The second and third forms for $\cos 2A$ are obtained by setting

$$\cos 2A = \cos^2 A - \sin^2 A$$
$$= (1 - \sin^2 A) - \sin^2 A$$
$$= 1 - 2 \sin^2 A,$$

and
$$\cos 2A = \cos^2 A - \sin^2 A$$
$$= \cos^2 A - (1 - \cos^2 A)$$
$$= 2\cos^2 A - 1.$$

Stated in words:

The sine of twice an angle is equal to 2 times the product of the sine and the cosine of the angle.

The cosine of twice an angle is equal to the square of the cosine of the angle minus the square of the sine of the angle.

The tangent of twice an angle is equal to twice the tangent of the angle divided by 1 minus the square of the tangent of the angle.

Example 1. Show that Identities (21), (22), and (23) are true for $A = 30°$.

Solution:

(21)
$$\sin 2A = 2\sin A \cos A;$$
$$\sin 60° = 2\sin 30° \cos 30°,$$
$$\frac{1}{2}\sqrt{3} = 2\left(\frac{1}{2}\right)\left(\frac{1}{2}\sqrt{3}\right),$$
$$\frac{1}{2}\sqrt{3} = \frac{1}{2}\sqrt{3}.$$

(22)
$$\cos 2A = \cos^2 A - \sin^2 A;$$
$$\cos 60° = \cos^2 30° - \sin^2 30°,$$
$$\frac{1}{2} = \left(\frac{1}{2}\sqrt{3}\right)^2 - \left(\frac{1}{2}\right)^2,$$
$$\frac{1}{2} = \frac{3}{4} - \frac{1}{4}.$$

(23)
$$\tan 2A = \frac{2\tan A}{1 - \tan^2 A};$$
$$\tan 60° = \frac{2\tan 30°}{1 - \tan^2 30°},$$
$$\sqrt{3} = \frac{2\left(\frac{1}{3}\sqrt{3}\right)}{1 - \left(\frac{1}{3}\sqrt{3}\right)^2} = \frac{\frac{2}{3}\sqrt{3}}{\frac{2}{3}} = \sqrt{3}.$$

Example 2. If $\tan \theta = \frac{4}{3}$ and θ is in the first quadrant, find $\sin 2\theta$, $\cos 2\theta$, and $\tan 2\theta$.

Solution: Sketch angle θ in standard position (Fig. 8-4); then, after finding $r = 5$, we get $\sin \theta = \frac{4}{5}$ and $\cos \theta = \frac{3}{5}$.

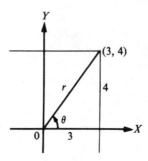

Figure 8-4

Then $\sin 2\theta = 2 \sin \theta \cos \theta = 2\left(\dfrac{4}{5}\right)\left(\dfrac{3}{5}\right) = \left(\dfrac{24}{25}\right),$

$$\cos 2\theta = \cos^2 \theta - \sin^2 \theta = \left(\dfrac{3}{5}\right)^2 - \left(\dfrac{4}{5}\right)^2 = -\dfrac{7}{25}.$$

By dividing the first of these two results by the second, we get

$$\tan 2\theta = \frac{\sin 2\theta}{\cos 2\theta} = -\frac{24}{7}.$$

The $\tan 2\theta$ could also have been obtained by use of Identity (23).

Example 3. Express $\cos^4 \theta$ in terms of first-degree functions of multiple angles.

Solution: From Identity (22c)

$$2 \cos^2 \theta - 1 = \cos 2\theta,$$

we get $\cos^2 \theta = \dfrac{1 + \cos 2\theta}{2}.$

Then $\cos^4 \theta = (\cos^2 \theta)^2 = \left(\dfrac{1 + \cos 2\theta}{2}\right)^2$

$$= \frac{1}{4}(1 + 2 \cos 2\theta + \cos^2 2\theta).$$

But $\cos^2 2\theta = \dfrac{1 + \cos 4\theta}{2}.$

Therefore, $$\cos^4 \theta = \frac{1}{4}\left(1 + 2\cos 2\theta + \frac{1 + \cos 4\theta}{2}\right)$$

$$= \frac{1}{8}(3 + 4\cos 2\theta + \cos 4\theta).$$

Example 4. Express $\sin 3\theta$ in terms of $\sin \theta$.

Solution:

$$\sin 3\theta = \sin (2\theta + \theta) = \sin 2\theta \cos \theta + \cos 2\theta \sin \theta$$

$$= (2 \sin \theta \cos \theta) \cos \theta + (1 - 2 \sin^2 \theta) \sin \theta$$

$$= 2 \sin \theta \cos^2 \theta + \sin \theta - 2 \sin^3 \theta$$

$$= 2 \sin \theta (1 - \sin^2 \theta) + \sin \theta - 2 \sin^3 \theta$$

$$= 3 \sin \theta - 4 \sin^3 \theta.$$

Example 5. Prove the identity $\dfrac{\sin 3x}{\sin x} - \dfrac{\cos 3x}{\cos x} = 2.$

Solution:

$$\frac{\sin 3x}{\sin x} - \frac{\cos 3x}{\cos x} = \frac{\sin 3x \cos x - \cos 3x \sin x}{\sin x \cos x}$$

$$= \frac{\sin (3x - x)}{\sin x \cos x} = \frac{\sin 2x}{\sin x \cos x}$$

$$= \frac{2 \sin x \cos x}{\sin x \cos x} = 2.$$

8-6 TRIGONOMETRIC FUNCTIONS OF HALF AN ANGLE

By using the last two forms of Identity (22) of Section 8-5 for the cosine of twice an angle, we can prove the following identities:

(24) $$\sin \frac{1}{2}\theta = \pm \sqrt{\frac{1 - \cos \theta}{2}},$$

(25) $$\cos \frac{1}{2}\theta = \pm \sqrt{\frac{1 + \cos \theta}{2}},$$

(26) $$\tan \frac{1}{2}\theta = \pm \sqrt{\frac{1 - \cos \theta}{1 + \cos \theta}} \quad \text{unless} \quad \cos \theta = -1.$$

The double sign in the right hand member of these identities is to be interpreted as either + or −; not both signs simultaneously. For example, if $90° < \frac{1}{2}\theta < 180°$, then $\sin\frac{1}{2}\theta$ is positive, $\cos\frac{1}{2}\theta$ is negative, and $\tan\frac{1}{2}\theta$ is negative.

As an aid in memorizing these half-angle identities it is helpful to remember: (1) that each identity is written in terms of the cosine; (2) that the sign preceding the cosine, under the radical, is minus for the sine of half an angle and plus for the cosine of half an angle; and (3) that the sign before the radical is determined by the quadrant in which the half angle lies.

Proof. To obtain Identity (24) we use Identity (22b)

$$\cos 2A = 1 - 2\sin^2 A,$$

and get

$$2\sin^2 A = 1 - \cos 2A.$$

Then by substituting $\frac{1}{2}\theta$ for A we obtain

$$2\sin^2\frac{1}{2}\theta = 1 - \cos 2\left(\frac{1}{2}\theta\right),$$

or

$$\sin\frac{1}{2}\theta = \pm\sqrt{\frac{1-\cos\theta}{2}}.$$

The derivations of Identities (25) and (26), from Identities (22c) and (24), are left as exercises for the student.

Example 1. Given that $90° < \theta < 180°$ and that $\sin\theta = \frac{3}{5}$, find $\sin\frac{1}{2}\theta$, $\cos\frac{1}{2}\theta$, and $\tan\frac{1}{2}\theta$.

Solution: In order to apply the half-angle identities we must find $\cos\theta$ and determine the quadrant in which the half angle lies. We draw a triangle of reference (Fig. 8-5), compute $x = -4$, and obtain $\cos\theta = -\frac{4}{5}$.

Since $90° < \theta < 180°$, $45° < \frac{1}{2}\theta < 90°$, so that the functions of $\frac{1}{2}\theta$ are positive, but $\cos\theta < 0$ and by Identity (24)

$$\sin\frac{1}{2}\theta = \sqrt{\frac{1-\left(-\frac{4}{5}\right)}{2}} = \sqrt{\frac{9}{10}} = \frac{3}{10}\sqrt{10}.$$

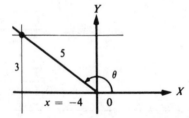

Figure 8-5

By making use of Identity (25) we obtain

$$\cos \frac{1}{2}\theta = \sqrt{\frac{1 + \left(-\frac{4}{5}\right)}{2}} = \sqrt{\frac{1}{10}} = \frac{1}{10}\sqrt{10}.$$

By making use of Identity (26), or $\tan \frac{1}{2}\theta = \dfrac{\sin \frac{1}{2}\theta}{\cos \frac{1}{2}\theta}$, we find

$$\tan \frac{1}{2}\theta = 3.$$

Example 2. Prove the identity

$$\cos^2 \frac{1}{2}x = \frac{\tan x + \sin x}{2 \tan x}.$$

Solution: We shall reduce the right-hand member of the identity to the form of the left-hand member.

$$\frac{\tan x + \sin x}{2 \tan x} = \frac{\dfrac{\sin x}{\cos x} + \sin x}{2\dfrac{\sin x}{\cos x}} = \frac{\sin x \,(1 + \cos x)}{2 \sin x} = \frac{1 + \cos x}{2}$$

$$= \left(\sqrt{\frac{1 + \cos x}{2}}\right)^2 = \cos^2 \frac{1}{2}x, \quad \text{by Identity (25)}.$$

Example 3. Solve the equation

$$3 \cos x - 1 = 2 \sin^2 \frac{1}{2}x$$

for all possible values of x in the interval $0° \le x < 360°$.

Solution: Express $\sin \frac{1}{2}x$ in terms of x and proceed as follows:

$$3 \cos x - 1 = 2 \sin^2 \frac{1}{2}x$$

$$= 2\left(\frac{1 - \cos x}{2}\right) = 1 - \cos x.$$

Thus, $\qquad\qquad 3 \cos x - 1 = 1 - \cos x,$

and $\qquad\qquad\qquad \cos x = \frac{1}{2}.$

Therefore, $\qquad\qquad x = 60° \text{ and } 300°.$

Exercises 8-6

In Exercises 1–4 use the tables to verify each statement.

1. $\sin 80° \neq 2 \sin 40.$

2. $\tan 86° \neq 2 \tan 43°.$

3. $\cos 70° \neq 2 \cos 35°.$

4. $\sin 70° \neq 2 \sin 35°.$

5. By substitution, verify Identities (21), (22), and (23) for $A = 60°.$

6. By substitution, verify Identities (24), (25), and (26) for $\theta = 60°.$

7. Find the values of sine, cosine, and tangent of $105°$ using the value of $\cos 210° = -\frac{1}{2}\sqrt{3}.$ Express these values in simplified radical form.

8. Find the values of sine, cosine, and tangent of $157\frac{1}{2}°$ using the value of $\cos 315° = \frac{1}{2}\sqrt{2}.$ Express these values in simplified radical form.

In Exercises 9–20 simplify each expression by reducing it to a single term involving only one function of an angle.

9. $2 \sin 40° \cos 40°.$

10. $\cos^2 40° - \sin^2 40°.$

11. $\cos^2 50° - \sin^2 50°. = cos100$

12. $2 \sin 20° \cos 20°.$

13. $\dfrac{2 \tan 10°}{1 - \tan^2 10°}.$

14. $\sqrt{\dfrac{1 - \cos 80°}{2}}.$

15. $\sqrt{\dfrac{1 + \cos 70°}{2}}.$

16. $\dfrac{2 \tan 3\alpha}{1 - \tan^2 3\alpha}.$

17. $6 \sin 3\alpha \cos 3\alpha.$

18. $\sin \frac{1}{2}\beta \cos \frac{1}{2}\beta.$

19. $5\sqrt{\dfrac{1 - \cos 5\beta}{1 + \cos 5\beta}}.$

20. $2 \cos^2 5\beta - 1.$

21. Given that $270° < x < 360°$ and that $\tan x = -\frac{5}{12}$, without using tables find $\sin 2x$, $\cos 2x$, $\tan 2x$, $\sin \frac{1}{2}x$, $\cos \frac{1}{2}x$, and $\tan \frac{1}{2}x$, and name the quadrants in which $2x$ and $\frac{1}{2}x$ lie.

22. Find the value of $z = 8 \cos^2 \theta - 24 \sin \theta \cos \theta + \sin^2 \theta$, when $\tan 2\theta = -\frac{24}{7}$ and $270° < 2\theta < 360°.$

In Exercises 23–40 prove the identities by operating on one member and leaving the other unchanged.

23. $\dfrac{2 \cot x}{\csc^2 x - 2} = \tan 2x.$

24. $\dfrac{1 - \tan^2 x}{1 + \tan^2 x} = \cos 2x.$

25. $\cos 3x = 4 \cos^3 x - 3 \cos x.$

26. $\dfrac{\sin 5x}{\sin x} - \dfrac{\cos 5x}{\cos x} = 4 \cos 2x.$

27. $\dfrac{1 - \tan \frac{1}{2}x}{1 + \tan \frac{1}{2}x} = \dfrac{1 - \sin x}{\cos x}.$

28. $\sin^4 x = \frac{1}{8}(3 - 4 \cos 2x + \cos 4x).$

29. $\dfrac{\cos^3 x + \sin^3 x}{\cos x + \sin x} = 1 - \frac{1}{2} \sin 2x.$

30. $\dfrac{1 - \tan \frac{1}{2}x}{1 + \tan \frac{1}{2}x} = \dfrac{\cos x}{1 + \sin x}.$

31. $\tan (x + 45°) + \tan (x - 45°) = 2 \tan 2x.$

32. $\tan (x + 45°) - \tan (x - 45°) = 2 \sec 2x.$

33. $2(\cos 3x \cos x + \sin 3x \sin x)^2 = 1 + \cos 4x.$

34. $2(\sin 2x \cos x - \cos 2x \sin x)^2 = 1 - \cos 2x.$

35. $\dfrac{\sin 2x}{2 \sin x} = \cos^2 \frac{1}{2}x - \sin^2 \frac{1}{2}x.$

36. $\cos^2 2x - \cos^2 x = \sin^2 x - \sin^2 2x.$

37. $\dfrac{1 + \sin 2x + \cos 2x}{1 + \sin 2x - \cos 2x} = \cot x.$

38. $\tan \frac{1}{2}x = \dfrac{\sec x - 1}{\tan x}.$

39. $\frac{1}{2}[\sin (x + y) + \sin (x - y)] = \sin x \cos y.$

40. $\frac{1}{2}[\cos (x + y) + \cos (x - y)] = \cos x \cos y.$

8-7 PRODUCTS OF SINES AND COSINES EXPRESSED AS SUMS

The identities that we wish to develop in this section are:

$$(27) \qquad \sin A \cos B = \frac{1}{2}[\sin (A + B) + \sin (A - B)],$$

$$(28) \qquad \cos A \sin B = \frac{1}{2}[\sin (A + B) - \sin (A - B)],$$

$$(29) \qquad \sin A \sin B = \frac{1}{2}[\cos (A - B) - \cos (A + B)],$$

$$(30) \qquad \cos A \cos B = \frac{1}{2}[\cos (A + B) + \cos (A - B)].$$

Examples.

1. $\sin 3\theta \cos 2\theta = \frac{1}{2}[\sin (3\theta + 2\theta) + \sin (3\theta - 2\theta)]$

$\qquad\qquad = \frac{1}{2}[\sin 5\theta + \sin \theta].$

2. $\cos 55° \cos 15° = \frac{1}{2}[\cos (55° + 15°) + \cos (55° - 15°)]$

$\qquad\qquad = \frac{1}{2}[\cos 70° + \cos 40°].$

These identities, and those of Section 8-8, are important in certain aspects of calculus and applied mathematics; however in this book we shall make little use of them, and it is not necessary to memorize them. The student should, however, be able to derive them. The derivation is not difficult.

To derive Identity (27), we add Identities (17) and (18) of Section 8-3.

(17) $\qquad\qquad\qquad \sin (A + B) = \sin A \cos B + \cos A \sin B,$

(18) $\qquad\qquad\qquad \underline{\sin (A - B) = \sin A \cos B - \cos A \sin B,}$

$\qquad\qquad \sin (A + B) + \sin (A - B) = 2 \sin A \cos B.$

Dividing by 2, we obtain Identity (27),

$$\sin A \cos B = \frac{1}{2}[\sin (A + B) + \sin (A - B)].$$

Subtracting Identity (18) from Identity (17), we get

$$\sin (A + B) - \sin (A - B) = 2 \cos A \sin B.$$

We divide by 2 to obtain Identity (28).

$$\cos A \sin B = \frac{1}{2}[\sin (A + B) - \sin (A - B)].$$

To derive Identity (30), we add Identities (15) and (16) of Section 8-2.

(15) $\qquad\qquad\qquad \cos (A + B) = \cos A \cos B - \sin A \sin B,$

(16) $\qquad\qquad\qquad \underline{\cos (A - B) = \cos A \cos B + \sin A \sin B,}$

$\qquad\qquad \cos (A + B) + \cos (A - B) = 2 \cos A \cos B.$

We divide by 2 to obtain Identity (30).

$$\cos A \cos B = \frac{1}{2}[\cos (A + B) + \cos (A - B)].$$

To derive Identity (29), we subtract Identity (16) from Identity (15) and obtain

$$\cos (A + B) - \cos (A - B) = -2 \sin A \sin B.$$

We divide by -2 to obtain Identity (29).

$$\sin A \sin B = \frac{1}{2}[\cos (A - B) - \cos (A + B)].$$

8-8 SUMS OF SINES AND COSINES EXPRESSED AS PRODUCTS

In Identities (27), (28), (29), and (30) we let $A + B = x$ and $A - B = y$. Then upon adding and subtracting these two equations

$$A + B = x \qquad\qquad A + B = x,$$
$$\underline{A - B = y} \qquad\qquad A - B = y,$$
$$2A \qquad = x + y \qquad\qquad 2B = x - y,$$

we get $\qquad A = \frac{1}{2}(x + y) \qquad$ and $\qquad B = \frac{1}{2}(x - y).$

Using these values for $A + B$, $A - B$, A, and B in the same identities, we get

$$(31) \qquad \mathbf{\sin x + \sin y = 2 \sin \frac{1}{2}(x + y) \cos \frac{1}{2}(x - y),}$$

$$(32) \qquad \mathbf{\sin x - \sin y = 2 \cos \frac{1}{2}(x + y) \sin \frac{1}{2}(x - y),}$$

$$(33) \qquad \mathbf{\cos x + \cos y = 2 \cos \frac{1}{2}(x + y) \cos \frac{1}{2}(x - y),}$$

$$(34) \qquad \mathbf{\cos x - \cos y = -2 \sin \frac{1}{2}(x + y) \sin \frac{1}{2}(x - y).}$$

Example 1. Prove the identity $\dfrac{\sin 4\theta - \sin 2\theta}{\cos 4\theta + \cos 2\theta} = \tan \theta.$

Solution: Applying Identity (32) to the numerator and Identity (33) to the denominator of the left-hand member, we find

$$\frac{\sin 4\theta - \sin 2\theta}{\cos 4\theta + \cos 2\theta} = \frac{2 \cos \frac{1}{2}(4\theta + 2\theta) \sin \frac{1}{2}(4\theta - 2\theta)}{2 \cos \frac{1}{2}(4\theta + 2\theta) \cos \frac{1}{2}(4\theta - 2\theta)}$$

$$= \frac{2 \cos 3\theta \sin \theta}{2 \cos 3\theta \cos \theta} = \tan \theta.$$

Example 2. Solve the equation $\sin 4x + \sin 2x = \cos x$ for all values of x in the interval $0° \le x < 360°$.

Solution: Applying Identity (31) to the left-hand member of the equation, we get

$$\sin 4x + \sin 2x = \cos x,$$

$$2 \sin 3x \cos x = \cos x.$$

Then $$2 \sin 3x \cos x - \cos x = 0$$

or $$\cos x(2 \sin 3x - 1) = 0.$$

This is equivalent to the two equations

$$\cos x = 0 \quad \text{and} \quad \sin 3x = \frac{1}{2}.$$

From $\cos x = 0$ we get $x = 90°$ and $270°$.

From $\sin 3x = \frac{1}{2}$ we get $3x = 30° + n360°$ and $3x = 150° + n360°$

or $$x = 10° + n120° \quad \text{and} \quad x = 50° + n120°.$$

Therefore,

$$x = 10°, 130°, 250° \quad \text{and} \quad x = 50°, 170°, 290°.$$

The values of x are

$$x = 10°, 50°, 90°, 130°, 170°, 250°, 270°, \text{ and } 290°,$$

all of which satisfy the original equation.

Exercises 8-8

In Exercises 1–6 transform each of the products into a sum or difference of sines or cosines.

1. $2 \sin 35° \cos 25°$.

2. $2 \cos 55° \cos 35°$.

3. $6 \cos 5\theta \cos 3\theta$.

4. $8 \sin 7\theta \cos \theta$.

5. $\frac{3}{2} \sin 2\beta \sin \beta$.

6. $\frac{1}{2} \cos 3\beta \sin 2\beta$.

In Exercises 7–12 transform each of the sums or differences into a product of sines and cosines.

7. $\sin 40° + \sin 30°$.

8. $\cos 50° + \cos 20°$.

9. $\cos 3\theta + \cos 5\theta$.

10. $\sin 5\theta - \sin 7\theta$.

11. $\sin \frac{7}{2}\alpha - \sin \frac{3}{2}\alpha$.

12. $\sin \frac{5}{2}\alpha + \sin \frac{1}{2}\alpha$.

In Exercises 13–30 prove the identities by operating on one member and leaving the other unchanged.

USE IDENTITY TABLES

13. $\sin (150° + x) + \sin (150° - x) = \cos x.$

14. $\cos (135° + x) - \cos (135° - x) = -\sqrt{2} \sin x.$

15. $\cos (45° + x) + \cos (45° - x) = \sqrt{2} \cos x.$

16. $\sin (30° + x) + \sin (30° - x) = \cos x.$

17. $\cos (x + 30°) - \cos (x - 30°) = -\sin x.$

18. $\sin 2x + 2 \sin 4x + \sin 6x = 4 \cos^2 x \sin 4x.$

19. $\sin^2 6x - \sin^2 4x = \sin 2x \sin 10x.$

20. $\cos^2 2x - \cos^2 6x = \sin 4x \sin 8x.$

21. $\sin (2x - y) \cos y + \cos (2x - y) \sin y = \sin 2x.$

22. $\cos (2x + y) \cos y + \sin (2x + y) \sin y = \cos 2x.$

23. $\dfrac{\cos 70° + \cos 50°}{\sin 280° + \sin 260°} = -\tfrac{1}{2}.$

24. $\dfrac{\cos 9x - \cos 5x}{\sin 17x - \sin 3x} = -\sin 2x \sec 10x.$

25. $\dfrac{\sin 5x + \sin 3x}{\cos 5x + \cos 3x} = \tan 4x.$

26. $\dfrac{\sin \alpha - \sin \theta}{\cos \alpha + \cos \theta} = \tan \tfrac{1}{2}(\alpha - \theta).$

27. $\dfrac{\sin 40° - \sin 20°}{\cos 220° - \cos 200°} = \sqrt{3}.$

28. $\dfrac{\sin x - \sin 3x}{\sin^2 x - \cos^2 x} = 2 \sin x.$

29. $\dfrac{\sin 5x - 2 \sin 3x + \sin x}{\cos 5x - \cos x} = \tan x.$

30. $\dfrac{(\sin 3x + \sin 9x) + (\sin 5x + \sin 7x)}{(\cos 3x + \cos 9x) + (\cos 5x + \cos 7x)} = \tan 6x.$

In Exercises 31–50 solve the equations for all values of x in the interval $0 \leqq x < 2\pi$.

31. $\sin 3x = 1.$

32. $\cos 3x = 1.$

33. $\cos \tfrac{1}{2}x = 0.$

34. $\sin \tfrac{1}{2}x = 1.$

35. $\cos 2x(2 \cos x - 1) = 0.$

36. $\sin 2x(2 \sin x + 1) = 0.$

37. $\sin 2x(\tan x - 1) = 0.$

38. $\cos x(\tan x + 1) = 0.$

39. $\sin 2x - \sin x = 0.$

40. $\cos 2x + \cos x = 0.$

41. $\cos 3x + \cos x = 0.$

42. $\sin 3x - \sin x = 0.$

43. $\sin 2x - \cos x = 0.$

44. $\sin 4x + \sin 2x = \cos x.$

45. $\sin 3x + 4 \sin^2 x = 0.$

46. $\cos 7x - \cos x = 0.$

47. $\cos 5x + \cos 3x = 2 \cos 4x.$

48. $\sin 4x + \sin 2x = \cos x.$

49. $\sin x + \sin 3x + \sin 5x = 0.$

50. $\cos x + \cos 2x + \cos 3x = 0.$

8-9 REDUCTION OF $a \cos \theta + b \sin \theta$ TO $c \cos (\theta + \alpha)$

It is often desirable to transform a sum of two terms of the form $a \cos \theta + b \sin \theta$ into a single term such as $c \cos (\theta - \alpha)$.

We let the real nonzero numbers a and b of the expression $a \cos \theta + b \sin \theta$ represent the coordinates of a point (a, b) on the rectangular coordinate plane. Let α be an angle in standard position with its terminal side passing through the point (a, b) (Fig. 8-6). Then

(1) $\sin \alpha = \dfrac{b}{\sqrt{a^2 + b^2}}, \qquad \cos \alpha = \dfrac{a}{\sqrt{a^2 + b^2}}, \qquad \tan \alpha = \dfrac{b}{a}.$

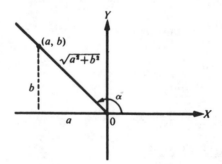

Figure 8-6

We multiply and divide the expression $a \cos \theta + b \sin \theta$ by $\sqrt{a^2 + b^2}$, substitute the values of $\sin \alpha$ and $\cos \alpha$ of (1) into the resulting expression, and apply Identity (16).

$$a \cos \theta + b \sin \theta = \sqrt{a^2 + b^2}\left(\frac{a}{\sqrt{a^2 + b^2}}\cos \theta + \frac{b}{\sqrt{a^2 + b^2}}\sin \theta\right)$$

$$= \sqrt{a^2 + b^2}(\cos \alpha \cos \theta + \sin \alpha \sin \theta)$$

$$= \sqrt{a^2 + b^2}(\cos \theta \cos \alpha + \sin \theta \sin \alpha)$$

$$= \sqrt{a^2 + b^2} \cos (\theta - \alpha).$$

Hence,

---

if θ is any angle,

(35) $\qquad\qquad a \cos \theta + b \sin \theta = c \cos(\theta - \alpha),$

where

$$c = \sqrt{a^2 + b^2}, \sin \alpha = \frac{b}{c}, \cos \alpha = \frac{a}{c}, \tan \alpha = \frac{b}{a}.$$

Example 1. Transform $4 \cos \theta + 3 \sin \theta$ to the form $c \cos(\theta - \alpha)$, and find c and α.

Solution: Plot the point $(4, 3)$ and place α in standard position with its terminal side passing through $(4, 3)$. (See Fig. 8-7a.) Then $c = \sqrt{4^2 + 3^2} = 5$, $\sin \alpha = \frac{3}{5}$, $\cos \alpha = \frac{4}{5}$, and $\tan \alpha = \frac{3}{4}$. Therefore, $\alpha \approx 36° 52'$ or any angle coterminal with $36° 52'$. By applying Formula (35) we obtain

$$4 \cos \theta + 3 \sin \theta \approx 5 \cos(\theta - 36° 52').$$

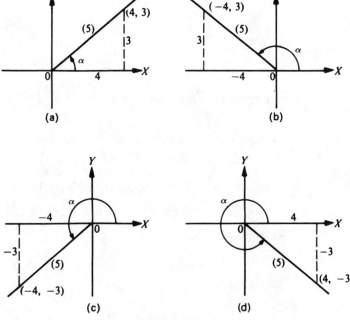

Figure 8-7

To further clarify what Formula (35) is doing for us in this problem, we show the following development. We plot the point (4, 3) and compute c; then

$$4 \cos \theta + 3 \sin \theta = 5 \left(\frac{4}{5} \cos \theta + \frac{3}{5} \sin \theta \right)$$

$$= 5(\cos \alpha \cos \theta + \sin \alpha \sin \theta)$$

$$= 5 \cos (\theta - \alpha)$$

$$= 5 \cos (\theta - 36° \, 52').$$

Example 2. Transform $-4 \cos \theta + 3 \sin \theta$ to the form $c \cos (\theta - \alpha)$, and find c and α.

Solution: Plot the point $(-4, 3)$, and place α in standard position with its terminal side passing through $(-4, 3)$. (See Fig. 8-7b.) Then we have $c = \sqrt{(-4)^2 + 3^2} = 5$, $\sin \alpha = \frac{3}{5}$, $\cos \alpha = -\frac{4}{5}$, and $\tan \alpha = -\frac{3}{4}$. Therefore, $\alpha \approx 143° \, 08'$, and by applying Formula (35) we obtain

$$-4 \cos \theta + 3 \sin \theta \approx 5 \cos (\theta - 143° \, 08').$$

As an alternate approach to this transformation, we plot the point $(-4, 3)$ and apply Formula (1) of Section 8-9 as follows:

$$-4 \cos \theta + 3 \sin \theta = 5 \left(-\frac{4}{5} \cos \theta + \frac{3}{5} \sin \theta \right)$$

$$= 5 \, (\cos \alpha \cos \theta + \sin \alpha \sin \theta)$$

$$= 5 \cos (\theta - \alpha)$$

$$\approx 5 \cos (\theta - 143° \, 08').$$

Example 3. Transform $-4 \cos \theta - 3 \sin \theta$ to the form $c \, \cos (\theta - \alpha)$, and find c and α.

Solution: Plot the point $(-4, -3)$, and place α in standard position with its terminal side passing through $(-4, -3)$. (See Fig. 8-7c.) Then $c = \sqrt{(-4)^2 + (-3)^2} = 5$, $\sin \alpha = -\frac{3}{5}$, $\cos \alpha = -\frac{4}{5}$, and $\tan \alpha = \frac{3}{4}$. Therefore, $\alpha \approx 216° \, 52'$, and by applying Formula (35) we obtain

$$-4 \cos \theta - 3 \sin \theta \approx 5 \cos (\theta - 216° \, 52').$$

An alternate solution:

$$-4 \cos \theta - 3 \sin \theta = 5 \left(-\frac{4}{5} \cos \theta - \frac{3}{5} \sin \theta \right)$$

$$= 5(\cos \theta \cos \alpha + \sin \theta \sin \alpha)$$

$$\approx 5 \cos (\theta - 216° \, 52')$$

Example 4. Transform $4 \cos \theta - 3 \sin \theta$ to the form $c \cos (\theta - \alpha)$, and find c and α.

Solution: Plot the point $(4, -3)$, and place α in standard position with its terminal side passing through $(4, -3)$. (See Fig. 8-7d.) Then $c = \sqrt{4^2 + (-3)^2} = 5$, $\sin \alpha = -\frac{3}{5}$, $\cos \alpha = \frac{4}{5}$, and $\tan \alpha = -\frac{3}{4}$. Therefore, $\alpha \approx 323° 08'$, and when we apply Formula (35) we obtain

$$4 \cos \theta - 3 \sin \theta \approx 5 \cos (\theta - 323° 08').$$

Example 5. Show that the maximum value of the function $a \cos \theta + b \sin \theta = c \cos (\theta - \alpha)$ is $\sqrt{a^2 + b^2}$ and occurs when $\theta = \alpha$.

Solution: The maximum value of $a \cos \theta + b \sin \theta = c \cos (\theta - \alpha)$ occurs when $\cos (\theta - \alpha)$ equals its greatest value, 1. When

$$\cos (\theta - \alpha) = 1,$$

$$\theta - \alpha = 0,$$

and $$\theta = \alpha.$$

Hence, the maximum value of the function $a \cos \theta + b \sin \theta$ is $\sqrt{a^2 + b^2} = c$, which occurs when $\theta = \alpha$. Of course, the function $c \cos (\theta - \alpha)$ is periodic and has an unlimited number of values of θ, all of which yield the maximum value of c for the function. These values of θ can be found as follows:

$$\cos (\theta - \alpha) = 1,$$

$$\theta - \alpha = n360°, \quad (n \text{ is any integer})$$

$$\theta = \alpha + n360°.$$

Example 6. Show that the minimum value of the function $a \cos \theta + b \sin \theta = c \cos (\theta - \alpha)$ is $-\sqrt{a^2 + b^2}$ and occurs when $\theta = \alpha + (2n + 1)180°$ (n is any integer).

Solution: The minimum value of the function

$$a \cos \theta + b \sin \theta = c \cos (\theta - \alpha)$$

occurs when $\cos (\theta - \alpha)$ has its least value, -1.
When

$$\cos (\theta - \alpha) = -1,$$

$$\theta - \alpha = 180° + n360°, \quad (n \text{ is any integer})$$

$$\theta = \alpha + 180° + n360°,$$

$$\theta = \alpha + (2n + 1)180°.$$

By applying Formula (35) we can quickly transform the function $a \cos \theta + b \sin \theta$ (a composite wave form with two terms) into an equivalent function with wave form and a single term, from which we can easily determine the maximum, minimum, and periodic properties of the function.

Exercises 8-9

In Exercises 1–10 transform each expression to the form $c \cos (\theta - \alpha)$, and determine c and a value of α in the interval $0 \leqq \alpha < 360°$.

1. $5 \cos \theta + 12 \sin \theta$.   2. $12 \cos \theta + 5 \sin \theta$.

3. $5 \cos \theta - 12 \sin \theta$.   4. $12 \cos \theta - 5 \sin \theta$.

5. $-12 \cos \theta - 5 \sin \theta$.   6. $-5 \cos \theta - 12 \sin \theta$.

7. $-12 \cos \theta + 5 \sin \theta$.   8. $-5 \cos \theta + 12 \sin \theta$.

9. $3 \sin \theta - 2 \cos \theta$.   10. $2 \sin \theta - 3 \cos \theta$.

11. Find the maximum value of the function $\cos \theta - 2 \sin \theta$ and a value of θ in the interval $0 \leqq \theta < 360°$ at which this maximum occurs.

12. Find the minimum value of the function $-24 \cos \theta + 7 \sin \theta$ and the smallest positive value of θ at which this minimum first occurs.

13. Find the smallest positive value of θ for which $-5 \cos \theta + 12 \sin \theta = 0$.

14. Find the smallest positive value of θ for which $5 \cos \theta - 12 \sin \theta = 0$.

15. Let the real nonzero numbers a and b of the expression $a \cos \theta + b \sin \theta$ represent the coordinates of the point (b, a) on the rectangular coordinate plane. Give the derivation of the formula

$$a \cos \theta + b \sin \theta = k \sin (\theta + \beta),$$

where $k = \sqrt{a^2 + b^2}$, $\sin \beta = a/k$, $\cos \beta = b/k$, and $\tan \beta = a/b$.

16. Express the right-hand member of the expression

$$V = 100 \sin 500t + 200 \cos 500t$$

as a single term, and give the maximum value of V.

REVIEW EXERCISES

1. Given $\sin A = \frac{4}{5}$, with A in Q I, and $\cos B = -\frac{12}{13}$, with B in Q III, find (a) $\sin (A + B)$ and (b) $\cos (A - B)$.

2. Given $\tan A = \frac{3}{4}$, with A in Q I, and $\sin B = \frac{4}{5}$, with B in Q II, find (a) $\sin (A - B)$ and (b) $\tan (A + B)$.

3. Use an identity to compute the value of $\tan 105°$ from the functions of $60°$ and $45°$. Express your answer in simplified radical form.

4. Use an identity to compute the value of $\tan 15°$ from the functions of $45°$ and $30°$. Express your answer in simplified radical form.

In Exercises 5–10 prove the identities.

5. $\tan (\pi - \theta) = -\tan \theta$.

6. $\cos \left(\dfrac{\pi}{2} + \theta \right) = -\sin \theta$.

7. $\sin (\pi + \theta) = -\sin \theta$.

8. $\sin (\pi - \theta) = \sin \theta$.

9. $\dfrac{\sin 2x}{\sin x} - \dfrac{\cos 2x}{\cos x} = \sec x$.

10. $\dfrac{\cos 4x}{\sin 2x} + \dfrac{\sin 4x}{\cos 2x} = \csc 2x$.

In Exercises 11–16 simplify each expression by reducing it to a single term involving only one function of an angle.

11. $2 \sin 25° \cos 25°$.

12. $\cos^2 50° - \sin^2 50°$.

13. $\dfrac{2 \tan 20°}{1 - \tan^2 20°}$.

14. $\sqrt{\dfrac{1 - \cos 70°}{2}}$.

15. $\sin (210° + x) + \sin (210° - x)$.

16. $\cos (x - 60°) - \cos (x + 60°)$.

17. Simplify $\sin (-x) + \cos (-x) + \sin x + \cos x$.

18. Simplify $\tan (-x) + \sec (-x) - \tan x - \sec x$.

In Exercises 19 and 20 solve for all the values of x in the interval $0 \leqq x < 2\pi$.

19. $\sin 2x + \cos x = 0$.

20. $\cos 2x - \cos x = 0$.

21. Find $\tan \theta$ if $\tan 2\theta = \frac{4}{3}$ and $0° < \theta < 90°$.

22. Transform $3 \cos \theta + 4 \sin \theta$ to the form $c \cos (\theta - \alpha)$, and find c and α.

CHAPTER 8: DIAGNOSTIC TEST

The purpose of this test is to see how well you understand the work covered in Chapter 8. We recommend that you work this test before your instructor tests you on this chapter. Allow yourself approximately 50 minutes to do the test.

Solutions to the problems, together with section references, are given in the Answer Section at the end of this book. We suggest that you study the sections referred to for the problems you do incorrectly.

---

1. Given $\tan \theta = \frac{4}{3}$ and $0° < \theta < 90°$, find the following:
    (a) $\sin 2\theta$.   (b) $\cos \frac{1}{2}\theta$.   (c) $\tan 2\theta$.

2. Given $\cos A = -\frac{3}{5}$, with A in Q II, and $\tan B = \frac{4}{3}$, with B in Q I, find the following:
    (a) $\sin (A - B)$.   (b) $\cos (A + B)$.   (c) $\tan (A - B)$.

3. Expand and simplify

    (a) $\tan (\pi + \theta)$.   (b) $\cos \left(\dfrac{\pi}{2} - \theta\right)$.

4. Reduce each expression to a single trigonometric function.

    (a) $\dfrac{2 \tan 40°}{1 - \tan^2 40°}$.   (b) $2 \sin 3x \cos 3x$.

    (c) $\sqrt{\dfrac{1 - \cos 4x}{1 + \cos 4x}}$.   (d) $\tan x + \tan (-x) + \cos (-x)$.

5. Transform the sum, $\sin 7\theta + \sin \theta$, into a product of sines and cosines.

    In Problems 6–8 solve each equation for all the values of x in the interval $0 \leq x < 2\pi$.

6. $\sin 2x + \sin x = 0$.

7. $\tan \frac{1}{2}x = \sqrt{3}$.

8. $\cos 2x = 1$.

9. Prove the identity $\dfrac{\sin 5\theta - \sin 3\theta}{\cos 5\theta + \cos 3\theta} = \tan \theta$.

10. Transform the expression $-3 \cos \theta + 4 \sin \theta$ to the form $c \cos (\theta - \alpha)$, and find c and α.

9

Graphical Methods

9-1 GENERAL TECHNIQUE

In Sections 2-17, 3-9, 4-8, 5-5, 5-6, and 5-7 we discussed the basic graphs of the sine, cosine, tangent, cotangent, secant, and cosecant functions; in Section 3-11 we discussed graphing by the addition of ordinates, sometimes called the composition of ordinates. We shall now deal with graphs of the more general forms of these functions.

Part of our previous method of graphing trigonometric functions was to assume arbitrary units of measure for the degree along the horizontal axis. Now that we have a knowledge of radian measure we shall, in general, use identical units of measure on both the horizontal and vertical axes.

In graphing the more general forms of the six basic trigonometric functions we shall in most cases (1) locate some of the consecutive *x*-intercepts (the points where the curve crosses the *x*-axis); (2) find the upper and lower bounds, if these exist, or the vertical lines, if they exist, which separate the branches of the curve; (3) locate a few points on the curve; and (4) then sketch the balance of the curve from our previous knowledge of the general shape of the curve in question.

175

9-2 ANALYSIS OF THE GENERAL SINE
AND COSINE CURVES

The graph of $y = a \sin bx$, where a and b are constants, is illustrated in the following example.

Example 1. Graph the function $y = 2 \sin 3x$, where x is expressed in radian measure, and units are the same on the horizontal and vertical axes.

Solution: We are familiar with the general shape of the graph of $y = \sin \theta$ (Figs. 2-26 and 2-27, Section 2-17); our concern here is to (1) locate some of the consecutive x-intercepts; (2) determine the upper and lower bounds; (3) locate a few points on the curve; and then (4) use this information to sketch the curve. We set $y = 0$ and find some of the consecutive x-intercepts.

$$2 \sin 3x = 0,$$

$$\sin 3x = 0,$$

$$3x = n\pi \quad (n \text{ an integer}),$$

$$x = n\frac{1}{3}\pi.$$

Therefore, some of the x-intercepts are

$$-\pi, \ -\frac{2}{3}\pi, \ -\frac{1}{3}\pi, 0, \frac{1}{3}\pi, \frac{2}{3}\pi, \pi, \text{ etc. (See Fig. 9.1.)}$$

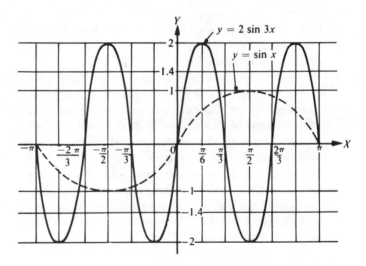

Figure 9-1

Since $\pi \approx 3.1416$, $\frac{1}{3}\pi \approx \frac{1}{3}(3.1416) = 1.0472 \approx 1$. Therefore, we shall make only a scarcely discernible distortion of the curves if, in graphing the trigonometric functions, we use one horizontal unit of distance to represent $\frac{1}{3}\pi$. We agree to make this convenient close approximation.

The amplitude (greatest value of $y = 2 \sin 3x$) is 2; therefore, this curve oscillates back and forth between $y = 2$ and $y = -2$. We know now where the curve crosses the x-axis and its upper and lower bounds, but we do not know when it is above or below the x-axis. To find this information we solve $y = 2 \sin 3x$ for y using some value of x midway between two consecutive x-intercepts. We set $x = \frac{1}{6}\pi$; then

$$y = 2 \sin 3x$$

$$= 2 \sin 3\left(\frac{1}{6}\pi\right)$$

$$= 2 \sin \left(\frac{1}{2}\pi\right) = 2.$$

Consequently, the curve is above the x-axis when $0 < x < \frac{1}{3}\pi$. Generally, one more point, such as the one for $x = \frac{1}{12}\pi$, $y = 2 \sin (3 \cdot \frac{1}{12}\pi) = \sqrt{2}$ ≈ 1.4, is sufficient for determining the shape of the arc. We draw this arc, and with our previous knowledge of the sine curve we know that congruent arcs lie below the x-axis between $x = -\pi$ and $-\frac{2}{3}\pi$, $-\frac{1}{3}\pi$ and 0, etc. The graphs of $y = \sin x$ and $y = 2 \sin 3x$ are shown together in Figure 9-1.

The function $y = \sin x$ has a period of 2π and an amplitude of 1; the function $y = 2 \sin 3x$ has a period of $\frac{1}{3}(2\pi)$ and an amplitude of 2. For any constant $b > 0$, the quantity bx may be said to increase "b times as fast" as x. Consequently, $\sin bx$ goes through b cycles, or periods, while $\sin x$ goes through one period. The period of $y = a \sin bx$, where $b > 0$, is therefore $2\pi/b$; its amplitude, or greatest value, is $|a|$.

If $b < 0$ in the function $y = a \sin bx$, then write $y = a \sin bx = a \sin [-(-bx)] = -a \sin (-bx)$. The coefficient of x is then $-b > 0$ and the period of $y = a \sin bx$ is $2\pi/-b$. In either case ($b < 0$ or $b > 0$) the period of $y = a \sin bx$ is $\dfrac{2\pi}{|b|}$.

To find the intercepts for $y = a \sin bx$, set $y = 0$ and solve for x.

$$a \sin bx = 0$$

$$\sin bx = 0$$

$$bx = 0 + n\pi$$

$$x = \frac{n}{b}\pi$$

where n is any integer.

The graph of $y = a \cos bx$ can be made in a similar way. It has a period of $2\pi/b$ and amplitude $|a|$. To find its x-intercepts set $y = 0$; then

$$a \cos bx = 0,$$

$$\cos bx = 0,$$

$$bx = \frac{1}{2}\pi + n\pi$$

Therefore

$$x = \frac{1}{2} \cdot \frac{\pi}{b} + n\frac{\pi}{b}$$

$$= (2n + 1)\frac{\pi}{2b}. \qquad (n \text{ any integer})$$

The graph of $y = a \sin(bx + c)$, where a, b, and c are constants, is illustrated in the next examples.

Example 2. Graph the function $y = \sin(x + \frac{1}{2}\pi)$.

Solution: Our first concern is to find the x-intecepts. To find these x-intercepts we set $y = 0$. Then

$$\sin\left(x + \frac{1}{2}\pi\right) = 0,$$

$$x + \frac{1}{2}\pi = n\pi,$$

$$x = n\pi - \frac{1}{2}\pi,$$

and

$$x = -\frac{1}{2}\pi, \frac{1}{2}\pi, \frac{3}{2}\pi, \frac{5}{2}\pi, \text{etc.}$$

The amplitude of this curve is 1. The graphs of $y = \sin x$ and $y = \sin(x + \frac{1}{2}\pi)$ are shown together in Figure 9-2.

It is evident from Example 2 that the graph of $y = \sin(x + \frac{1}{2}\pi)$ may be obtained from that of $y = \sin x$ by shifting the graph of $y = \sin x$ a distance of $\frac{1}{2}\pi$ to the left, as shown in Figure 9-2. It is important to note that when we apply Identity (17) to the right-hand member of $y = \sin(x + \frac{1}{2}\pi)$, the function reduces to $y = \cos x$, the graph of which is shifted a distance of $\frac{1}{2}\pi$ to the left of the graph of $y = \sin x$. We say that the graph of $y = \sin(x + \frac{1}{2}\pi)$ *differs in phase* by $\frac{1}{2}\pi$ from the graph of $y = \sin x$.

The period of $y = a \sin(bx + c)$ is $\dfrac{2\pi}{|b|}$, and its amplitude is $|a|$. A zero of y comes where $(bx + c) = 0$ or at $x = -c/b$, instead of at $x = 0$

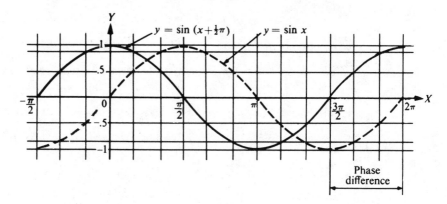

Figure 9-2

as it does for the graph of $y = a \sin bx$. Hence, the graph of $y = a \sin (bx + c)$ is the same as that of $y = a \sin bx$ except that the whole curve is displaced $\left|\dfrac{c}{b}\right|$ units (Fig. 9-3). The amount of such a displacement is called the *phase difference* between the two curves.

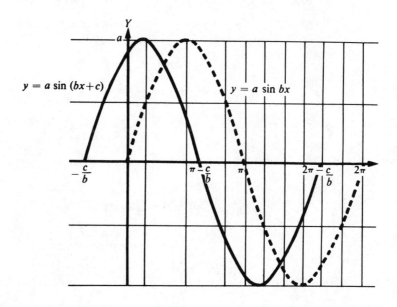

Phase difference $= \dfrac{c}{b}$

Figure 9-3

The graph of $y = a \sin(bx + c)$ *has a simple wave form with amplitude* $|a|$ *and period* $\dfrac{2\pi}{|b|}$ *differing in phase from that of* $y = \sin x$ *by* $\left|\dfrac{c}{b}\right|$. *The phase shift is left when* c/b *is positive and right when* c/b *is negative.*

The graph of $y = a \cos(bx + c)$ *has a simple wave form with amplitude* $|a|$ *and period* $\dfrac{2\pi}{|b|}$ *differing in phase from that of* $y = \cos x$ *by* $\left|\dfrac{c}{b}\right|$. *The phase shift is left when* c/b *is positive and right when* c/b *is negative.*

Example 3. Find the period and amplitude of each of the following functions:

(a) $3 \sin \pi x$.
(b) $-2 \sin\left(\frac{2}{5}x - \frac{1}{3}\pi\right)$.

Solution:

(a) Period $= 2\pi/b = 2\pi/\pi = 2$. The amplitude $= |a| = |3| = 3$.
(b) Period $= 2\pi/\frac{2}{5} = 5\pi$. The amplitude $= |-2| = 2$.

Example 4. Graph $y = 3 \cos\left(\frac{1}{2}x - \pi\right)$ for one complete wave.

Solution: First we set $y = 0$ and find three consecutive x-intercepts.
When $y = 0$,

$$\cos\left(\frac{1}{2}x - \pi\right) = 0,$$

$$\frac{1}{2}x - \pi = \frac{\pi}{2} + n\pi,$$

$$\frac{1}{2}x = \frac{3\pi}{2} + n\pi,$$

$$x = 3\pi + 2n\pi,$$

$$x = (3 + 2n)\pi.$$

We give n values and find x-intercepts.

Points

n	x	y
-2	$-\pi$	0
-1	π	0
0	3π	0

Next we find a point on the curve when $x = 0$. ($x = 0$ is midway between the two consecutive x-intercepts, $-\pi$ and π.)

$$y = 3 \cos \left(\frac{1}{2} \cdot 0 - \pi \right)$$

$$= 3 \cos (-\pi)$$

$$= -3.$$

(This point $(0, -3)$ is a minimum point on the curve which when combined with the three points already found enables us to complete the curve in Figure 9-4.)

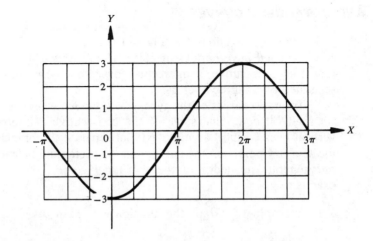

Figure 9-4

Exercises 9-2

In Exercises 1–10 find the period and amplitude of each function.

1. $3 \sin 2x$.
2. $2 \cos 3x$.
3. $5 \cos \frac{2}{3}x$.
4. $3 \sin \frac{3}{4}x$.
5. $-2 \sin \frac{1}{6}\pi x$.
6. $-5 \cos \frac{1}{3}\pi x$.
7. $4 \cos \left(x + \frac{1}{2}\pi \right)$.
8. $4 \sin \left(x - \frac{1}{3}\pi \right)$.
9. $-5 \sin \left(\frac{1}{5}\pi x - \frac{1}{6}\pi \right)$.
10. $-3 \sin \left(\frac{1}{2}\pi x + \frac{1}{3}\pi \right)$.

In Exercises 11–22 graph each function for one complete wave.

11. $y = 3 \sin 2x$.
12. $y = 2 \cos 3x$.
13. $y = \cos \frac{1}{2}x$.
14. $y = \sin \frac{1}{2}x$.
15. $y = -2 \cos 2x$.
16. $y = -2 \sin 2x$.

17. $y = \frac{1}{2}\sin \pi x.$

18. $y = \frac{2}{3}\cos \pi x.$

19. $y = 2\sin\left(\frac{2}{3}x + \pi\right).$

20. $y = 2\cos\left(\frac{1}{3}x - \pi\right).$

21. $y = 3\cos\left(\pi x - \frac{1}{4}\pi\right).$

22. $y = 3\sin\left(\frac{1}{2}\pi x + \frac{1}{4}\pi\right).$

In Exercises 23–26 graph the equations after changing them to the form $y = c\cos(x - \alpha)$. See Section 8–9.

23. $y = \cos x + 3\sin x.$

24. $y = 3\cos x - \sin x.$

25. $y = 3\sin x - 4\cos x.$

26. $y = 2\cos x + \sin x.$

9-3 *ANALYSIS OF THE GENERAL TANGENT AND COTANGENT CURVES*

It was shown in Sections 4-8 (Fig. 4-19) and 5-5 (Fig. 5-6) that both $y = \tan\theta$ and $y = \cot\theta$ have periods of 180°. Then (using 180° $= \pi$ radians) it follows that the consecutive x-intercepts for either $y = \tan x$ or $y = \cot x$ are separated by an interval of π.

To graph curves of the type $y = a\tan bx$, where a and b are constants, we shall (1) determine some of the consecutive x-intercepts; (2) locate the vertical lines (vertical asymptotes) that separate the branches of the curve; (3) locate a few points on the curve; and (4) sketch the balance of the curve from our general knowledge of the $y = \tan\theta$ curve. We illustrate this with an example.

Example 1.   Graph the function $y = 2\tan\frac{1}{2}x$.

Solution:

(1) Set $y = 0$ and find some of the consecutive x-intercepts. If $2\tan\frac{1}{2}x = y$, and if $y = 0$,

$$2\tan\frac{1}{2}x = 0, \tan\frac{1}{2}x = 0,$$

$$\frac{1}{2}x = n\pi, x = 2n\pi. \qquad n \text{ is any integer}$$

Thus, some of the consecutive x-intercepts are

$$x = -2\pi, 0, 2\pi, \text{etc.}$$

We plot these points (Fig. 9-5).

(2) In Section 4-8 (Fig. 4-19) we found that $y = \tan\theta$ is undefined for values of θ midway between the consecutive θ-intercepts. This is indeed the case with all curves of the type $y = a\tan bx$. For example, for $y = 2\tan\frac{1}{2}x$ when $x = \pi$,

$$y = 2\tan\frac{1}{2}x = 2\tan\frac{1}{2}\pi,$$

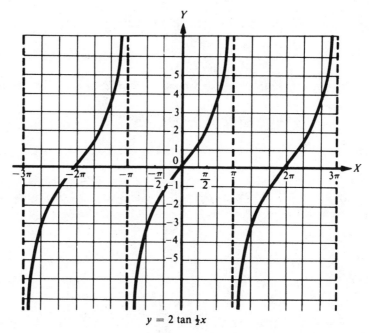

$$y = 2 \tan \tfrac{1}{2}x$$

Figure 9-5

which is undefined. Therefore, some of the vertical asymptotes which separate the branches of the curve are $x = -3\pi, x = -\pi, x = \pi$, *and* $x = 3\pi$.

(3) We locate three points on the curve between $x = 0$ and $x = \pi$. It is always desirable to locate the point for which θ is midway between the x-intercept and the vertical asymptotes which separates the branches of the curve. A brief table of points on this curve is given:

x	$\tfrac{1}{3}\pi$	$\tfrac{1}{2}\pi$	$\tfrac{2}{3}\pi$
y	$\tfrac{2}{3}\sqrt{3} \approx 1.15$	2	$2\sqrt{3} \approx 3.46$

With these points, the x-intercepts, the vertical asymptotes which separate the branches of the curve, and our knowledge of the $y = \tan \theta$ curve, we are able to sketch the balance of the curve for $y = 2 \tan \tfrac{1}{2}x$.

The function $y = a \cot bx$ can be graphed in a similar way by considering the graph of $y = \cot \theta$.

The functions $y = a \tan (bx + c)$ and $y = a \cot (bx + c)$, where a, b, and c are constants, have periods of $\dfrac{\pi}{|b|}$.

We have seen that $y = \tan x$ has the period π. That is, for any x_1,

(1) $\tan (x_1 + \pi) = \tan x_1.$

Therefore, the period of $\tan(bx + c)$ is $\dfrac{\pi}{|b|}$, since, if $x = x_1 + \pi/b$,

$$\tan(bx + c) = \tan(bx_1 + \pi + c) = \tan(bx_1 + c)$$

by Equation (1). Similarly, $\cot(bx + c)$ has the period $\dfrac{\pi}{|b|}$.

Example 2. Graph the function

$$y = 2\cot\left(\frac{1}{3}x - \frac{1}{6}\pi\right).$$

Solution:

(1) We set $y = 0$ and find some of the consecutive x-intercepts.

$$2\cot\left(\frac{1}{3}x - \frac{1}{6}\pi\right) = y,$$

$$\cot\left(\frac{1}{3}x - \frac{1}{6}\pi\right) = 0,$$

$$\left(\frac{1}{3}x - \frac{1}{6}\pi\right) = \frac{1}{2}\pi + n\pi,$$

$$x = 2\pi + 3n\pi. \qquad (n \text{ is any integer})$$

Therefore, some of the consecutive x-intercepts are

$$x = -4\pi, -\pi, 2\pi, \text{ and } 5\pi.$$

(See Fig. 9-6.)

(2) We draw the vertical asymptotes which separate the branches of the curve midway between the x-intercepts. These are

$$x = -2\frac{1}{2}\pi, x = \frac{1}{2}\pi, \text{ and } x = 3\frac{1}{2}\pi.$$

(3) We locate a few points as follows: For $x = \frac{3}{4}\pi$, we have

$$y = 2\cot\left(\frac{1}{3} \cdot \frac{3}{4}\pi - \frac{1}{6}\pi\right) = 2\cot\left(\frac{1}{12}\pi\right) = 2\cot\left(\frac{1}{12} \cdot 180°\right)$$

$$= 2\cot 15° \approx 2 \times 3.73 = 7.46 \text{ from Table 1.}$$

x	$\frac{3}{4}\pi$	π	$1\frac{1}{4}\pi$	2π	$2\frac{3}{4}\pi$	3π
y	7.46	3.46	2	0	−2	−3.46

(4) With the above information and our knowledge of the graph of $y = \cot\theta$, we sketch the curve for $y = 2\cot\left(\frac{1}{3}x - \frac{1}{6}\pi\right)$.

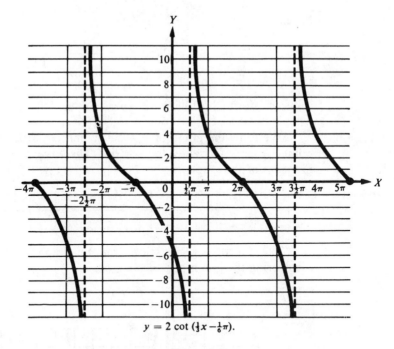

$$y = 2 \cot \left(\tfrac{1}{3}x - \tfrac{1}{6}\pi\right).$$

Figure 9-6

9-4 ANALYSIS OF THE GENERAL SECANT
AND COSECANT CURVES

The graphs for $y = \sec \theta$ and $y = \csc \theta$ were explained in Sections 5-6 and 5-7.

The graph of $y = a \csc bx$, where a and b are constants, is illustrated in the following example.

Example. Graph the function $y = 2 \csc \tfrac{1}{2}x$.

Solution: The graph of $y = 2 \csc \tfrac{1}{2}x$ can be obtained from that of $y = \sin \tfrac{1}{2}x$:

$$y = 2 \csc \frac{1}{2}x = \frac{2}{\sin \frac{1}{2}x}.$$

When $x = \pi$,

$$y = 2 \csc \frac{1}{2}\pi = \frac{2}{\sin \frac{1}{2}\pi} = \frac{2}{1} = 2.$$

When $x = \frac{1}{3}\pi$,

$$y = 2 \csc \frac{1}{6}\pi = \frac{2}{\sin \frac{1}{6}\pi} = \frac{2}{\frac{1}{2}} = 4.$$

Thus, we form the following table, noting that $\csc \frac{1}{2}x$ is undefined (U) when $x = 2n\pi$.

We draw the graph of $y = \sin \frac{1}{2}x$ and construct vertical lines at the points where it crosses the x-axis. These vertical lines separate the branches of the graph of $y = 2 \csc \frac{1}{2}x$. (See Fig. 9-7.)

x	-2π	$-1\frac{2}{3}\pi$	$-\pi$	$-\frac{1}{3}\pi$	0	$\frac{1}{3}\pi$	π	$1\frac{2}{3}\pi$	2π	$2\frac{1}{3}\pi$	3π	$3\frac{2}{3}\pi$	4π
y	U	-4	-2	-4	U	4	2	4	U	-4	-2	-4	U

The function $y = a \sec bx$ can be graphed in a similar way by considering the graph of $y = a \cos bx$.

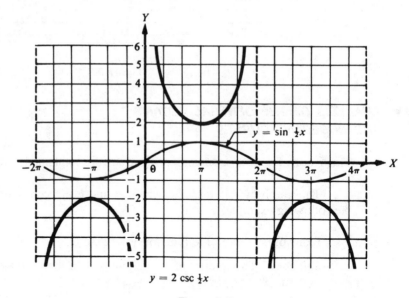

$$y = 2 \csc \tfrac{1}{2}x$$

Figure 9-7

9-5 GRAPHING BY THE ADDITION OF ORDINATES

If a function is in the form of the sum of two or more simpler functions, its graph can be drawn by the method known as the *addition of ordinates*, or as the *composition of ordinates*. This method was discussed in Section 3-10, but at that time, without the knowledge of radian measure, we could not graph functions of the type $y = ax + b \sin cx$. We demonstrate the graph of this kind of a function, by the method of the addition of ordinates, with an example.

Example. Graph the function $y = \frac{1}{2}x - \sin x$.

Solution: Rather than work with the function $y = \frac{1}{2}x - \sin x$ as a difference between two functions, it may be easier for the student to think of it as a sum such as $y = \frac{1}{2}x + (-\sin x)$. We first graph the two functions $y = \frac{1}{2}x$ and $y = -\sin x$. The graph of $y = \frac{1}{2}x$ is a straight line passing through the points $(0, 0)$, $(1, \frac{1}{2})$, $(\pi, \frac{1}{2}\pi)$, etc. Some of the x-intercepts of the function $y = -\sin x$ are -2π, $-\pi$, 0, π, 2π, etc. When $x = \frac{1}{2}\pi$, $y = -1$; therefore, the arc for the graph of $y = -\sin x$, between $x = 0$ and $x = \pi$, lies below the x-axis. (See Fig. 9-8.)

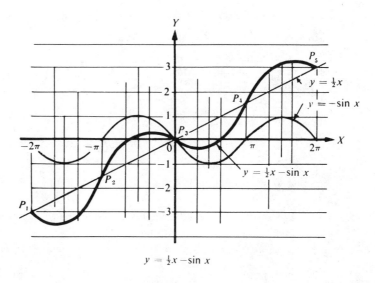

$$y = \tfrac{1}{2}x - \sin x$$

Figure 9-8

Then, for any value of x, the height, or ordinate, of the graph of $y = \frac{1}{2}x - \sin x$ can be found by adding the ordinates of the other two functions for that value of x. In this way we can quickly find as many points as we wish for the desired curve and then draw it. It is helpful to draw vertical lines from the x-intercepts, of the $y = -\sin x$ curve, to the line $y = \frac{1}{2}x$. The points of intersection of these vertical lines and the line $y = \frac{1}{2}x$ (P_1, P_2, P_3, P_4, and P_5) are points on the composite curve where it crosses the line $y = \frac{1}{2}x$. The composite curve lies below the line $y = \frac{1}{2}x$ between P_1 and P_2, and between P_3 and P_4; it lies above the line $y = \frac{1}{2}x$ between P_2 and P_3, and between P_4 and P_5.

Exercises 9-5

In Exercises 1–10 graph each function for one complete wave.

1. $y = \tan \frac{1}{3}x.$ 2. $y = \cot \frac{1}{2}x.$ 3. $y = \sec 2x.$

4. $y = \csc 2x.$ 5. $y = 2 \cot \frac{1}{4}x.$ 6. $y = 2 \tan \frac{1}{4}x.$

7. $y = 3 \csc x.$ 8. $y = 2 \sec x.$ 9. $y = 2 \tan (x - \frac{1}{4}\pi).$

10. $y = \frac{2}{3} \tan (\frac{1}{2}x + \frac{1}{3}\pi).$

In Exercises 11–20 use the method of composition of ordinates to graph each of the functions on the specified intervals.

11. $y = x + \cos x,$ $= \frac{1}{2}\pi \leq x \leq 2\frac{1}{2}\pi.$

12. $y = x + \sin x,$ $-\pi \leq x \leq 2\pi.$

13. $y = 1 + \sin x,$ $0 \leq x \leq 2\pi.$

14. $y = 1 - \cos x,$ $-\frac{1}{2}\pi \leq x \leq 2\frac{1}{2}\pi.$

15. $y = 2 \sin \frac{1}{2}x - 1,$ $0 \leq x \leq 4\pi.$

16. $y = 2 \sin \frac{1}{2}x - x,$ $0 \leq x \leq 4\pi.$

17. $y = \tan x + \sin x,$ $-\frac{1}{3}\pi \leq x \leq \frac{1}{3}\pi.$

18. $y = \cos x + \cot x,$ $\frac{1}{6}\pi \leq x \leq \frac{5}{6}\pi.$

19. $y = x + \sin^2 x,$ $0 \leq x \leq 3\pi.$

20. $y = x + \cos^2 x,$ $-\frac{1}{2}\pi \leq x \leq \frac{3}{2}\pi.$

REVIEW EXERCISES

In Exercises 1–4 find the amplitude and period of each function. Do not draw the graphs for these four exercises.

1. $y = 3 \cos 2x.$ 2. $y = 2 \sin (3x + \frac{1}{2}\pi).$

3. $y = \frac{1}{2} \tan 3x.$ 4. $y = 4 \cot 2x.$

In Exercises 5–16 graph each function on the indicated interval and give the amplitude and period of each.

5. $y = \frac{1}{2} \sin 2x,$ $0 \leq x \leq \pi.$

6. $y = 2 \cos 2x,$ $0 \leq x \leq \pi.$

7. $y = 3 \sin \frac{1}{2}\pi x,$ $0 \leq x \leq 4.$

8. $y = \frac{1}{2} \tan \pi x,$ $0 \leq x \leq 2.$

9. $y = 3 \cos \left(\frac{1}{3}x + \frac{\pi}{2}\right),$ $-3\pi \leq x \leq 3\pi.$

10. $y = 3 \sin (2x - \pi),$ $\dfrac{\pi}{2} \leq x \leq \dfrac{\pi}{2}.$

11. $y = \frac{1}{2} \cot 2x,$ $0 \leqq x \leqq \dfrac{\pi}{2}.$

12. $y = 2 \sec x,$ $-\dfrac{\pi}{2} \leqq x \leqq \dfrac{3\pi}{2}.$

13. $y = 2 \csc x,$ $0 \leqq x \leqq 2\pi.$

14. $y = 3 \cos \left(2x - \dfrac{\pi}{2} \right),$ $0 \leqq x \leqq \pi.$

15. $y = \frac{1}{2}x + \cos 2x,$ $0 \leqq x \leqq \dfrac{3\pi}{4}.$

16. $y = \frac{1}{2} + \sin x,$ $0 \leqq x \leqq 2\pi.$

CHAPTER 9: DIAGNOSTIC TEST

The purpose of this test is to see how well you understand the work covered in Chapter 9. We recommend that you work this test before your instructor tests you on this chapter. Allow yourself approximately 60 minues to do the test.

Solutions to the problems, together with section references, are given in the Answer Section at the end of this book. We suggest that you study the sections referred to for the problems you do incorrectly.

---

1. Find the amplitude and period of the curve $y = 4 \sin \left(\dfrac{2}{3}x + \dfrac{\pi}{2} \right).$

2. Find the amplitude and period of the curve $y = 2 \tan 3x.$

3. Find the period of the curve $y = 3 \sec 2x.$

4. Find the period of the curve $y = 2 \csc 3x.$

In Problems 5–10 graph each function on the indicated interval.

5. $y = 4 \sin \frac{1}{2}x,$ $0 \leqq x \leqq 4\pi.$

6. $y = \tan \frac{1}{3}x,$ $-\dfrac{3\pi}{2} \leqq x \leqq \dfrac{3\pi}{2}.$

7. $y = 2 \sec 2x,$ $0 \leqq x \leqq \dfrac{3\pi}{4}.$

8. $y = \cos x - x,$ $0 \leqq x \leqq 2\pi.$

9. $y = \cot \pi x,$ $0 \leqq x \leqq 2.$

10. $y = \csc \dfrac{\pi}{2}x.$ $-2 \leqq x \leqq 2.$

10

Logarithms

10-1 USE OF LOGARITHMS

Many applications of mathematics make use of exponential and logarithmic functions. They are used, for example, in calculations of growth and decay of bacteria, population, and radium as well as compound interest, continuous compound amount of principal and interest, and probability.

Logarithms were invented in the seventeenth century, and until recent years they were an important tool in all kinds of arithmetic calculations. Now, with the influx of inexpensive calculators, this application of logarithms is less important. However, neither logarithmic theory nor the use of logarithms in expressions and equations has diminished in importance. Certain kinds of equations are solved most easily by the use of logarithms.

An **algebraic function** is a function made up of algebraic expressions. In this chapter we shall be working with transcendental functions. A **transcendental function** is any function that is not an algebraic function. Trigonometric functions and logarithmic functions are examples of transcendental functions.

10-2 LAWS OF EXPONENTS

As we shall see, *logarithms are exponents*. In this section we recall certain laws of exponents. These laws of exponents, which are discussed in courses in algebra, hold for all types of real exponents for a positive real base. For the purposes of this course we shall restrict the exponents to rational numbers, and we shall make the following assumptions.

Let a be any positive number and x and y be any two rational numbers. Then

I. $a^x \cdot a^y = a^{x+y}$.   *Example:* $2^3 \cdot 2^4 = 2^7$.

II. $\dfrac{a^x}{a^y} = a^{x-y}$.   *Example:* $\dfrac{2^7}{2^2} = 2^5$.

III. $(a^x)^y = a^{xy}$.   *Example:* $(2^3)^4 = 2^{12}$.

IV. $\sqrt[y]{a^x} = a^{x/y}$.   *Example:* $\sqrt[3]{2^5} = 2^{5/3}$.

V. $a^0 = 1$.   *Example:* $2^0 = 1$.

VI. $a^{-x} = \dfrac{1}{a^x}$.   *Example:* $2^{-5} = \dfrac{1}{2^5} = \dfrac{1}{32}$.

VII. If a is any number greater than 1, and if y is positive, it is always possible to find a real exponent x such that
$$a^x = y.$$

But if y is negative or zero, no real value of x exists satisfying this relationship.

Definition. *If a is any real number and n is a positive integer, then*
$$a^n = a \cdot a \cdots a, \qquad n \text{ factors of } a;$$

*n is called the **exponent**, a the **base**, and a^n the **nth power** of a.*

10-3 DEFINITION OF A LOGARITHM

If b, x, and N are three numbers such that

(1)
$$N = b^x,$$

then we say that x is the logarithm of N to the base b, and we write

(2)
$$x = \log_b N.$$

Definition. *The **logarithm** of a number to a given base is the exponent to which the base must be raised to give the number.*

We assume in this book that $\log_b N$ is used only for $N > 0$ and $b > 0 \neq 1$, which implies that $\log_b N$ is a real number.

Equations (1) and (2) are different ways of expressing the same relation among the three numbers b (the *base*), N (the *number*), and x (the *exponent*, or *logarithm*). Equation (1) states the relation in *exponential form*, and Equation (2) says the same thing in *logarithmic form*. If x *is the logarithm of N to a given base, then N is called the* **antilogarithm** *of x to that base.*

The following values are very important and should be memorized. Since $b^1 = b$ and $b^0 = 1$, for any base, b,

$$(3) \quad \log_b b = 1 \qquad \text{and} \qquad (4) \quad \log_b 1 = 0.$$

In each of Examples 1, 2, and 3 we express the same relation between the three respective numbers, first in exponential form and then in logarithmic form.

	EXPONENTIAL FORM	LOGARITHMIC FORM

Example 1. $\qquad 4^3 = 64. \qquad\qquad \log_4 64 = 3.$

4 is the base.
3 is the exponent of the power.
4^3 is the third power of 4, and 64 is the third power of 4.

4 is the base.
3 is the logarithm of 64 to the base 4.

Example 2. $\qquad 4^{2.5} = 32. \qquad\qquad \log_4 32 = 2.5.$

4 is the base.
2.5 is the exponent of the power.
$4^{2.5}$ is the 2.5 power of 4, and 32 is the 2.5 power of 4.

4 is the base.
2.5 is the logarithm of 32 to the base 4.

Example 3. $\qquad e^2 = 7.39. \qquad\qquad \log_e 7.39 = 2.$

e is the base. ($e \approx 2.72$)
2 is the exponent of the power.
e^2 is the second power of e, and 7.39 is the second power of e.

e is the base.
2 is the logarithm of 7.39 to the base e.

Example 4. Find the value of $\log_{32} 8$.

Solution: Let

$$x = \log_{32} 8.$$

Write in exponential form:

$$32^x = 8.$$

Express 32 and 8 as powers of 2:

$$(2^5)^x = 2^3,$$
$$2^{5x} = 2^3.$$

Hence, $\qquad\qquad\qquad 5x = 3, x = \dfrac{3}{5}.$

Therefore, $\qquad\qquad \log_{32} 8 = \dfrac{3}{5}.$

Example 5. Find the number N if $\log_{81} N = \frac{3}{4}$.

Solution:

Write in exponential form: $\qquad N = (81)^{3/4}.$

Write 81 as a power of 3: $\qquad N = (3^4)^{3/4}.$

Apply Law III of Exponents: $\qquad N = 3^4 \cdot \frac{3}{4} = 3^3 = 27.$

Exercises 10-3

In Exercises 1–16 express each equation in logarithmic form.

1. $2^3 = 8.$
2. $3^2 = 9.$
3. $10^2 = 100.$
4. $10^3 = 1000.$
5. $25^{1/2} = 5.$
6. $16^{1/2} = 4.$
7. $8^{-2/3} = \frac{1}{4}.$
8. $4^{-3/2} = \frac{1}{8}.$
9. $5^0 = 1.$
10. $7^0 = 1.$
11. $3^0 = 1.$
12. $6^0 = 1.$
13. $5^1 = 5.$
14. $3^1 = 3.$
15. $10^1 = 10.$
16. $4^1 = 4.$

In Exercises 17–28 express each equation in exponential form.

17. $\log_4 64 = 3.$
18. $\log_8 64 = 2.$
19. $\log_2 16 = 4.$
20. $\log_5 125 = 3.$
21. $\log_{16} 4 = \frac{1}{2}.$
22. $\log_9 3 = \frac{1}{2}.$
23. $\log_{10} 10 = 1.$
24. $\log_{10} .01 = -2.$
25. $\log_7 7 = 1.$
26. $\log_{10} 100 = 2.$
27. $\log_b b = 1.$
28. $\log_{10} .001 = -3.$

In Exercises 29–37 find the value of each of the logarithms.

29. $\log_3 27.$
30. $\log_4 64.$
31. $\log_4 8.$
32. $\log_{27} 9.$
33. $\log_5 1.$
34. $\log_6 1.$
35. $\log_9 (\frac{1}{3}).$
36. $\log_8 (\frac{1}{64}).$
37. $\log_4 32.$

In Exercises 38–46 find the unknown b, N, or x.

38. $\log_2 N = 3.$
39. $\log_3 N = 2.$
40. $\log_{10} 1000 = x.$

41. $\log_{10} 10,000 = x.$ **42.** $\log_b 32 = 2.5.$ **43.** $\log_b 27 = 1.5.$

44. $10^{\log_{10} 3} = x.$ **45.** $3^{\log_3 2} = x.$ **46.** $2^{\log_2 3} = x.$

10-4 FUNDAMENTAL LAWS OF LOGARITHMS

Since logarithms are exponents, the laws of exponents enable us to derive certain laws of logarithms that are used in solving logarithmic equations and in computation. Any base, b, greater than 0, $\neq 1$ may be used. We shall assume that M and N are positive numbers.

I. *Multiplication.   The logarithm of the product of two or more numbers is equal to the sum of the logarithms of the numbers.*

$$\log_b (M \cdot N) = \log_b M + \log_b N,$$

$$\log_b (M \cdot N \cdot P \cdots R) = \log_b M + \log_b N + \log_b P + \cdots + \log_b R.$$

Proof. *Let*

$$x = \log_b M \qquad \text{and} \qquad y = \log_b N.$$

Write in exponential form:

$$M = b^x \qquad \text{and} \qquad N = b^y.$$

Multiply M by N:

$$MN = b^x \cdot b^y = b^{x+y}.$$

Write in logarithmic form:

$$\log_b MN = x + y.$$

Replace x and y by their values:

$$\log_b MN = \log_b M + \log_b N.$$

Example 1. $\log_{10} (289 \times 356) = \log_{10} 289 + \log_{10} 356.$

II. *Division.   The logarithm of the quotient of two numbers is equal to the logarithm of the dividend minus the logarithm of the divisor.*

$$\log_b \left(\frac{M}{N} \right) = \log_b M - \log_b N.$$

Proof. *Let*

$$x = \log_b M \qquad \text{and} \qquad y = \log_b N.$$

Write in exponential form:

$$M = b^x \qquad \text{and} \qquad N = b^y.$$

Divide M by N:

$$\frac{M}{N} = \frac{b^x}{b^y} = b^{x-y}.$$

Write in logarithmic form:

$$\log_b\left(\frac{M}{N}\right) = x - y.$$

Replace x and y by their values:

$$\log_b\left(\frac{M}{N}\right) = \log_b M - \log_b N.$$

Example 2. $\log_{10}\left(\frac{125}{346}\right) = \log_{10} 125 - \log_{10} 346.$

III. *The logarithm of a power of a number is equal to the logarithm of the number multiplied by the exponent of the power.*

$$\log_b (M^n) = n \log_b M.$$

Proof. Let

$$x = \log_b M.$$

Write in exponential form:

$$M = b^x.$$

Raise to the nth power:

$$M^n = (b^x)^n = b^{nx}.$$

Write in logarithmic form:

$$\log_b M^n = nx.$$

Replace x by its value:

$$\log_b M^n = n \log_b M.$$

Example 3. $\log_{10} (3.51)^7 = 7 \log_{10} 3.51.$

IV. *The logarithm of a root of a number is equal to the logarithm of the number divided by the index of the root.*

$$\log_b (\sqrt[n]{M}) = \frac{\log_b M}{n}.$$

Proof. Since by Law IV of Exponents, $\sqrt[n]{M} = M^{1/n}$, by the above law of logarithms we have

$$\log_b (\sqrt[n]{M}) = \log_b (M^{1/n}) = \frac{1}{n} \log_b M.$$

Example 4. $\log_{10} \sqrt[5]{146} = \log_{10} (146)^{1/5} = \frac{1}{5} \log_{10} 146.$

Example 5.

(a) $\log_{10} \dfrac{345 \times 561}{284} = \log_{10} 345 + \log_{10} 561 - \log_{10} 284;$

(b) $\log_{10} \dfrac{156 \times (4.15)^3}{(1.05)^{12}} = \log_{10} 156 + 3 \log_{10} 4.15 - 12 \log_{10} 1.05;$

(c) $\log_{10} \dfrac{68.4 \times \sqrt{7.61}}{\sqrt[4]{3.48}} = \log_{10} 68.4 + \frac{1}{2} \log_{10} 7.61 - \frac{1}{4} \log_{10} 3.48.$

Example 6. Given $\log_{10} 2 = .301$, $\log_{10} 3 = .477$, and $\log_{10} 7 = .845$, find $\log_{10} \frac{54}{7}$.

Solution:

$$\log_{10} \tfrac{54}{7} = \log_{10} \frac{2 \times 3^3}{7}$$

$$= \log_{10} 2 + 3 \log_{10} 3 - \log_{10} 7$$

$$= .301 + 3(.477) - .845$$

$$= .887.$$

Example 7. Express $3 \log_b x + 5 \log_b y - \frac{1}{2} \log_b z$ as a single logarithm.

Solution:

$$3 \log_b x + 5 \log_b y - \tfrac{1}{2} \log_b z = \log_b x^3 + \log_b y^5 - \log_b z^{1/2}$$

$$= \log_b \frac{x^3 y^5}{\sqrt{z}}.$$

Exercises 10-4

Given $\log_{10} 2 = .301$, $\log_{10} 3 = .477$, $\log_{10} 7 = .845$, find the following logarithms. (Recall that $\log_{10} 10 = 1$.)

1. $\log_{10} 14.$ 2. $\log_{10} 21.$ 3. $\log_{10} \frac{18}{7}.$

4. $\log_{10} \frac{27}{7}.$ 5. $\log_{10} 5.$ 6. $\log_{10} 50.$

7. $\log_{10} \sqrt{21}$.   8. $\log_{10} \sqrt{27}$.   9. $\log_{10} 30$.

10. $\log_{10} 60$.   11. $\log_{10} 98$.   12. $\log_{10} 72$.

13. $\log_{10} \sqrt[5]{2}$.   14. $\log_{10} \sqrt[10]{2}$.   15. $\log_{10} 6000$.

Write each expression as a single logarithm. (Assume that all of the logarithms have the same base.)

16. $\log x + \log y$.   17. $\log x - \log y$.

18. $2 \log x - 3 \log y$.   19. $3 \log x + 2 \log y$.

20. $\frac{1}{2} \log x + \log y - \log z$.   21. $\frac{1}{4} \log x - \log y + \log z$.

22. $\log (x^2 - y^2) - \log (x + y)$.   23. $\log (x^2 - y^2) - \log (x - y)$.

In each of the following solve for y in terms of x.

24. $\log_2 y = 3x$.   25. $\log_3 y = 2x$.

26. $\log_{10} y = 3 \log_{10} x$.   27. $\log_{10} y = -2 \log_{10} x$.

28. $\log_{10} y = \log_{10} x^2 - \log_{10} x$.   29. $\log_{10} y = \log_{10} x^3 - 2 \log_{10} x$.

30. $\frac{1}{3} \log_{10} y = 3 \log_{10} x - \log_{10} x^2$.   31. $\frac{1}{3} \log_{10} y = \log_{10} x^4 - 3 \log_{10} x$.

Solve the following equations for x.

32. $\log_{10} (x + 3) + \log_{10} 4 = \log_{10} 100 + \log_{10} 2$.

33. $\log_{10} (x - 5) + \log_{10} 5 = \log_{10} 10 + \log_{10} 3$.

34. $\log_{10} (x^2 - 4) = 1 + \log_{10} (x + 2)$.

35. $\log_{10} (x^2 - 1) = 2 + \log_{10} (x - 1)$.

10-5 SYSTEMS OF LOGARITHMS

There are two important systems of logarithms. The *natural*, or *Napierian*, system uses the base e, where e is approximately 2.71828. This system is used in most advanced applications of mathematics. Many authors use the notation $\ln x$ instead of $\log_e x$. The system that is used in the solution of triangles and that is more convenient for computation is the *common*, or *Briggs*, system which uses the base 10. *Hereafter, when we speak of the logarithm of a number without specifying the base, it is to be understood that the base of the logarithm is* 10—and instead of writing $\log_{10} N$, we shall write $\log N$.

10-6 LOGARITHMS TO THE BASE 10

The logarithms of integral powers of $(\cdots, 10^{-2}, 10^{-1}, 10^0, 10^1, 10^2, \cdots)$ are integers.

Example 1.

	EXPONENTIAL FORM	LOGARITHMIC FORM
(a)	$10^3 = 1000$	$\log 1000 = 3.$
(b)	$10^2 = 100$	$\log 100 = 2.$
(c)	$10^1 = 10$	$\log 10 = 1.$
(d)	$10^0 = 1$	$\log 1 = 0.$
(e)	$10^{-1} = .10$	$\log .10 = -1.$
(f)	$10^{-2} = .01$	$\log .01 = -2.$

Most logarithms are irrational numbers that are given with an accuracy specified to a certain number of decimal places. A collection of such numbers is given in Table 2. These decimal approximations are found by methods beyond the level of this book.

Example 2. The following logarithms are given to four decimal accuracy:

	EXPONENTIAL FORM	LOGARITHMIC FORM
(a)	$10^{2.3010} = 200$	$\log 200 = 2.3010.$
(b)	$10^{1.6990} = 50$	$\log 50 = 1.6990.$
(c)	$10^{0.3010} = 2$	$\log 2 = .3010.$

Any positive number written in ordinary decimal notation can be written in scientific notation, that is, in the form $M \times 10^n$ where $1 \leqq M < 10$. (See Section 2-12 which includes a discussion of how to write a number in scientific notation.)

Example 3.

(a)	$2714 = 2.714 \times 10^3.$
(b)	$315.6 = 3.156 \times 10^2.$
(c)	$21.59 = 2.159 \times 10^1.$
(d)	$7.187 = 7.187 \times 10^0.$
(e)	$.8570 = 8.570 \times 10^{-1}.$
(f)	$.01444 = 1.444 \times 10^{-2}.$
(g)	$.00009889 = 9.889 \times 10^{-5}.$
(h)	$1000 = 1.000 \times 10^3.$

Writing $1000 = 1.000 \times 10^3$ indicates that the zeros are significant digits.

In scientific notation, two numbers that have the same sequence of significant digits can differ only in the power of 10 that is used.

Example 4.

(a) $$314 = 3.14 \times 10^2.$$

(b) $$31.4 = 3.14 \times 10^1.$$

(c) $$.314 = 3.14 \times 10^{-1}.$$

(d) $$.00000314 = 3.14 \times 10^{-6}.$$

The relation $y = \log x$ is a function since for each $x > 0$, there corresponds exactly one value for y. Furthermore, this function is "increasing" in the sense that if $x_1 < x_2$, then $\log x_1 < \log x_2$. Therefore, if $1 < N < 10$, then $\log 1 < \log N < \log 10$ and $0 < \log N < 1$. See Example 5.

Example 5.

(a) $$\log 1 = 0.$$

(b) $$\log 1.10 = .0414.$$

(c) $$\log 2.00 = .3010.$$

(d) $$\log 3.14 = .4969.$$

(e) $$\log 9.99 = .9996.$$

(f) $$\log 10.0 = 1.0000.$$

The logarithms of the numbers of Example 4 can be expressed as follows:

Example 6.

(a)
$$\begin{aligned}
\log 314 &= \log (3.14 \times 10^2) \\
&= \log 3.14 + \log 10^2 \\
&= \log 3.14 + 2 \log 10 \\
&= \log 3.14 + 2(1) \\
&= .4969 + 2 = 2.4969.
\end{aligned}$$

(b)
$$\begin{aligned}
\log 31.4 &= \log (3.14 \times 10^1) \\
&= \log 3.14 + \log 10^1 \\
&= .4969 + 1 = 1.4969.
\end{aligned}$$

(c)
$$\log .314 = \log (3.14 \times 10^{-1})$$
$$= \log 3.14 + \log 10^{-1}$$
$$= \log 3.14 + (-1) \log 10$$
$$= .4969 - 1.$$

By adding 10 and then subtracting 10 from the number $(.4969 - 1)$, we can write
$$.4969 - 1 = .4969 - 1 + 10 - 10$$
$$= .4969 + 9 - 10$$
$$= 9.4969 - 10.$$

(d)
$$\log .00000314 = \log (3.14 \times 10^{-6})$$
$$= \log 3.14 + \log 10^{-6}$$
$$= .4969 - 6$$
$$= 4.4969 - 10.$$

Hence, the common logarithm of any positive number can be written as the sum of two parts: (1) a non-negative decimal less than 1 and (2) the exponent applied to 10 when the number is written in scientific notation.

The decimal part of the common logarithm of a number is called the **mantissa** of the logarithm. As we saw in Example 6, the *mantissa depends only on the particular sequence of significant digits in the number.* It is independent of the position of the decimal point.

The exponent of 10, which is the whole part of the logarithm, is called the **characteristic**. *The characteristic depends only on the position of the decimal point in the number.* It is independent of the sequence of significant digits.

10-7 RULE FOR THE CHARACTERISTIC OF A COMMON LOGARITHM

As we stated in the previous section, the characteristic of the logarithm of a given number is equal to the exponent of 10 when that number is written in scientific notation.

Example 1. The characteristic and the exponent of 10 is the same number.

	Number	The number in scientific notation	Characteristic of the logarithm of the number
(a)	2045	2.045×10^3	3
(b)	204.5	2.045×10^2	2
(c)	20.45	2.045×10^1	1
(d)	2.045	2.045×10^0	0
(e)	.2045	2.045×10^{-1}	−1
(f)	.02045	2.045×10^{-2}	−2
(g)	.002045	2.045×10^{-3}	−3
(h)	.0002045	2.045×10^{-4}	−4

Some students find the ∧ symbol, called **caret**, an aid in converting a number to its equivalent form in scientific notation. We use the caret to show the position where the decimal point of the transformed number is to be placed. Some of the examples above are repeated below, showing the use of the caret.

Examples 2. $2045 = 2_\wedge 045. = 2.045 \times 10^3$.

3. $204.5 = 2_\wedge 04.5 = 2.045 \times 10^2$.

Notice that the caret is placed immediately to the right of the first nonzero digit of the number. Then the number is rewritten with the decimal point in the position where the caret was used. This equivalent number is then multiplied by 10^n, where n is the number of places the decimal point is to the right or left of the caret, and is positive when the decimal point is to the right of the caret, as in Examples 2 and 3 above, and negative when the decimal point is to the left of the caret, as in Examples 4 and 5 below.

Examples 4. $.2045 = .2_\wedge 045 = 2.045 \times 10^{-1}$.

5. $.02045 = .02_\wedge 045 = 2.045 \times 10^{-2}$

For computational purposes, when using logarithms, it is better to write negative characteristics as the difference of two numbers. For example,

the characteristic −1 is written 9 − 10,
the characteristic −2 is written 8 − 10,

the characteristic −4 is written 6 − 10,
the characteristic −12 is written 8 − 20.

Exercises 10-7

Write the characteristic of the logarithm of each of the following numbers.

1. 542.	2. 125.	3. 93,000,000.	4. 186,000.
5. .0745.	6. .0815.	7. .000904.	8. .000516.
9. 8056.	10. 2018.	11. .00634.	12. .808.
13. 3612.	14. 54,000.	15. .0684.	16. .1006.
17. .9008.	18. 3047.	19. 5,880,000,000,000.	

20. .000000514.

In Exercises 21–30 rewrite each number and place its decimal point according to the given characteristic of its logarithm.

Number	Characteristic	Number	Characteristic
21. 1278	2	22. 2354	1
23. 5061	5	24. 6043	4
25. 9455	0	26. 8544	0
27. 1008	8—10	28. 2009	7—10
29. 7599	5—10	30. 4988	4—10

10-8 TABLES OF LOGARITHMS

The mantissa of the logarithm of a number is very often a never-ending decimal. **Tables of common logarithms** contain the *mantissas* of logarithms of numbers, to a specified number of decimal places. In Table 2 there are listed, to four decimal places, the mantissas of the logarithms of all whole numbers from 1 to 999. In this table of mantissas the decimal point which belongs in front of each mantissa is omitted; it should be supplied when the mantissa is taken from the table. This table may be used to find the logarithm of a given number or to find the number corresponding to a given logarithm.

To find the logarithm of a given number: Determine the characteristic, by the methods given in Section 10-7, and then find the mantissa from Table 2.

To find the mantissa for the logarithm of a number, the first two digits of the number are given in the column at the left under the caption "N." The third digit of the number occurs at the top of the table in the same line as N. For example, on page 278 we find log 1.50 = .1761. The next entry in the same line is log 1.51 = .1790. In each case, the mantissa obtained directly from the table is the entire logarithm, since the characteristic is zero.

To find the number corresponding to a given logarithm: The number that corresponds to a given logarithm is often referred to as the **antilogarithm** of the logarithm. Thus in the statement log 1.50 = .1761, the number 1.50 is the antilogarithm of the number .1761. The method of finding the antilogarithm is just the reverse of that explained above for finding the logarithm of a given number.

Since the characteristic of the logarithm depends only on the position of the decimal point in the number, and since the mantissa depends only on the succession of digits in the number, in order to find the antilogarithm of a given logarithm we first disregard the characteristic and look for the given mantissa in the table.

In order to find the number N when log N = 1.2455, we first disregard the characteristic 1 and then follow down the columns of mantissas (Table 2) until 2455 is found. An examination of this section of the table shows 2455 in the line with N = 17 and in the column under "6." Hence, if

$$\log N = 1.2455, \quad \text{then} \quad N = 17.6.$$

10–9 INTERPOLATION

Interpolation is the process of approximating a number that is not an entry in the table but lies between two consecutive entries. Interpolation as applied to numbers involved in the table of trigonometric functions was explained in Section 3-6. The same type of linear interpolation is used here with our four-place table of logarithms.

The logarithms of numbers with three significant digits are read directly from Table 2. By interpolation we can find the logarithm of a number with four significant digits. If we wish to find the logarithm of a number having more than four significant digits, we usually round it off to four places before using the table, because the table is not designed to give greater accuracy than this.

Example 1.   Use Table 2 to find log 183.2.

Solution: The characteristic is 2. The logarithm of 183.2 lies between log 183.0 and log 184.0. We interpolate as follows:

$$1\left\{.2\left\{\begin{matrix}\log 183.0 = 2.2625 \\ \log 183.2 = \quad ?\end{matrix}\right\}d\right\}.0020$$
$$\left.\log 184.0 = 2.2645\right\}$$

Therefore, $\log 183.2 = 2.2625 + .2(.0020)$

$$= 2.2629$$

The product $.2 \times 20$ can be found by the tables of proportional parts, which are on the pages of the table. The numbers above the short columns are the tabular differences that occur for the page. The nine numbers in the column are, in order, .1, .2, .3, etc., of the number above the column. Thus, in Example 1 above $.2 \times 20 = 4$. The student will find the proportional parts tables more helpful when the tabular difference is a larger number.

Example 2.   Given $\log N = 8.2110 - 10$, find N.

Solution: We first disregard the characteristic and seek only the succession of digits in the number N which correspond to the mantissa .2110. We find that this mantissa lies between the two table entries 2095 and 2122.

$$1\left\{\begin{matrix}\log 162 = 2095 \\ \log N \quad = 2110 \\ \log 163 = 2122\end{matrix}\right\}15\left.\right\}27.$$

Therefore, the succession of digits for N is

$$162 + \frac{15}{27}(1) \approx 1626. \qquad \left(\frac{15}{27} \approx .6\right)$$

The characteristic is $(8 - 10)$, that is, -2. Therefore,

$$N = .01626 \text{ (accurate to four significant digits).}$$

The table of proportional parts (of the tables) can be used to find $\frac{15}{27} \approx .7$ as follows: Look down the column having 27 at the top for the number nearest 15. It is 16.2, which corresponds to .6.

Exercises 10-9

In Exercises 1–12 assume the following numbers to be exact numbers and find the logarithm of each.

1. 125.	2. 161.	3. .537.
4. .562.	5. 67.35.	6. 71.66.
7. .0008.	8. .0040.	9. 93,000,000.
10. 186,000.	11. .05013.	12. .06767.

In Exercises 13–20 find the antilogarithms of each of the given numbers.

13. .7889.   **14.** .8215.   **15.** 3.6314.

16. 2.4378.   **17.** 7.0792 − 10.   **18.** 6.0569 − 10.

19. 1.2460.   **20.** 1.7120.

10-10 COMPUTATIONS WITH LOGARITHMS

Before any computation is done in a problem, it is important to analyze the problem and make an outline of the procedure to be followed. Without an outline the student is likely to waste time and possibly omit some important detail. This is especially true in solving triangles by the use of logarithms. The characteristics of the logarithms of each given number should be written in the proper location in the outline before going to the tables for the mantissas.

In the five following examples the numbers in the examples are assumed to be exact, and the answer for each example is given correct to four significant figures.

Example 1. Find the value of $x = 21.7 \times 3.14 \times .896$.

Solution: By Law I, Section 10-4,

$$\log x = \log 21.7 + \log 3.14 + \log .896.$$

The blank outline, with the characteristics of the numbers written in, is shown at the left. The same outline is also shown at the right after it has been filled in from the tables.

$$\begin{aligned}
\log 21.7 &= 1. & \log 21.7 &= 1.3365 \\
\log 3.14 &= 0. & \log 3.14 &= .4969 \\
\log .896 &= 9. \qquad -10(+) & \log .896 &= 9.9523 - 10(+) \\
\hline
\log x &= & \log x &= 11.7857 - 10, \\
x &= & x &= 61.06.
\end{aligned}$$

Example 2. Find the value of $x = \dfrac{9.041}{.06544}$.

Solution: Take the logarithms of both sides of the equation.

$$\log x = \log 9.041 - \log .06544.$$

$$\begin{aligned}
\log 9.041 &= 0. & \log 9.041 &= 10.9562 - 10 \\
\log .06544 &= 8. \qquad -10(-) & \log .06544 &= 8.8158 - 10(-) \\
\hline
\log x &= & \log x &= 2.1404 \\
x &= & x &\approx 138.2.
\end{aligned}$$

Notice that 10 was added to and then subtracted from the logarithm of 9.041 in order to keep the decimal part of $\log x$ positive.

Example 3.   Find the value of $x = \dfrac{(1.50)^4}{(3.46)(.896)}$.

Solution: Take the logarithm of both sides of the equation.

$$\log x = 4 \log 1.50 - (\log 3.46 + \log .896)$$

$$= 4(. \quad\quad) - [(. \quad\quad) + (9. \quad\quad - 10)].$$

We let the above be our outline and proceed to find the mantissas from the tables.

$$\log x = 4(.1761) - [.5391 + (9.9523 - 10)]$$

$$= .7044 - (10.4914 - 10)$$

$$= .2130.$$

Then

$$x = 1.633.$$

Example 4.   Find the value of $x = \sqrt[3]{.716}$.

Solution: Rewrite the equation in the form

$$x = (.716)^{1/3}.$$

Then take the logarithm of both sides of this equation.

$$\log x = \log (.716)^{1/3}$$

$$= \frac{1}{3}\log .716$$

$$= \frac{1}{3}(9.8549 - 10).$$

We are about to divide the logarithm $(9.8549 - 10)$ by 3. In order to make the negative part of the characteristic evenly divisible by 3, we both add and subtract 20 in the characteristic. Thus, we rewrite the logarithm $(9.8549 - 10)$ as

$$9.8549 - 10 = 9.8549 + 20 - 20 - 10$$

$$= 29.8549 - 30.$$

Then

$$\log x = \frac{1}{3}(29.8549 - 30)$$

$$= 9.9516 - 10.$$

Therefore,

$$x = .8946.$$

Example 5. Find the value of $x = \dfrac{75.0 \times (-3.85)}{5.79}$.

Solution: As stated in Section 10-3, we are restricted to the logarithms of positive numbers, whereas the factor (-3.85) is negative. By inspection we observe that

$$x = \frac{75.0 \times (-3.85)}{5.79} = -\frac{75.0 \times 3.85}{5.79} = -F,$$

where $F = \dfrac{75.0 \times 3.85}{5.79}$. Then we proceed as follows:

$$\log 75.0 = 1. \qquad\qquad \log 75.0 = 1.8751$$
$$\log 3.85 = 0. \quad (+) \qquad \log 3.85 = \underline{\quad .5855(+)}$$
$$\qquad\qquad\qquad\qquad\qquad\qquad 2.4606$$
$$\log 5.79 = 0. \quad (-) \qquad \log 5.79 = \underline{\quad .7627(-)}$$
$$\log F = \qquad\qquad\qquad \log F = 1.6979,$$
$$F = \qquad\qquad\qquad\quad F \approx 49.9.$$

Therefore, $x = -F \approx -49.9.$

Exercises 10-10

In Exercises 1–36 use logarithms (Table 2) to obtain the results correct to four significant figures. Assume the given numbers to be exact numbers.

The answers given for these exercises have been obtained by using the four place tables. If you use a calculator, your answers will not all agree with the given answers. A calculator will give more accurate answers.

1. 28.4×3.05.
2. 48.0×7.61.
3. $609 \div 25.4$.
4. $711 \div 49.1$.
5. $1.91 \div 54.7$.
6. $2.83 \div 65.7$.
7. $(1.16)^5$.
8. $(2.10)^6$.
9. $\sqrt[4]{75.6}$.
10. $\sqrt[5]{80.9}$.
11. $\sqrt[3]{.856}$.
12. $\sqrt[3]{.199}$.
13. $\sqrt[5]{.036}$.
14. $\sqrt[6]{.084}$.
15. $(18.5)^{2/3}$.
16. $(59.4)^{3/4}$.
17. $\sqrt{\dfrac{850}{1.46 \times 7.84}}$.
18. $\sqrt{\dfrac{386}{7.15 \times 25.8}}$.
19. $1.46 + \log 27.3$.
20. $7.86 + \log 14.9$.
21. $38.6 \times \log 187$.
22. $58.6 \times \log 784$.

23. $\dfrac{85 \times (-6.75)}{38.4}$.

24. $\dfrac{28,600}{7.15 \times (-38.7)}$.

25. $(2.81)^{-2.5}$.

26. $(18.1)^{-3.5}$.

27. $\dfrac{\log 18.5}{1.17}$.

28. $\dfrac{\log 197}{2.86}$.

29. $\dfrac{\log 175}{\log 284}$.

30. $\dfrac{\log 198}{\log 574}$.

31. $\dfrac{.184 \times 87.1}{35.1 \times 2.17}$.

32. $\dfrac{.387 \times 94.1}{21.1 \times 3.00}$.

33. $\sqrt{\dfrac{75 \times (3.5)^4}{\sqrt{15.6} \times 2.45}}$.

34. $\sqrt[3]{\dfrac{(1.16)^5 \times 31.7}{\sqrt{18.5} \times 7.75}}$.

35. $(\log 186)(\log .065)$.

36. $(\log 175)(\log .0187)$.

37. Calculate the volume of a sphere of radius $r = 2.62$ from the formula $V = \frac{4}{3}\pi r^3$. Use $\pi = 3.14$.

38. Calculate the volume of a right-circular cylinder whose altitude is $h = 7.25$ and the radius of whose base is $r = 1.75$ from the formula $V = \pi r^2 h$. Use $\pi = 3.14$.

10-11 THE GRAPH OF $y = \log_b x$

For a given value of the base b we can obtain the graph of the function

$$y = \log_b x$$

by plotting positive values of x as the abscissas and the corresponding values of $\log_b x$ as the ordinates of points on the graph.

We show graphs for the functions $y = \log_{10} x$ and $y = \log_2 x$ (Fig. 10-1). Coordinates of points on the graph of $y = \log_{10} x$ can be obtained either from a table of logarithms or from the exponential form of $y = \log_{10} x$, namely, $x = 10^y$. In the exponential form we give y values and solve for x. The coordinates of some of these points are given below.

x	.01	.1	1	10	100
y	-2	-1	0	1	2

To find coordinates of points on the graph of the function $y = \log_2 x$, we change the function to exponential form, give y values, and solve for x.

$$y = \log_2 x,$$

$$x = 2^y.$$

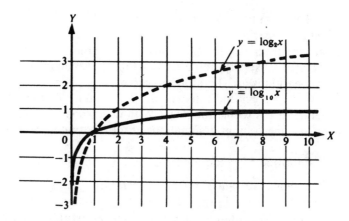

Figure 10-1

The coordinates of some of the points on the graph of the function $y = \log_2 x$ are given below.

x	$\frac{1}{4}$	$\frac{1}{2}$	1	2	4	8
y	-2	-1	0	1	2	3

Several important properties of logarithms are illustrated by the graph:

1. The logarithm of a number greater than 1 is positive.
2. The logarithm of 1 is zero.
3. The logarithms of numbers between 0 and 1 are negative.
4. As x approaches zero as a limit from the positive side of 0, $\log x$ decreases through negative values, and its numerical value increases beyond all bounds.
5. The logarithm of 0 does not exist.
6. The logarithm of a negative number is not on the graph of real numbers (our definition of the logarithm excludes the logarithm of a negative number).
7. As x increases through positive values, $\log x$ increases more and more slowly.

10-12 LOGARITHMS OF TRIGONOMETRIC FUNCTIONS

In Table 1 there are listed the logarithms of the sine, cosine, tangent, and cotangent functions of angles from $0°$ to $90°$ inclusive.

The student will observe that the -10, following the mantissa, is omitted in the tables. Omitting the -10 in tables of trigonometric functions is done to save space.

The quantity −10 is to be appended to all logarithms of the sine and cosine, to logarithms of the tangent from 0° to 45°, and to logarithms of the cotangent from 45° to 90°. This results from the fact that for the intervals mentioned these trigonometric functions have values that are less than 1, and their logarithms are therefore negative.

With degrees indicated in the left column of the page, use the column headings at the top of the page. With degrees given in the right column of the page, use the column designations at the bottom of the page.

Example 1.   Find log tan 3° 40′.

Solution: Locate 3° 40′ in the "Degrees" column of Table 1. Then to the right of 3° 40′ in the "Tangent Log" column read 8.8067. The minus 10 of this logarithm is understood but not written. Therefore,

$$\log \tan 3° \, 40' = 8.8067 - 10.$$

Example 2.   Find log cos 62° 27′.

Solution: We know that log cos 62° 27′ must lie between log cos 62° 20′ and log cos 62° 30′. In Table 1 we find

$$10'\left\{\begin{matrix}7'\left\{\begin{matrix}\log \cos 62° \, 20' = 9.6668\\ \log \cos 62° \, 27' = \quad ?\end{matrix}\right.\\ \log \cos 62° \, 30' = 9.6644\end{matrix}\right\}.0024$$

$$\log \cos 62° \, 27' = 9.6668 - .7(.0024) - 10$$

$$= 9.6651 - 10$$

Since 62° 27′ is .7 of the way from 62° 20′ to 62° 30′, we assume that log 62° 27′ is .7 of the way from 9.6668 − 10 to 9.6644 − 10. The tabular difference is .0024; therefore we subtract .7(.0024) from the 9.6668 − 10 entry. We subtract instead of add because for acute angles cos θ decreases, instead of increasing as θ increases.

Example 3.   Find $x = 37.1 \sin 27° \, 50'$.

Solution: We use Table 2 to find log 3.71; Table 1 for log sin 27° 50′; and then Table 2 to find the antilogarithm of log x.

$$\log 37.1 = 1. \qquad\qquad\qquad \log 37.1 = 1.5694$$

$$\log \sin 27° \, 50' = 9. \quad\underline{\quad\quad} - 10(+) \quad \log \sin 27° \, 50' = \underline{9.6692 - 10(+)}$$

$$\log x = \qquad\qquad\qquad\qquad \log x = 11.2386 - 10,$$

$$x = \qquad\qquad\qquad\qquad\qquad x = 17.3.$$

Exercises 10-12

Assume the given numbers to be exact and round off answers to table accuracy. In Exercises 1–10 find the logarithms.

1. $\log \sin 1° 20'$.

2. $\log \cos 2° 40'$.

3. $\log \tan 38° 10'$.

4. $\log \sin 75° 50'$.

5. $\log \cos 25° 15'$.

6. $\log \tan 31° 20'$.

7. $\log \sin 157° 40'$.

8. $\log \cos 151° 14'$.

9. $\log \cot 135° 45'$.

10. $\log \cot 171° 10'$.

In Exercises 11–20 solve for the acute angle θ.

11. $\log \sin \theta = 9.6418 - 10$.

12. $\log \tan \theta = 9.6946 - 10$.

13. $\log \cos \theta = 9.7222 - 10$.

14. $\log \cos \theta = 9.5576 - 10$.

15. $\log \tan \theta = 8.7429 - 10$.

16. $\log \sin \theta = 8.9104 - 10$.

17. $\log \sin \theta = 9.2100 - 10$.

18. $\log \cos \theta = 9.2215 - 10$.

19. $\log \tan \theta = 1.3494$.

20. $\log \tan \theta = 1.3599$.

In Exercises 21–26 solve for x. Use Tables 1 and 2.

21. $x = 1.355 \sin 45° 15'$.

22. $x = 205.4 \cos 15° 27'$.

23. $x = \dfrac{25.6 \sin 51° 10'}{\sin 35° 25'}$.

24. $x = \dfrac{3.84 \sin 78° 15'}{\sin 41° 55'}$.

25. $x = \dfrac{3.85 \tan 75° 10'}{\cos 155°}$.

26. $x = \dfrac{7.84 \tan 81° 44'}{\cos 175°}$.

REVIEW EXERCISES

Assume the given numbers to be exact and round off answers to table accuracy. In Exercises 1 and 2 express each equation in logarithmic form.

1. $9^2 = 81$.

2. $2^4 = 16$.

In Exercises 3 and 4 express each equation in exponential form.

3. $\log_4 16 = 2$.

4. $\log_{25} 5 = \frac{1}{2}$.

In Exercises 5–10 find the unknown b, N, or x.

5. $\log_3 N = 2$.

6. $\log_{10} N = 3$.

7. $\log_b \frac{9}{4} = 2$.

8. $\log_b 8 = 3$.

9. $\log_{10} 1000 = x$.

10. $\log_{10} .01 = x$.

In Exercises 11–14 write each expression as a single logarithm. (Assume that all the logarithms have the same base.)

11. $2 \log x + 3 \log y$.

12. $3 \log x - 2 \log y$.

13. $\log (x - y) - \log (x^2 - y^2)$.   14. $\log (x^2 - y^2) - 2 \log (x - y)$.

In Exercises 15–18 solve each equation for x.

15. $\log x + \log (7 - x) = 1$.   16. $\log (11 - x) + \log x = 1$.

17. $\log (2x + 1) = \log 1 + \log (x + 2)$.

18. $\log (1 + 3x) = \log 1 + \log (x + 4)$.

In Exercises 19 and 20 find θ accurate to the nearest minute.

19. $\sin \theta = .2513$.   20. $\cos \theta = .3350$.

In Exercises 21–24 assume the given numbers to be exact and solve for x. Use Tables 1 and 2.

21. $x = 2.567 \sin 25° 37'$.   22. $x = 45.61 \cos 42° 15'$.

23. $\dfrac{x}{\sin 27° 15'} = \dfrac{15.7}{\sin 102° 37'}$.   24. $\dfrac{x}{\sin 34° 16'} = \dfrac{23.4}{\sin 115° 24'}$.

CHAPTER 10: DIAGNOSTIC TEST

The purpose of this test is to see how well you understand the work covered in Chapter 10. We recommend that you work this test before your instructor tests you on this chapter. Allow yourself approximately 60 minutes to do the test.

Solutions to the problems, together with section references, are given in the Answer Section at the end of this book. We suggest that you study the sections referred to for the problems you do incorrectly.

1.   Express $3^4 = 81$ in logarithmic form.

2.   Express $\log_9 3 = \frac{1}{2}$ in exponential form.

3.   Give the number value for each of the following expressions:
     (a) $\log_{10} 10,000$.   (b) $\log_2 \frac{1}{4}$.   (c) $\log_5 5$.   (d) $\log_6 1$.

4.   Write $4 \log x - 3 \log y$ as a single logarithm.

5.   Solve for x.
     $2 \log (x + 3) = \log (7x + 1) + \log 2$.

6.   Find θ accurate to the nearest minute when $\log \tan \theta = 9.9904 - 10$.

In Problems 7–10 solve for x. Consider the given numbers to be exact and round your answers to table accuracy.

7. $x = \log \sin 130° 45'$.

8. $x = \cos 23° 40'$.

9. $x = 452 \tan 35° 10'$.

10. $x = \dfrac{27.4 \sin 54° 20'}{\cos 25° 40'}$.

11

Solution of Oblique Triangles

11-1 *SOLUTION OF OBLIQUE TRIANGLES*

Any triangle that is not a right triangle is an *oblique triangle*.

The solution of oblique triangles is not limited to this chapter. The Law of Sines (Section 2-14) and the Law of Cosines (Section 3-8) were used previously to solve oblique triangles. In this chapter we shall add other formulas for finding unknown parts of triangles and for finding the area of an oblique triangle. For oblique triangles, the necessary three parts can be given in four ways, and solutions for these ways will be discussed under the following four cases:

Case 1. One side and two angles.
Case 2. Two sides and the angle opposite one of them.
Case 3. Two sides and the included angle.
Case 4. Three sides.

As stated in Section 2-13, we shall denote the vertices and corresponding angles of a triangle in general by capital letters, A, B, and C, and the lengths of the opposite sides by the corresponding small letters, a, b, and c. The formulas developed for solving oblique triangles apply to right triangles but are usually less desirable for right triangles than those given in Chapters 2, 3, and 4.

In solving a problem, a sketch is helpful both as an aid in determining which formula is to be used and as a check for the detection of gross errors in the solution.

In all cases, before the numerical calculations are carried out, a blank computation form, such as was used in Section 10-10, should be made out completely before turning to the tables.

The given parts should be used as much as possible in the calculations, so that possible errors in computed parts may not be carried on into further calculations.

Before going on we recall the Law of Sines,

$$\frac{a}{\sin A} = \frac{b}{\sin B} = \frac{c}{\sin C},$$

and the Law of Cosines,

$$a^2 = b^2 + c^2 - 2bc \cos A,$$
$$b^2 = a^2 + c^2 - 2ac \cos B,$$
$$c^2 = a^2 + b^2 - 2ab \cos C.$$

11-2   THE AREA OF A TRIANGLE

Corresponding to different sets of data (Cases 1, 2, 3, and 4) there are several formulas for the area of a triangle. We shall denote the area of a triangle by K. If a, b, c are the lengths of the sides in inches, for example, K will be expressed in square inches.

> I. *Given two sides and the included angle (Case 3). The area of a triangle is equal to one-half the product of any two sides times the sine of the included angle*:
>
> (1)
> $$K = \begin{cases} \dfrac{1}{2}ab \sin C, \\[2mm] \dfrac{1}{2}ac \sin B, \\[2mm] \dfrac{1}{2}bc \sin A. \end{cases}$$

Proof. We construct triangle ABC showing A first as an acute and then as an obtuse angle (Fig. 11-1). Let h be the altitude drawn from vertex

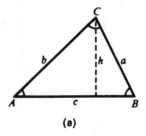 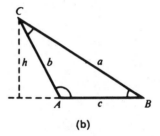

(a)   (b)

Figure 11-1

C. Then

$$K = \frac{1}{2}ch \quad \text{and} \quad h = b \sin A.$$

Therefore, $$K = \frac{1}{2}cb \sin A.$$

II. *Given one side and two angles (Case 1).* The area of a triangle is equal to one-half the square of any side times the product of the sines of the adjacent angles divided by the sine of the third angle:*

(2)
$$K = \begin{cases} \dfrac{1}{2}a^2\dfrac{\sin B \, \sin C}{\sin A}, \\[2ex] \dfrac{1}{2}b^2\dfrac{\sin A \, \sin C}{\sin B}, \\[2ex] \dfrac{1}{2}c^2\dfrac{\sin A \, \sin B}{\sin C}. \end{cases}$$

Proof. Let *b* be any side of a triangle. From the Law of Sines we have

$$\frac{b}{\sin B} = \frac{a}{\sin A}, \quad \text{or} \quad b = \frac{a \sin B}{\sin A}.$$

Substituting this value of *b* in the formula $K = \frac{1}{2}ab \sin C$, we have

$$K = \frac{1}{2}a\left(\frac{a \sin B}{\sin A}\right) \sin C = \frac{1}{2}a^2\frac{\sin B \, \sin C}{\sin A}.$$

---

* When two angles of a triangle are given, the third angle is also considered as known.

III. *Given the three sides (Case 4). The area of a triangle is found by the formula*

(3)
$$K = \sqrt{s(s - a)(s - b)(s - c)},$$

where $s = \dfrac{a + b + c}{2}.$

This formula is known as **Hero's formula**.

Proof. We square both members of Formula (1), $K = \frac{1}{2}bc \sin A$ and obtain

(a)
$$K^2 = \frac{1}{4}b^2c^2 \sin^2 A = \frac{1}{4}b^2c^2(1 - \cos^2 A)$$

$$= \frac{1}{4}b^2c^2(1 + \cos A)(1 - \cos A).$$

From the Law of Cosines,

(b)
$$\cos A = \frac{b^2 + c^2 - a^2}{2bc}.$$

We first add and then subtract the right member of (b) to and from 1 and obtain

(c)
$$1 + \cos A = 1 + \frac{b^2 + c^2 - a^2}{2bc} = \frac{2bc + b^2 + c^2 - a^2}{2bc}$$

$$= \frac{(b + c)^2 - a^2}{2bc} = \frac{(b + c + a)(b + c - a)}{2bc},$$

(d)
$$1 - \cos A = 1 - \frac{b^2 + c^2 - a^2}{2bc} = \frac{2bc - b^2 - c^2 + a^2}{2bc}$$

$$= \frac{a^2 - (b - c)^2}{2bc} = \frac{(a + b - c)(a - b + c)}{2bc}.$$

We substitute the results of (c) and (d) into (a) and get

(e)
$$K^2 = \frac{1}{16}(a + b + c)(b + c - a)(a + b - c)(a - b + c).$$

When we set $(a + b + c) = 2s$ and subtract $2a$ from both members of the equation, we get $(b + c - a) = 2(s - a)$. In a similar way,

$$(a + b - c) = 2(s - c), \quad \text{and} \quad (a - b + c) = 2(s - b).$$

When these expressions are substituted in (e) we get

$$K^2 = s(s - a)(s - b)(s - c).$$

Taking the square root of both members, we obtain

$$K = \sqrt{s(s - a)(s - b)(s - c)}.$$

IV. *Given two sides and the angle opposite one of them (Case* 2). *Proceed as follows:*

(i) *Determine whether one or two triangles are possible (See Section 2-15.)*
(ii) *Use the Law of Sines to find another side or angle.*
(iii) *Then use Formula* (1): $K = \frac{1}{2}bc \sin A$.

Example 1. Find the area of the triangle ABC if $a = 75.4$, $b = 39.7$, and $C = 42° 40'$. We assume the data to be exact.

Solution: Use $K = \frac{1}{2}ab \sin C$ and Tables 1 and 2.

$$\log 75.4 = 1. \qquad\qquad \log 75.4 = 1.8774.$$
$$\log 39.7 = 1. \qquad\qquad \log 39.7 = 1.5988.$$
$$\log \sin 42° 40' = 9. \quad - 10(+) \quad \log \sin 42° 40' = 9.8311 - 10(+)$$
$$\qquad\qquad\qquad\qquad\qquad\qquad\qquad\qquad 3.3073$$
$$\log 2 = \underline{\qquad\quad} (-) \qquad \log 2 = .3010 \qquad (-)$$
$$\log K = \qquad\qquad\qquad\qquad \log K = 3.0063,$$
$$K = \qquad\qquad\qquad\qquad\quad K = 1015 \text{ square units.}$$

Example 2. Find the area of the triangle ABC if $a = .816$, $b = .315$, and $c = .717$. We assume the given numbers to be exact.

Solution: Use Formula (3)

$$K = \sqrt{s(s - a)(s - b)(s - c)},$$

where

$$s = \frac{a + b + c}{2} = \frac{.816 + .315 + .717}{2} = .924.$$

$$s - a = .108, \qquad s - b = .609, \qquad s - c = .207.$$

Taking the logarithm of both members of Formula (3), we get

$$\log K = \frac{1}{2}\log s(s-a)(s-b)(s-c).$$

$$\log s = 9.9657 - 10$$

$$\log(s-a) = 9.0334 - 10$$

$$\log(s-b) = 9.7846 - 10$$

$$\log(s-c) = \underline{9.3160 - 10(+)}$$

$$\log K = \frac{1}{2}(38.0997 - 40)$$

$$= 19.0498 - 20,$$

$$K = .1122 \text{ square units.}$$

Exercises 11-2

(Assume the numbers given in the following exercises to be exact and round off the answers to table accuracy. Use Tables 1 or 2.) In Exercises 1–16 find the areas of the triangles ABC with the given parts.

1. $a = 100, b = 20, C = 41° 40'$.

2. $a = 80, b = 20, C = 38° 20'$.

3. $A = 22° 20', B = 48° 20', c = 10$.

4. $A = 30°, B = 41° 50', c = 10$.

5. $a = 7, b = 8, c = 9$.

6. $a = 9, b = 10, c = 11$.

7. $a = 3.85, b = 5.05, C = 74° 10'$.

8. $a = 1.75, b = 3.19, C = 61° 20'$.

9. $b = .760, c = .415, A = 157° 50'$.

10. $a = .875, c = .391, B = 137° 10'$.

11. $b = 1.88, C = 39° 10', A = 105° 50'$.

12. $a = 1.05, B = 41° 10', C = 110° 40'$.

13. $a = 18.7, b = 9.1, c = 10.4$.

14. $a = 15.8, b = 8.1, c = 11.5$.

15. $c = 15.87, A = 41° 16', C = 89° 43'$.

16. $a = 17.71, A = 55° 47', B = 110° 16'$.

17. A triangular corner lot is at the intersection of two streets which meet at an oblique angle. The frontages on the streets are 125 feet and 85 feet, and the line across the back of the lot measures 160 feet. Find the area of the lot.

18. A triangular lot has one side 125 feet long, another side 112 feet long, and the included angle is 80°. Find the area of the lot.

19. How many acres are there in a triangular field whose sides measure 430 feet, 780 feet, and 550 feet? (One acre equals 43,560 square feet.)

20. How many acres are there in a triangular lot whose sides measure 125 feet, 95 feet, and 160 feet?

21. A farmer wishes to buy a triangular field on the corner formed by two roads which meet at an angle of 115°. The frontage on one road is 44.3 rods and on the other 62.5 rods. What will the field cost him at $350 per acre? (One acre equals 160 square rods.)

22. Find the area of a hexagon inscribed in a circle whose radius is 2.625 inches.

23. A bar of steel with uniform triangular cross section is 12 feet long. The cross-sectional triangle has angles of 50°, 50°, and 80°. The side between the 50° angles measures 6.5 inches. The ends of the bar are perpendicular to its longitudinal axis. (a) Calculate the number of cubic feet in the bar. (b) Steel is approximately 7.8 times as heavy as water, and 1 cubic foot of water weighs 62.4 pounds; find the weight of the bar.

24. The area of a triangle is 468 square units, and two of its angles are 58° and 76°. Find the length of the longest side.

25. Prove that the area of a parallelogram is equal to the product of the lengths of a pair of adjacent sides and the sine of their included angle.

26. Two adjacent sides of a parallelogram have lengths of 3.75 and 1.89 and include an angle of 115° 40'. Find the area of the parallelogram.

11-3 THE LAW OF TANGENTS

In the solution of an oblique triangle when two sides and the included angle are known (Case 3), either the Law of Cosines or the Law of Tangents can be used. When the lengths of the sides of the triangle are not easily squared and logarithms are to be used, then the Law of Tangents is preferred, for it is well suited to logarithmic computation.

The *Law of Tangents*:

In any triangle the difference between any two sides is to their sum as the tangent of one-half of the difference between the opposite angles is to the tangent of one-half of the sum of those angles:

$$\frac{a - b}{a + b} = \frac{\tan\frac{1}{2}(A - B)}{\tan\frac{1}{2}(A + B)}, \qquad \frac{b - c}{b + c} = \frac{\tan\frac{1}{2}(B - C)}{\tan\frac{1}{2}(B + C)},$$

$$\frac{c - a}{c + a} = \frac{\tan\frac{1}{2}(C - A)}{\tan\frac{1}{2}(C + A)}.$$

Proof.   From the Law of Sines we have

(1)
$$\frac{a}{c} = \frac{\sin A}{\sin C}$$

and

(2)
$$\frac{b}{c} = \frac{\sin B}{\sin C}.$$

Subtracting Equation (2) from Equation (1) and adding Equations (1) and (2) yields

(3)
$$\frac{a - b}{c} = \frac{\sin A - \sin B}{\sin C}$$

and

(4)
$$\frac{a + b}{c} = \frac{\sin A + \sin B}{\sin C}.$$

Dividing each member of (3) by the corresponding member of (4) yields

(5)
$$\frac{a - b}{a + b} = \frac{\sin A - \sin B}{\sin A + \sin B}.$$

Applying Identities (32) and (31) to the right-hand member of (5), we obtain

$$\frac{a - b}{a + b} = \frac{2 \cos \frac{1}{2}(A + B) \sin \frac{1}{2}(A - B)}{2 \sin \frac{1}{2}(A + B) \cos \frac{1}{2}(A - B)}$$

$$= \frac{\sin \frac{1}{2}(A - B)}{\cos \frac{1}{2}(A - B)} \cdot \frac{\cos \frac{1}{2}(A + B)}{\sin \frac{1}{2}(A + B)}.$$

Hence,

$$\frac{a - b}{a + b} = \frac{\tan \frac{1}{2}(A - B)}{\tan \frac{1}{2}(A + B)}.$$

In a similar manner we can derive

$$\frac{b - c}{b + c} = \frac{\tan \frac{1}{2}(B - C)}{\tan \frac{1}{2}(B + C)}, \qquad \frac{c - a}{c + a} = \frac{\tan \frac{1}{2}(C - A)}{\tan \frac{1}{2}(C + A)}.$$

11-4 APPLICATIONS OF THE LAW OF TANGENTS

When two sides and the included angle of an oblique triangle are known, we use the Law of Tangents to find the other two angles and then apply the Law of Sines to find the third side of the triangle.

Suppose that a and b and the included angle C are given. If $a > b$, we use the Law of Tangents in the form involving $(a - b)$ and $\frac{1}{2}(A - B)$; if $b > a$, we use a form containing $(b - a)$ and $\frac{1}{2}(B - A)$, to avoid negative factors. Let us assume that $a > b$; then the unknown in the Law of Tangents is the expression $\frac{1}{2}(A - B)$. With C known we find $\frac{1}{2}(A + B)$ as follows:

$$A + B + C = 180°.$$

Then

$$A + B = 180° - C,$$

and

$$\frac{1}{2}(A + B) = \frac{1}{2}(180° - C).$$

We use the law of tangents to find

$$\frac{1}{2}(A - B) = \underline{\hspace{2cm}}.$$

The angles A and B can now be found from the relations $A = \frac{1}{2}(A + B) + \frac{1}{2}(A - B)$ and $B = \frac{1}{2}(A + B) - \frac{1}{2}(A - B)$; side c can be found by using the Law of Sines.

Example. Assume the given numbers to be exact and solve the triangle ABC, given

$$a = 591, \quad b = 786, \quad C = 37° \, 20'.$$

Solution: Since $b > a$, we use the form

$$\frac{b - a}{b + a} = \frac{\tan \frac{1}{2}(B - A)}{\tan \frac{1}{2}(B + A)}.$$

$$b = \; 786, \qquad \frac{1}{2}(B + A) = \frac{1}{2}(180° - 37° \, 20')$$

$$a = \; 591, \qquad\qquad\qquad = \frac{1}{2}(179° \, 60' - 37° \, 20')$$

$$\overline{}$$

$$b + a = 1377, \qquad\qquad\qquad = \frac{1}{2}(142° \, 40') = 71° \, 20'.$$

$$b - a = \; 195.$$

$$\tan \frac{1}{2}(B - A) = \frac{(b - a) \tan \frac{1}{2}(B + A)}{b + a}$$

$$= \frac{195 \tan 71° \, 20'}{1377}$$

$$\log 195 = 2.2900$$

$$\log \tan 71° \, 20' = \underline{10.4713 - 10} \, (+)$$

$$12.7613 - 10$$

$$\log 1377 = \underline{\; 3.1389 \qquad} (-)$$

$$\log \tan \frac{1}{2}(B - A) = \; 9.6224 - 10.$$

Therefore,

$$(1) \quad \frac{1}{2}(B - A) = 22° 45'.$$

But

$$(2) \quad \frac{1}{2}(B + A) = 71° 20'(+)$$

which gives

$$B = 94° 05' \text{ to the nearest minute.}$$

When we subtract (1) from (2), we find $A = 48° 35'$, to the nearest minute. We use the Law of Sines to find c.

$$\frac{c}{\sin C} = \frac{a}{\sin A}.$$

$$c = \frac{a \sin C}{\sin A} = \frac{591 \sin 37° 20'}{\sin 48° 35'}.$$

$$\log 591 = 2.7716$$

$$\log \sin 37° 20' = 9.7828 - 10(+)$$

$$\overline{12.5544 - 10}$$

$$\log \sin 48° 35' = 9.8750 - 10(-)$$

$$\overline{\log c = 2.6794.}$$

Therefore,

$$c = 478.0 \text{ to four significant digits.}$$

Exercises 11-4

Solve the following triangles. Assume the given numbers to be exact and round off the answers to table accuracy. Use Tables 1 and 2.

1. $b = 23.4, c = 28.9, A = 35° 40'$.
2. $a = 15.8, c = 24.9, B = 68° 00'$.
3. $a = 4.15, c = 5.97, B = 58° 20'$.
4. $b = 2.81, c = 3.06, A = 41° 40'$.
5. $a = 65.1, b = 50.9, C = 52° 40'$.
6. $a = 63.8, b = 49.6, C = 73° 20'$.
7. $b = 184, c = 172, A = 105° 40'$.
8. $a = .391, c = .515, B = 115° 20'$.
9. $a = .865, b = .631, C = 127° 20'$.
10. $b = 53.17, c = 33.65, A = 79° 50'$.
11. Two streets meet at an angle of $96° 30'$. A triangular lot at their intersection has a frontage of 158.6 feet on one street and 127.7 feet

on the other. Find the angles the third side makes with the two streets and the length of the third side.

12. In the parallelogram $ABCD$, $AB = 375$, $AD = 186$, and angle $BAD = 75°\ 10'$. Find the length of the diagonal AC.

11-5 THE HALF-ANGLE FORMULAS IN TERMS OF THE SIDES OF A TRIANGLE

Let ABC be any triangle with sides a, b, and c. Let AO, BO, and CO be the bisectors of angles A, B, and C, respectively (Fig. 11-2). The bisectors

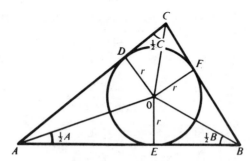

Figure 11-2

of the angles of a triangle meet at a point that is equidistant from the sides of the triangle; O is therefore the center of the circle inscribed in the triangle. We draw OE, OF, and OD perpendicular to the respective sides of the triangle. Hence, $OE = OF = OD = r$, the radius of the inscribed circle. Then

$$\text{area } \triangle ABC = \text{area } \triangle ABO + \text{area } \triangle BCO + \text{area } \triangle AOC,$$

and it follows that

(1) $$K = \frac{1}{2}cr + \frac{1}{2}ar + \frac{1}{2}br = r\left(\frac{a + b + c}{2}\right) = rs.$$

By Formula (3) of Section 11-2

(2) $$K = \sqrt{s(s - a)(s - b)(s - c)}.$$

Equating (1) and (2), we get

$$rs = \sqrt{s(s - a)(s - b)(s - c)},$$

and

$$r = \sqrt{\frac{(s - a)(s - b)(s - c)}{s}}.$$

> *The radius of the inscribed circle in any triangle ABC is given by*
>
> (3)
> $$r = \sqrt{\frac{(s - a)(s - b)(s - c)}{s}},$$
>
> *where* $s = \frac{1}{2}(a + b + c)$.

By elementary geometry, in Figure 11-2

$$AE = AD, \qquad BE = BF, \qquad CF = CD.$$

Then
$$AB + BC + AC = 2s,$$

$$AE + BF + CF = s,$$

$$AE = s - (BF + CF) = s - a.$$

Thus, in the right triangle AEO

$$\tan \frac{1}{2}A = \frac{r}{s - a}.$$

When we apply a similar procedure to triangles BOE and CDO, we obtain

$$\tan \frac{1}{2}B = \frac{r}{s - b} \qquad \text{and} \qquad \tan \frac{1}{2}C = \frac{r}{s - c}.$$

Hence:

> *In any triangle ABC,*
>
> (4) $\quad \tan \dfrac{1}{2}A = \dfrac{r}{s - a}, \qquad \tan \dfrac{1}{2}B = \dfrac{r}{s - b}, \qquad \tan \dfrac{1}{2}C = \dfrac{r}{s - c},$
>
> *where* $\quad r = \sqrt{\dfrac{(s - a)(s - b)(s - c)}{s}} \qquad$ *and* $\qquad s = \dfrac{1}{2}(a + b + c).$

Sines of the half-angles. If A, B, C are the angles of a triangle, then $\frac{1}{2}A$, $\frac{1}{2}B$, and $\frac{1}{2}C$ are acute angles, and their functions are all positive. Therefore, by Identity (24), Section 8-6,

(5) $\quad \sin \dfrac{1}{2}A = \sqrt{\dfrac{1 - \cos A}{2}}, \qquad$ and $\qquad \sin^2 \dfrac{1}{2}A = \dfrac{1}{2}(1 - \cos A).$

By the Law of Cosines,

$$\cos A = \frac{b^2 + c^2 - a^2}{2bc}.$$

We substitute this expression for $\cos A$ in (5) and obtain

$$\sin^2 \frac{1}{2}A = \frac{1}{2}\left(1 - \frac{b^2 + c^2 - a^2}{2bc}\right) = \frac{a^2 - b^2 + 2bc - c^2}{4bc}$$

$$= \frac{a^2 - (b^2 - 2bc + c^2)}{4bc} = \frac{a^2 - (b - c)^2}{4bc}.$$

We factor the numerator as the difference of two squares, getting

$$\sin^2 \frac{1}{2}A = \frac{[a - (b - c)][a + (b - c)]}{4bc} = \frac{a - b + c}{2} \cdot \frac{a + b - c}{2} \cdot \frac{1}{bc}$$

$$= \frac{(s - b)(s - c)}{bc}.$$

In a similar way formulas can be obtained for $\sin^2 \frac{1}{2}B$ and $\sin^2 \frac{1}{2}C$. Hence:

In any triangle ABC,

$$(6) \quad \sin \frac{1}{2}A = \sqrt{\frac{(s - b)(s - c)}{bc}}, \qquad \sin \frac{1}{2}B = \sqrt{\frac{(s - a)(s - c)}{ac}},$$

$$\sin \frac{1}{2}C = \sqrt{\frac{(s - a)(s - b)}{ab}},$$

where $s = \frac{1}{2}(a + b + c).$

If we wish to find all of the angles of a triangle, when the sides are known, it is best to use the half-angle tangent formulas. If only one of the angles is desired, the formula for the sine of the half-angle is satisfactory unless the angle is very large or very small.

Example. Find the radius of the inscribed circle and the angles of triangle ABC, given

$$a = 31.4,$$
$$b = 27.1,$$
$$c = 35.1.$$

Solution: We assume the given numbers to be exact and round answers to table accuracy.

By addition, $2s = a + b + c = 93.6.$

$$s = 46.8 \qquad r = \sqrt{\frac{(s - a)(s - b)(s - c)}{s}}.$$

$$s - a = 15.4, \qquad \log(s - a) = 1.1875$$
$$s - b = 19.7, \qquad \log(s - b) = 1.2945$$
$$s - c = 11.7, \qquad \log(s - c) = \underline{1.0682}\,(\,+\,)$$
$$3.5502$$

$$\log s = \underline{1.6702}\,(\,-\,)$$
$$\log r^2 = 1.8800\,(\,\div\,2\,)$$

$$\log r = .9400.$$
$$r = 8.710.$$

$$\tan\frac{1}{2}A = \frac{r}{s - a}, \qquad\qquad \tan\frac{1}{2}B = \frac{r}{s - b}.$$

$$\log r = 10.9400 - 10 \qquad\qquad \log r = 10.9400 - 10$$
$$\log(s - a) = \underline{1.1875}\quad(\,-\,) \qquad \log(s - b) = \underline{1.2945}\quad(\,-\,)$$
$$\log\tan\frac{1}{2}A = 9.7525 - 10. \qquad \log\tan\frac{1}{2}B = 9.6455 - 10.$$

$$\frac{1}{2}A = 29°\,20' + \frac{28}{29}(10) \qquad\qquad \frac{1}{2}B = 23°\,50' + \frac{3}{34}(10')$$

$$A = 58°\,40' + 19' \qquad\qquad\qquad B = 46°\,100' + 02'$$

$$A = 58°\,59' \qquad\qquad\qquad\qquad B = 47°\,42'.$$

$$\tan\frac{1}{2}C = \frac{r}{s - c}.$$

$$\log r = 10.9400 - 10$$
$$\log(s - c) = \underline{1.0682}\quad(\,-\,)$$
$$\log\tan\frac{1}{2}C = 9.8718 - 10.$$

$$\frac{1}{2}C = 36°\,40',$$

$$C = 73°\,20'.$$

Check.

$$A = 58° 59',$$
$$B = 47° 42',$$
$$C = 73° 20',$$
$$\overline{A + B + C = 180° 01'.}$$

We solve the above example for angle A using the formula

$$\sin \frac{1}{2}A = \sqrt{\frac{(s - b)(s - c)}{bc}}.$$

$\log (s - b) = 1.2945$		$\log b = 1.4330$	
$\log (s - c) = 1.0682 \ (+)$		$\log c = 1.5453 \ (+)$	
(1)	$\overline{2.3627.}$	(2)	$\overline{2.9783.}$

Subtracting (2) from (1)

$$12.3627 - 10$$
$$\underline{2.9783 \qquad (-)}$$

we get
$$\log \sin^2 \frac{1}{2}A = 9.3844 - 10$$
$$= \overline{19.3844 - 20(\div 2)}$$
$$\log \sin \frac{1}{2}A = 9.6922 - 10.$$
$$\frac{1}{2}A = 29° 20' + \frac{21}{22}(10'),$$
$$A = 58° 59'.$$

Exercises 11-5

Use the formula for tangent of half an angle to solve Exercises 1–4 and the formula for sine of half an angle to solve Exercises 5–8. Assume the given numbers to be exact and round to table accuracy.

1. $a = 116, b = 95, c = 127.$
2. $a = 209, b = 184, c = 239.$
3. $a = .356, b = .751, c = .609.$
4. $a = .475, b = .385, c = .208.$
5. $a = 75.6, b = 49.7, c = 38.5.$
6. $a = 37.1, b = 41.8, c = 19.5.$

7. $a = 182, b = 293, c = 161.$

8. $a = 512, b = 409, c = 749.$

9. What is the radius of the largest circular track that can be laid out within a triangular field whose sides are 275 yards, 198 yards, and 345 yards long?

10. A motorboat race follows a triangular course. The first leg of 1760 yards is due north, and the other two legs of 1320 yards and 2200 yards lie to the east of the first one. Find the bearing of the third leg.

REVIEW EXERCISES

In all the exercises of this set assume the given numbers to be exact and round your answers to table accuracy.

In Exercises 1–6 find the areas of the triangles ABC having the given parts.

1. $A = 39° 47', b = 120, c = 100.$

2. $a = 100, B = 47° 21', c = 110.$

3. $A = 31° 14', B = 84° 52', c = 10.$

4. $a = 20, B = 75° 11', C = 54° 37'.$

5. $a = 10, b = 13, c = 15.$

6. $a = 11, b = 10, c = 9.$

In Exercises 7 and 8 solve for the missing parts of each triangle.

7. $a = 161, c = 111, B = 53° 10'.$

8. $a = 180, b = 130, C = 75° 20'.$

9. Given triangle ABC with $a = 25, b = 18,$ and $c = 21,$ solve for angle B.

10. Given triangle ABC with $a = 45, b = 27,$ and $c = 32,$ solve for angle $C.$

11. Find the radius of the largest circle that can be inscribed in the triangle whose sides are 150, 125, and 117.

12. Find the radius of the largest circle that can be inscribed in the triangle whose sides are 181, 173, and 140.

CHAPTER 11: DIAGNOSTIC TEST

The purpose of this test is to see how well you understand the work covered in Chapter 11. We recommend that you work this test before your instructor tests you on this chapter. Allow yourself approximately 60 minutes to do the test.

Solutions to the problems, together with section references, are given in the Answer Section at the end of this book. You should study the sections referred to for the problems you do incorrectly.

Assume that the numbers given in the problems to be exact and round your answers to table accuracy. Express degree measure accurate to the nearest minute.

1. Find the area of triangle ABC if $a = 100$, $b = 140$, and $C = 81° 14'$.

2. Find the area of triangle ABC if $a = 15$, $b = 10$, and $c = 11$.

3. Find angle A of triangle ABC if $a = 21$, $b = 32$, and $c = 27$.

4. Find the area of triangle ABC if $A = 75° 14'$, $B = 61° 53'$, and $a = 12$.

5. Solve for angles A and C and side b of triangle ABC if $a = 25$, $B = 53° 10'$, and $c = 20$.

12

Inverse Trigonometric Functions

12-1 INVERSE TRIGONOMETRIC RELATIONS

We often need a notation to express an angle in terms of one of its trigonometric functions.

> If $x = \sin \theta$, then θ is an angle whose sine is x. We then say that θ is *an inverse sine of x*, and we express this by the notation
>
> (1) $\qquad\qquad\qquad \theta = \textbf{arcsin } \textbf{\textit{x}}.$

This notation is an abbreviation for "θ is an angle whose sine is x."*

Example 1. If $\sin \theta = \frac{1}{2}$, then we can say that "$\theta$ is the angle whose sine is $\frac{1}{2}$," or

$$\theta = \arcsin \frac{1}{2}.$$

---

* Another form in common use is $\theta = \sin^{-1} x$. When this form is used, it must be understood that the -1 is not an exponent.

Thus, θ can be equal to $\frac{1}{6}\pi$, $\frac{5}{6}\pi$, $\frac{13}{6}\pi$, $-\frac{7}{6}\pi$, and so forth, or, in general, $\theta = \frac{1}{6}\pi + 2n\pi$, or $\frac{5}{6}\pi + 2n\pi$, where n is an integer.

From the observation of Example 1 we see that the relation $y = \arcsin x$ is infinitely many-valued. Since y, in the equation $y = \arcsin x$, is not uniquely determined when x is given, we refrain from calling it a function (if y is a function of x, at most one value of y corresponds to any one value of x). Such a pairing in which to each x there can correspond none, one, or several values of y is called a *relation*. This leads to the need for defining the so-called *principal values* of the inverse relations, in terms of which all the related inverse functions can be expressed. This is done in Section 12-3.

The remaining inverse trigonometric relations are defined in a corresponding way. The expression arccos x denotes an angle whose cosine is x, arctan x denotes an angle whose tangent is x, and so forth. The other three inverse trigonometric relations, arccot x, arcsec x, and arccsc x, are of less importance. The arccsc x and arcsec x can be expressed in terms of arcsin x and arccos x respectively. For example, the expression arccsc x can be considered equivalent to arcsin $1/x$. More specifically,

$$\text{arccsc } 2 = \arcsin \frac{1}{2} = \frac{\pi}{6}.$$

Example 2. Let θ be an angle placed in standard position with its terminal side passing through the point $P(4, 3)$. (See Fig. 12-1.) We describe the relations shown in Figure 12-1 in the following different but equivalent ways. Each statement applies to Figure 12-1, in which θ is an acute angle.

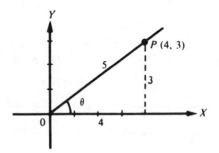

Figure 12-1

1. $\sin \theta = \frac{3}{5}$; θ is the acute angle whose sine is $\frac{3}{5}$;

$$\theta = \arcsin \tfrac{3}{5}.$$

2. $\cos \theta = \frac{4}{5}$; θ is the acute angle whose cosine is $\frac{4}{5}$;

$$\theta = \arccos \tfrac{4}{5}.$$

3. $\tan \theta = \frac{3}{4}$; θ is the acute angle whose tangent is $\frac{3}{4}$;

$$\theta = \arctan \tfrac{3}{4}.$$

4. $\cot \theta = \frac{4}{3}$; θ is the acute angle whose cotangent is $\frac{4}{3}$;

$$\theta = \operatorname{arccot} \tfrac{4}{3}.$$

5. $\sec \theta = \frac{5}{4}$; θ is the acute angle whose secant is $\frac{5}{4}$;

$$\theta = \operatorname{arcsec} \tfrac{5}{4}.$$

6. $\csc \theta = \frac{5}{3}$; θ is the acute angle whose cosecant is $\frac{5}{3}$;

$$\theta = \operatorname{arccsc} \tfrac{5}{3}.$$

Example 3. Solve the equation $y = \sin 2x$ for x in terms of y.

Solution: Since $\sin 2x = y$,

$$2x = \arcsin y,$$

and

$$x = \frac{1}{2} \arcsin y.$$

Example 4. Solve the equation $x = \frac{1}{3} \arctan 2y$ for y in terms of x.

Solution: Since $x = \frac{1}{3} \arctan 2y$,

$$3x = \arctan 2y,$$

and

$$2y = \tan 3x;$$

therefore,

$$y = \frac{1}{2} \tan 3x.$$

Example 5. Restricting $\arctan \frac{5}{12}$ to the first quadrant, find the value of each of the following expressions:

(a) $$\sin \left(\arctan \frac{5}{12} \right),$$

(b) $$\cos \left(\arctan \frac{5}{12} \right),$$

(c) $$\tan \left(\arctan \frac{5}{12} \right).$$

Solution: (a) This part of the example merely asks for the sine of the acute angle whose tangent is $\frac{5}{12}$. Let θ be this angle. Sketch the angle (Fig. 12-2); let $P(12, 5)$ be a point on the terminal side of θ; find the radius 13 for P; then read the answer for part (a) of the problem from the

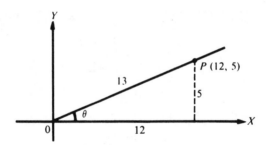

Figure 12-2

figure. Answers for parts (b) and (c) are obtained in the same manner from the figure.

(a) $$\sin\left(\arctan\frac{5}{12}\right) = \frac{5}{13}.$$

(b) $$\cos\left(\arctan\frac{5}{12}\right) = \frac{12}{13}.$$

(c) $$\tan\left(\arctan\frac{5}{12}\right) = \frac{5}{12}.$$

Exercises 12-1

In Exercises 1–12 find all values of each expression in the interval $0 \leqq \theta < 2\pi$. Do not use tables.

1. arcsin 1.

2. arccos 1.

3. arccos 0.

4. arcsin 0.

5. arctan 1.

6. arccot $\sqrt{3}$.

7. arcsin $\left(-\frac{1}{2}\right)$.

8. arccos $\left(-\frac{1}{2}\right)$.

9. arccos (-1).

10. arctan (-1).

11. arccsc 2.

12. arcsec (-2).

In Exercises 13–20 use Table 1 to find all values of each expression in the interval $0° \leqq \theta < 360°$.

13. arccos .8746.

14. arcsin .5150.

15. arctan .3906.

16. arctan .9380.

17. arcsin $(-.0581)$.

18. arccos $(-.9848)$.

19. arctan $(-.2126)$.

20. arcsin $(-.9983)$.

Solve each of the following Exercises for x in terms of y.

21. $y = 2\sin 3x$.

22. $y = \frac{1}{2}\cos 2x$.

23. $y = \frac{1}{3}\cos 2x$.

24. $y = 3\sin 4x$.

25. $y = \tan \pi x$.

26. $y = \sin \pi x$.

27. $y = 1 + \cos 3x$.

28. $y = 2 - \sin 2x$.

29. $y = \frac{1}{2}\pi - \arccos 2x$.

30. $y = \pi - \arctan 3x$.

In Exercises 31–40 find the value of each expression. Restrict the arc function expressions to acute angles.

31. $\tan (\arcsin \frac{4}{5})$.

32. $\tan (\arcsin \frac{3}{5})$.

33. $\sec (\arccos \frac{2}{3})$.

34. $\csc (\arcsin \frac{1}{5})$.

35. $\sin (\arcsin \frac{3}{7})$.

36. $\cos (\arccos \frac{5}{7})$.

37. $\cos (\arcsin \frac{2}{5})$.

38. $\sin (\arccos \frac{1}{3})$.

39. $\sin (\arcsin \frac{3}{5} + \arccos \frac{3}{5})$.

40. $\cos (\arctan \frac{5}{12} + \arcsin \frac{3}{5})$.

12-2 GRAPHS OF THE INVERSE TRIGONOMETRIC RELATIONS

The graphs of the inverse trigonometric relations show their behavior very clearly.

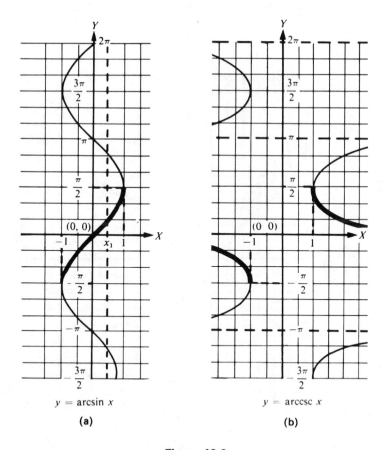

$y = \arcsin x$

(a)

$y = \text{arccsc } x$

(b)

Figure 12-3

To obtain the graph of $y = \arcsin x$, write the equivalent direct form of this relation, $x = \sin y$, and proceed as in Section 2-17. The coordinates of some of the points on the graph are given in the table below. (See Fig. 12-3a.)

y	$-\pi$	$-\frac{1}{2}\pi$	0	$\frac{1}{2}\pi$	π	$\frac{3}{2}\pi$	2π
x	0	-1	0	1	0	-1	0

The graphs (Figs. 12-3, 12-4, 12-5) of the remaining inverse trigonometric relations can be obtained in a corresponding way.

Since the sine function only takes on values from -1 to 1, the domain of $\arcsin x$ is $-1 \leqq x \leqq 1$. For example, $\arcsin 2$ is not defined in real numbers, since there is no angle whose sine is 2. On the other hand, $\arctan x$ (Fig. 12-5a) is defined for all real values of x, since the tangent can assume any real value. We can say that:

The domain of arcsin x and arccos x is $-1 \leqq x \leqq 1$, the domain of arctan x and arccot x is $-\infty < x < +\infty$, the domain of arcsec x and arccsc x is $x \geqq 1$ and $x \leqq -1$.

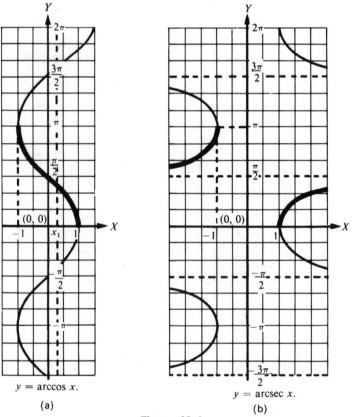

$y = \arccos x.$

(a)

$y = \arcsec x.$

(b)

Figure 12-4

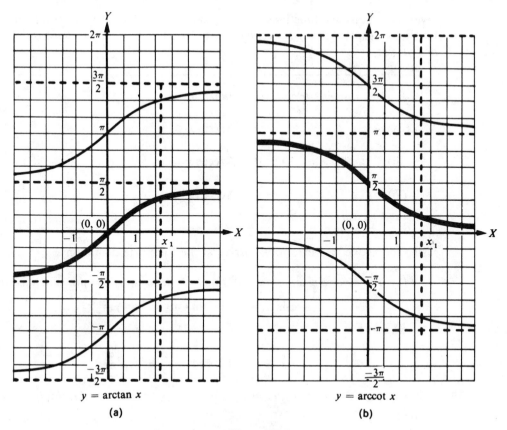

$y = \arctan x$

(a)

$y = \text{arccot } x$

(b)

Figure 12-5

A vertical line drawn at any point in the interval $-1 \leqq x \leqq 1$ on the graphs for $y = \arcsin x$ and $y = \arccos x$ (Figs. 12-3a and 12-4a) meets the graphs in infinitely many points; the ordinates of these points are the infinitely many values of arcsin x and arccos x for the corresponding value of x.

12-3 PRINCIPAL VALUES OF THE INVERSE TRIGONOMETRIC FUNCTIONS

In our discussion we have emphasized that the trigonometric inverse relations are many-valued. However, under suitable restrictions these inverse relations can be made one-to-one, and in this way we obtain the important **inverse trigonometric**, or **arc**, **functions**. As is seen by the graph of $y = \arcsin x$ (Fig. 12-3a), if the range of the relation is limited to $-\pi/2 \leqq \arcsin x \leqq \pi/2$, there is a unique value of y for each x of the domain. In a similar manner, the range for each of the other arc functions is restricted to distinguish between the inverse relations and functions. The capital letter "A" is used in writing the arc functions. Thus, *the range of each of the inverse trigonomeric functions is defined as follows*:

Range of inverse trigonometric functions:

(i) $$-\frac{\pi}{2} \leqq \mathbf{Arcsin}\, x \leqq \frac{\pi}{2},$$

(ii) $$0 \leqq \mathbf{Arccos}\, x \leqq \pi,$$

(iii) $$-\frac{\pi}{2} < \mathbf{Arctan}\, x < \frac{\pi}{2}.$$

Using the above restrictions, we arrive at the following definitions:

(1) *The inverse sine, or arcsine, function, denoted by Arcsin, is defined by*

$$\mathbf{Arcsin}\, x = y \quad \text{if and only if} \quad \sin y = x,$$

where

$$-1 \leqq x \leqq 1 \quad \text{and} \quad -\frac{\pi}{2} \leqq y \leqq \frac{\pi}{2}.$$

(2) *The inverse cosine, or arccosine, function, denoted by Arccos, is defined by*

$$\mathbf{Arccos}\, x = y \quad \text{if and only if} \quad \cos y = x,$$

where

$$-1 \leqq x \leqq 1 \quad \text{and} \quad 0 \leqq y \leqq \pi.$$

(3) *The inverse tangent, or arctangent, function, denoted by Arctan, is defined by*

$$\mathbf{Arctan}\, x = y \quad \text{if and only if} \quad \tan y = x,$$

where

$$-\infty < x < \infty \quad \text{and} \quad -\frac{\pi}{2} < y < \frac{\pi}{2}.$$

The graphs of these functions are indicated in the figures by heavy lines.

Authors sometimes differ in their choice of principal values for the other three inverse trigonometric functions, Arccot x, Arcsec x, and Arccsc x. Since these are of less importance in elementary work, we shall not be concerned with them in this course. However, the principal values commonly used for these are shown by heavy lines on the graphs of $y = \text{arccot } x$ (Fig. 12-5b), $y = \text{arcsec } x$ (Fig. 12-4b), and $y = \text{arccsc } x$ (Fig. 12-3b).

Example 1. In this example we illustrate the difference between the principal and general values of the expressions. The number n is any integer.

(a) $\qquad\qquad\qquad$ Arcsin $1 = \pi/2$.

(b) $\qquad\qquad\qquad$ arcsin $1 = \pi/2 + 2n\pi$.

(c) $\qquad\qquad\qquad$ Arctan $\sqrt{3} = \pi/3$.

(d) $\qquad\qquad\qquad$ arctan $\sqrt{3} = \pi/3 + n\pi$.

(e) $\qquad\qquad\qquad$ Arccos $(-1) = \pi$.

(f) $\qquad\qquad\qquad$ arccos $(-1) = \pi + 2n\pi$.

(g) $\qquad\qquad\qquad$ Arcsin $(-1) = -\pi/2$.

(h) $\qquad\qquad\qquad$ arcsin $(-1) = -\pi/2 + 2n\pi$.

Example 2. Find the value of $\sin (2 \text{ Arctan } \tfrac{1}{3})$.

Solution: Let $\theta = \text{Arctan } \tfrac{1}{3}$.
Then $\tan \theta = \tfrac{1}{3}$ and θ is an acute angle. Sketch θ in standard position (Fig. 12-6). By Identity (21), Section 8-5,

$$\sin \left(2 \text{ Arctan } \frac{1}{3} \right) = \sin 2\theta = 2 \sin \theta \cos \theta = 2\left(\frac{1}{\sqrt{10}}\right)\left(\frac{3}{\sqrt{10}}\right) = \frac{3}{5}.$$

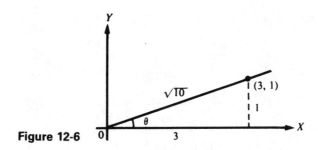

Figure 12-6

Example 3. Find the value of $\sin (\text{Arcsin } \tfrac{5}{13} + \text{Arccos } \tfrac{4}{5})$.

Solution: Set Arcsin $\tfrac{5}{13} = \alpha$ and Arccos $\tfrac{4}{5} = \beta$; then α and β are positive acute angles, and $\sin \alpha = \tfrac{5}{13}$, $\cos \beta = \tfrac{4}{5}$. Sketch angles α and β in standard position and compute the missing parts of the reference

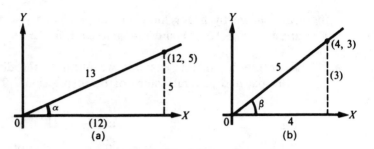

Figure 12-7

triangles (Fig. 12-7). Then, by applying the sin $(\alpha + \beta)$ identity, we have

$$\sin (\alpha + \beta) = \sin \alpha \cos \beta + \cos \alpha \sin \beta$$

$$= \frac{5}{13} \cdot \frac{4}{5} + \frac{12}{13} \cdot \frac{3}{5} = \frac{56}{65}.$$

Therefore, $\sin (\text{Arcsin } \frac{5}{13} + \text{Arccos } \frac{4}{5}) = \frac{56}{65}$.

Example 4.   Assume $0 < u < 5$ and show that

$$\tan \left(\text{Arcsin } \frac{u}{5} \right) = \frac{u}{\sqrt{25 - u^2}}.$$

Solution: Set Arcsin $u/5 = \theta$ and then $\sin \theta = u/5$.
Sketch θ in standard position and complete the reference triangle as shown in Figure 12-8. Compute the third side of the reference triangle.

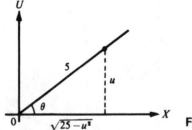

Figure 12-8

Then, from the figure,

$$\tan \left(\text{Arcsin } \frac{u}{5} \right) = \tan \theta = \frac{u}{\sqrt{25 - u^2}}.$$

Example 5.   Use inverse trigonometric notation to represent θ where $\frac{1}{2}\pi < \theta < \pi$ and $\tan \theta = -\frac{3}{4}$.

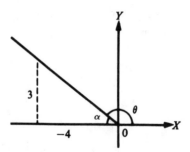

Figure 12-9

Solution: Sketch the angle in standard position (Fig. 12-9). Let α be the related angle of θ; then

$$\alpha = \text{Arctan}\,\frac{3}{4}.$$

But $$\theta = \pi - \alpha.$$

Therefore, $$\theta = \pi - \text{Arctan}\,\frac{3}{4}.$$

Example 6.   Solve $\text{Arctan}\,(3x^2 + 1) = 2\,\text{Arctan}\,\frac{1}{2}$.

Solution: Take the tangents of both members of the equation and equate them.

$$\tan\,[\text{Arctan}\,(3x^2 + 1)] = \tan\left(2\,\text{Arctan}\,\frac{1}{2}\right),$$

(1) $$3x^2 + 1 = \tan\left(2\,\text{Arctan}\,\frac{1}{2}\right).$$

Set $\text{Arctan}\,\frac{1}{2} = \theta$ and then $\tan \theta = \frac{1}{2}$. Apply the double-angle identity to the right member of Equation (1).

$$\tan\left(2\,\text{Arctan}\,\frac{1}{2}\right) = \tan 2\theta$$

$$= \frac{2\tan\theta}{1 - \tan^2\theta} = \frac{2 \cdot \frac{1}{2}}{1 - \left(\frac{1}{2}\right)^2} = \frac{4}{3}.$$

Therefore, Equation (1) becomes

$$3x^2 + 1 = \frac{4}{3}$$

and

$$x = \pm\frac{1}{3}.$$

Check: For $x = \pm\frac{1}{3}$, $(3x^2 + 1) = \frac{4}{3}$.
Then

$$\text{Arctan}\,(3x^2 + 1) = 2\,\text{Arctan}\,\frac{1}{2},$$

$$\text{Arctan}\,\frac{4}{3} = 2\,\text{Arctan}\,\frac{1}{2},$$

$$\tan\left(\text{Arctan}\,\frac{4}{3}\right) = \tan\left(2\,\text{Arctan}\,\frac{1}{2}\right),$$

$$\frac{4}{3} = \frac{4}{3}.$$

Exercises 12-3

Find the value of each of the following expressions without using tables.

1. Arctan 1.
2. $\text{Arccos}\,\frac{1}{2}$.
3. $\text{Arcsin}\,\frac{1}{2}\sqrt{2}$.
4. $\text{Arcsin}\,\frac{1}{2}$.
5. $\text{Arccos}\,(-\frac{1}{2})$.
6. $\text{Arcsin}\,(-\frac{1}{2})$.
7. $\text{Arctan}\,(-\sqrt{3})$.
8. $\text{Arctan}\,(-1)$.
9. $\text{Arcsin}\,(-\frac{1}{2})$.
10. $\text{Arccos}\,(-\frac{1}{2}\sqrt{2})$.
11. $\text{Arccos}\,\frac{1}{2}\sqrt{3}$.
12. $\text{Arctan}\,\frac{1}{3}\sqrt{3}$.
13. $\sin\,(\text{Arccos}\,\frac{1}{3})$.
14. $\cos\,(\text{Arctan}\,\frac{3}{5})$.
15. $\tan\,(\text{Arcsin}\,\frac{5}{13})$.
16. $\sin\,(\text{Arcsin}\,\frac{7}{8})$.
17. $\cos\,(\text{Arccos}\,\frac{5}{7})$.
18. $\tan\,(\text{Arccos}\,\frac{3}{5})$.
19. $\cos\,(2\,\text{Arctan}\,\frac{3}{5})$.
20. $\sin\,(2\,\text{Arccos}\,\frac{1}{3})$.
21. $\sin\,(2\,\text{Arcsin}\,\frac{7}{8})$.
22. $\tan\,(2\,\text{Arcsin}\,\frac{5}{13})$.
23. $\tan\,(2\,\text{Arccos}\,\frac{3}{5})$.
24. $\cos\,(2\,\text{Arccos}\,\frac{5}{7})$.
25. $\sin\,[\text{Arcsin}\,(-\frac{1}{2}) + \text{Arccos}\,(-\frac{1}{2})]$.
26. $\sin\,(\text{Arccos}\,\frac{1}{3} + \text{Arcsin}\,\frac{1}{3})$.
27. $\cos\,(\text{Arcsin}\,\frac{5}{13} - \text{Arccos}\,\frac{4}{5})$.
28. $\sin\,(\text{Arcsin}\,\frac{12}{13} - \text{Arcsin}\,\frac{4}{5})$.
29. $\tan\,(\text{Arctan}\,\frac{9}{40} + \text{Arctan}\,\frac{3}{4})$.
30. $\tan\,(\text{Arctan}\,\frac{4}{5} + \text{Arctan}\,\frac{1}{4})$.
31. $\tan\,(\text{Arctan}\,2 - \text{Arctan}\,\frac{1}{2})$.
32. $\tan\,(\text{Arctan}\,3 - \text{Arctan}\,2)$.

In Exercises 33–41 use inverse trigonometric notation to represent θ under the specified conditions.

33. $\sin\theta = \frac{3}{5}, \frac{1}{2}\pi < \theta < \pi$.
34. $\cos\theta = \frac{3}{5}, \frac{1}{2}\pi < \theta < \pi$.
35. $\cos\theta = -\frac{5}{13}, \pi < \theta < \frac{3}{2}\pi$.
36. $\sin\theta = -\frac{5}{13}, \pi < \theta < \frac{3}{2}\pi$.

37. $\tan \theta = -\frac{3}{4}, \frac{3}{2}\pi < \theta < 2\pi.$

38. $\tan \theta = -\frac{5}{12}, \frac{3}{2}\pi < \theta < 2\pi.$

39. $\sin \theta = \frac{4}{5}, 2\pi < \theta < \frac{5}{2}\pi.$

40. $\sin \theta = \frac{3}{5}, 2\pi < \theta < \frac{5}{2}\pi.$

41. $\tan 2\theta = \frac{24}{7}, \frac{1}{2}\pi < \theta < \pi.$

In Exercises 42–52 show that each of the given statements is true.

42. $\text{Arcsin} \frac{4}{5} + \text{Arctan} \frac{3}{4} = \frac{1}{2}\pi.$

43. $\text{Arccos} \frac{12}{13} + \text{Arctan} \frac{1}{4} = \text{Arccot} \frac{43}{32}.$

44. $\text{Arctan} \frac{3}{4} = 2 \text{ Arctan} \frac{1}{3}.$

45. $\frac{1}{2} \text{Arctan} \frac{7}{24} = \text{Arccos} \frac{7}{10}\sqrt{2}.$

46. $\text{Arctan} \frac{1}{2} - \text{Arctan} \frac{1}{3} = \text{Arctan} \frac{1}{7}.$

47. $\text{Arcsin } x + \text{Arccos } x = \frac{1}{2}\pi,$ if $-1 \leq x \leq 1.$

48. $\cos (\text{Arcsin } x) = \sqrt{1 - x^2},$ if $-1 \leq x \leq 1.$

49. $2 \text{ Arcsin } x = \text{Arccos} (1 - 2x^2),$ if $0 \leq x \leq 1.$

50. $\text{Arctan } x = \text{Arcsin} \dfrac{x}{\sqrt{1 + x^2}}$ for all values of $x.$

51. $\cos (\frac{1}{2} \text{Arccos } x) = \sqrt{\dfrac{1 + x}{2}},$ if $-1 \leq x \leq 1.$

52. $\tan 2(\text{Arccos } x) = \dfrac{2x\sqrt{1 - x^2}}{2x^2 - 1},$ if $-1 \leq x \leq 1.$

In Exercises 53-64 solve each equation for $x.$

53. $\text{Arctan } x + 2 \text{ Arctan } 1 = \frac{3}{4}\pi.$

54. $\text{Arctan } x = \text{Arcsin} \frac{12}{13}.$

55. $\text{Arccos } 2x = \text{Arcsin } x.$

56. $\text{Arccos } x + 2 \text{ Arcsin } 1 = \pi.$

57. $\text{Arcsin } 2x = \frac{1}{6}\pi + \text{Arccos } x.$

58. $\text{Arcsin } x + \text{Arcsin } 2x = \frac{1}{3}\pi.$

59. $\text{Arcsin } x + \text{Arccos } 2x = \frac{1}{6}\pi.$

60. $\text{Arccos } 2x - \text{Arccos } x = \frac{1}{3}\pi.$

61. $\text{Arcsin } x + \text{Arccos} (1 - x) = 0.$

62. $\text{Arctan} (x + 1) + \text{Arctan} (x - 1) = \text{Arctan} \frac{8}{31}.$

63. $\text{Arctan } 2x + \text{Arctan } 3x = \frac{3}{4}\pi.$

64. $\text{Arctan } x + \text{Arctan} (1 - x) = \text{Arctan} \frac{4}{3}.$

REVIEW EXERCISES

In Exercises 1–4 find all values of each expression in the interval $0 \leqq \theta < 2\pi$.

1. $\arccos \frac{1}{2}$.

2. $\arcsin (-1)$.

3. $\arctan \sqrt{3}$.

4. $\text{arcsec } 2$.

In Exercises 5–16 find the value of each expression.

5. $\text{Arcsin} (-1)$.

6. $\text{Arccos} (-1)$.

7. $\text{Arccos} (-\frac{1}{2})$.

8. $\text{Arctan} (-1)$.

9. $\sin (\text{Arcsin } \frac{2}{3})$.

10. $\cos (\text{Arccos } \frac{3}{5})$.

11. $\cos (\text{Arctan } \frac{4}{3})$.

12. $\sin (\text{Arctan } \frac{3}{4})$.

13. $\sin (2 \text{ Arcsin } \frac{3}{5})$.

14. $\cos (2 \text{ Arccos } \frac{3}{5})$.

15. $\tan (\text{Arctan } \frac{3}{2} + \text{Arctan } \frac{1}{2})$.

16. $\sin (\text{Arcsin } \frac{1}{5} + \text{Arccos } \frac{5}{7})$.

In Exercises 17–20 solve each exercise for x in terms of y.

17. $y = 3 \csc 2x$.

18. $y = 2 \cot 3x$.

19. $2y = \sec 3x - 1$.

20. $3y = 2 - \csc 4x$.

In Exercises 21–24 solve each equation for x.

21. $\text{Arctan } x + \text{Arctan } 1 = \pi/2$.

22. $\text{Arcsin } x + 2 \text{ Arccos } \frac{1}{2} = \pi$.

23. $\text{Arctan } (x - 1) - \text{Arctan } x = \text{Arctan } (-1)$.

24. $\text{Arctan } x + \text{Arctan } (x + 1) = \text{Arctan } (-1)$.

25. Show that $\text{Arccos } a = \text{Arcsin } \sqrt{1 - a^2}$, for $-1 \leqq a \leqq 1$.

26. Show that $\text{Arctan } a = \text{Arccos } \dfrac{1}{\sqrt{a^2 + 1}}$, for $a > 0$.

CHAPTER 12: DIAGNOSTIC TEST

The purpose of this test is to see how well you understand the work covered in Chapter 12. We recommend that you work this test before your instructor tests you on this chapter. Allow yourself approximately 40 minutes to do the test.

Solutions to the problems, together with section references, are given in the Answer Section at the end of this book. You should study the sections referred to for the problems you do incorrectly.

---

1. Find all values of each expression in the interval $0 \leqq \theta < 2\pi$.

   (a) $\arcsin \dfrac{\sqrt{3}}{2}$.

   (b) $\arccos (-1)$.

   (c) $\arctan (-1)$.

2. Find the value of each expression.
   (a) Arctan 1.   (b) Arccos $\frac{1}{2}$.   (c) Arcsin $(-\frac{1}{2})$.
   (d) tan (Arcsin $\frac{4}{5}$).   (e) sin (Arcsin $\frac{5}{13}$ + Arccos $\frac{4}{5}$).

3. Solve the equation cot $2x = y + 3$ for x in terms of y.

4. Show that Arcsin a = Arctan $\dfrac{a}{\sqrt{1 - a^2}}$ for $-1 < a < 1$.

5. Solve the following for x.

$$\text{Arctan } (x + 1) - \text{Arctan } (x - 1) = \text{Arctan } 2.$$

13

Complex Numbers

13-1 PURE IMAGINARY NUMBERS

The student will recall from algebra that a good deal of imagination, in the sense of inventiveness, has been required to construct the real number system. The further invention of a complex number system does not then seem so strange. It is fitting for us to study such a system, since modern engineering has found that it provides a convenient language for expressing vibratory motion, harmonic oscillation, damped vibrations, alternating currents, and other wave phenomena.

The square root of a negative number (for example, $\sqrt{-1}$, $\sqrt{-7}$, $\sqrt{-4}$) is called a **pure imaginary number**. To operate with such numbers a new symbol $i = \sqrt{-1}$ has been invented; then

$$\sqrt{-7} = \sqrt{7} \cdot \sqrt{-1} = \sqrt{7}i,$$

and

$$\sqrt{-4} = \sqrt{4} \cdot \sqrt{-1} = 2i.$$

$$-\sqrt{-4} = -\sqrt{4}\sqrt{-1} = -2i.$$

Accordingly, the symbol $\sqrt{-1}$, called the **imaginary unit** and usually designated by i, represents a new kind of number with the property that

(1)
$$\sqrt{-1} = i,$$
$$i^2 = (\sqrt{-1})^2 = -1.$$

From Property (1) it follows that

$$i^1 = i, \qquad\qquad i^2 = -1,$$
$$i^3 = i^2 \cdot i = -i, \qquad i^4 = (i^2)^2 = (-1)^2 = 1,$$
$$i^5 = i^4 \cdot i = i, \qquad i^6 = (i^2)^3 = (-1)^3 = -1,$$
$$i^7 = i^6 \cdot i = -i, \qquad i^8 = (i^2)^4 = (-1)^4 = 1.$$

Therefore, the integral powers of i have only four different values: i, -1, $-i$, 1.

13-2 COMPLEX NUMBERS

If a and b are real numbers, $a + bi$ is called a **complex number**, of which a is the **real part** and bi is the **imaginary part**. Complex numbers can be thought of as including all real numbers and all pure imaginary numbers. For example, if $b = 0$, the complex number $a + bi$ is *real*; if $b \neq 0$, the complex number is said to be *imaginary*; in particular, if $a = 0$ and $b \neq 0$, the complex number is a *pure imaginary* number.

    I. Fundamental principle of operations. *Algebraic operations involving only whole powers of complex numbers can be performed by writing i in place of $\sqrt{-1}$, then operating with i as if it were a real number, and, finally, replacing i^2 by -1.*
    II. Definition of equality. *Two complex numbers are said to be equal if and only if their real parts are equal and their imaginary parts are equal:*

$$x + yi = a + bi, \quad \textit{if and only if} \quad x = a \textit{ and } y = b.$$

    III. Conjugate of a complex number. *The conjugate of a complex number $a + bi$ is the complex number $a - bi$, which is obtained by changing the sign of the coefficient of the imaginary part.*

1. *Addition*: To add two complex numbers, add the real parts and add the pure imaginary parts.

$$(a + bi) + (c + di) = (a + c) + (b + d)i.$$

Example 1.

$$(3 + 2i) + (4 - 5i) = (3 + 4) + (2 - 5)i = 7 - 3i.$$

2. *Subtraction*: To subtract one complex number from another, subtract its real and pure imaginary parts from the corresponding parts of the other number.

$$(a + bi) - (c + di) = (a - c) + (b - d)i.$$

Example 2.

$$(3 + 2i) - (4 - 5i) = (3 - 4) + [2 - (-5)]i = -1 + 7i.$$

3. *Multiplication*: To multiply two complex numbers, multiply as you would in multiplying two binomials, and replace i^2 by -1.

$$(a + bi)(c + di) = ac + (ad + bc)i + bdi^2$$
$$= (ac - bd) + (ad + bc)i.$$

Example 3.

$$(3 + 2i)(4 - 5i) = 12 - 7i - 10i^2$$
$$= 12 - 7i - 10(-1) = 22 - 7i.$$

In multiplying numbers like $\sqrt{-a} \cdot \sqrt{-b}$, where $a > 0$ and $b > 0$, first rewrite each factor in the form

$$\sqrt{-a} = \sqrt{(-1)(a)} = \sqrt{-1} \cdot \sqrt{a} = i\sqrt{a},$$
$$\sqrt{-b} = \sqrt{(-1)(b)} = \sqrt{-1} \cdot \sqrt{b} = i\sqrt{b},$$

and then multiply the numbers as follows:

$$(i\sqrt{a})(i\sqrt{b}) = i^2\sqrt{ab} = -\sqrt{ab}.$$

The student is warned that

$$\sqrt{-a} \cdot \sqrt{-b} \neq \sqrt{(-a)(-b)}, \quad \text{or} \quad \sqrt{ab}.$$

---

4. *Division*: To divide one complex number by another, multiply both dividend and divisor by the conjugate of the divisor.

$$\frac{a + bi}{c + di} = \frac{(a + bi)(c - di)}{(c + di)(c - di)} = \frac{(ac + bd) + (bc - ad)i}{c^2 + d^2}.$$

Example 4.

$$\frac{3 + 2i}{4 - 3i} = \frac{(3 + 2i)(4 + 3i)}{(4 - 3i)(4 + 3i)} = \frac{6 + 17i}{16 + 9}$$

$$= \frac{6}{25} + \frac{17}{25}i.$$

Example 5. Solve $y + 2i = (2 + xi)(3 - 2i)$ for the real numbers x and y.

Solution: Express the product $(2 + xi)(3 - 2i)$ in the form $a + bi$.

$$(1) \qquad\qquad y + 2i = (2 + xi)(3 - 2i)$$

$$= 6 - 4i + 3xi - 2xi^2$$

$$= (6 + 2x) + (3x - 4)i.$$

Apply the definition of equality (II of Section 13-2) to (1):

$$y + 2i = (6 + 2x) + (3x - 4)i.$$

Therefore,

$$y = 6 + 2x, \quad \text{and} \quad 2 = 3x - 4 \quad \text{or} \quad x = 2,$$

and $\qquad y = 6 + 2x = 6 + 4 = 10.$

Example 6. Solve $9x^2 - 6x + 5 = 0$.

Solution: Apply the quadratic formula $x = (-b \pm \sqrt{b^2 - 4ac})/2a$.

$$x = \frac{6 \pm \sqrt{36 - 180}}{18} = \frac{6 \pm \sqrt{-144}}{18} = \frac{6 \pm 12i}{18} = \frac{1}{3} \pm \frac{2}{3}i.$$

Example 7. Expand $(3 + 2i)^3$, simplify, and write the result in the form $a + bi$.

Solution:

$$(3 + 2i)^3 = 3^3 + 3 \cdot 3^2(2i) + 3 \cdot 3(2i)^2 + (2i)^3$$
$$= 27 + 54i + 36i^2 + 8i^3$$
$$= 27 + 54i - 36 - 8i$$
$$= -9 + 46i.$$

Exercises 13-3

In Exercises 1–16 perform the indicated operations, simplify, and write the result in the form $a + bi$.

1. $(2 + 3i) + (5 - 2i)$.

2. $(1 - i) + (7 + 3i)$.

3. $(4 - 2i) - (7 - i)$.

4. $(5 + i) - (2 - 3i)$.

5. $(\frac{1}{3} + i) + (2 - \frac{1}{3}i)$.

6. $(\frac{2}{5} - 2i) + (\frac{1}{2} + \frac{1}{5}i)$.

7. $(2 + 3i)^2$.

8. $(3 - 2i)^2$.

9. $(3 + i)(3 - i)$.

10. $(7 - 2i)(7 + 2i)$.

11. $(2 + i)^3$.

12. $(2 - \frac{1}{2}i)^3$.

13. $\dfrac{1 - i}{1 + i}$.

14. $\dfrac{1 + i}{1 - i}$.

15. $\dfrac{4 + \sqrt{2}i}{4 - \sqrt{2}i}$.

16. $\dfrac{3 + \sqrt{5}i}{3 - \sqrt{5}i}$.

In Exercises 17–24 solve each equation for the real number x and y.

17. $3x + 4yi = 6 + 8i$.

18. $2x - yi = 4 + 3i$.

19. $y + 2i = (4 - 2i)(3 + xi)$.

20. $(x + y) + (x - 4y)i = 4 + 19i$.

21. $(2x - 3y) + (x + y)i = 3 + 4i$.

22. $(x + yi)^2 = 5 + 12i$.

23. $(x^2 - 6y) + 6i = -(2x + y)i$.

24. $(2 - xi)(-y + 3i) = 15 + 27i$.

In Exercises 25–28 solve each equation for x.

25. $x^2 - 4x + 5 = 0$.

26. $x^2 + x + 1 = 0$.

27. $9x^2 - 12x + 53 = 0$.

28. $\sqrt{5}x^2 + 4x + \sqrt{5} = 0$.

29. Show that the conjugate of the product of two conjugate numbers is equal to the product of their respective conjugates.

30. Show that the conjugate of the quotient of two conjugate numbers is equal to the quotient of their respective conjugates.

13-4 *GRAPHIC REPRESENTATION OF COMPLEX NUMBERS*

In algebra, real numbers are represented by points on a straight line. Since a complex number $x + yi$ depends on two independent real numbers x and y, and since (x, y) can be plotted as a point on the rectangular coordinate system, every complex number $x + yi$ can be associated with some point on the plane, and every point on the plane can be associated with some complex number. Thus, in Figure 13-1 the point P, whose rectangular coordinates are $(2, 3)$, represents the complex number $2 + 3i$, and the point P' represents the number $-2 + 3i$. In such connections, we shall speak of the point which represents the number $x + yi$ as "the point $x + yi$."

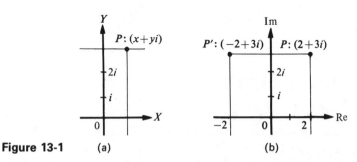

Figure 13-1   (a)   (b)

As thus used, the plane in Figure 13-1 is called the **complex plane**, and the number $(x + yi)$ is said to be the **rectangular**, or **algebraic**, **form** for a complex number. Also from this representation, the x-axis is often called the **axis of reals** (or the **real axis**), and the y-axis is called the **axis of imaginaries** (or **imaginary axis**). On this latter axis, the point a unit distance above O represents the imaginary unit i.

Since a real number is regarded as a complex number whose imaginary part is zero, it is represented by a point on the real axis (the x-axis). A pure imaginary number is a complex number whose real part is zero, and it is represented by a point on the imaginary axis (the y-axis). Some representative points (complex numbers) are shown in Figure 13-2.

Exercises 13-4

Represent graphically each of the following complex numbers and their conjugates.

1. $3 + i$.
2. $-2 + i$.
3. $-2 + 3i$.
4. $3 + 2i$.
5. $2 - 2i$.
6. $3 - 3i$.
7. $3i$.
8. 4.
9. -3.
10. $2i$.

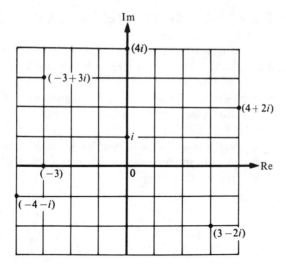

Figure 13-2

13-5 POLAR COORDINATES

The position of a point in a plane is determined if its rectangular coordinates are known. An alternative method of fixing the position of a point is to give its **polar coordinates**. These are r, the distance of the point from the origin, or **pole**, and θ, the angle which the line joining the point and the pole makes with the positive x-axis, or **polar axis**. The coordinates are written (r, θ); r is called the **radius** of the point; θ is its **angle**.

The radius is expressed in any convenient unit of length, and the angle is commonly measured in radians, though degree measure can be used. In previous chapters the radius of a point has always been taken as positive. In connection with polar coordinates, however, it is often desirable to let r be negative. To plot a point when its polar coordinates (r, θ) are given, we first place the angle θ in standard position, that is, with its vertex at the pole and its initial side on the polar axis. Then, if r is positive, the point (r, θ) lies on the terminal side of θ and at the distance r from the pole; if r is negative, (r, θ) lies

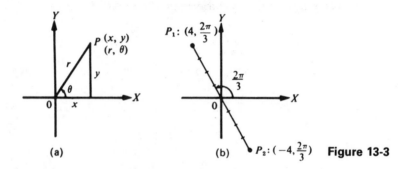

Figure 13-3

at the distance $|r|$ measured from the pole in the opposite direction, that is, along the line made by extending the terminal side of θ through the pole. Thus, in Figure 13-3b the two points P_1: $(4, \frac{2}{3}\pi)$ and P_2: $(-4, \frac{2}{3}\pi)$ are shown. If $r = 0$, the point (r, θ) is the pole, regardless of the value of θ. Unless there is some contrary indication, we shall assume that $r \neq 0$.

It is often desirable to use rectangular and polar coordinates together, the polar axis of the polar system coinciding with the positive x-axis, and x, y, and r being measured in the same units. We can always convert from one system to the other by means of the following relations (See Fig. 13-3a):

Polar coordinates:

(1) $\begin{cases} x = r \cos \theta, \\ y = r \sin \theta. \end{cases}$
(2) $\begin{cases} r = \sqrt{x^2 + y^2}, \\ \theta = \arctan \dfrac{y}{x}. \end{cases}$

In the equations of (2), r is taken as positive, and θ is any one of the infinitely many values of arctan (y/x) whose terminal side lies in the same quadrant (or on the same axis) as the point (x, y). If we wish to do so, we can make $r = -\sqrt{x^2 + y^2}$ and take θ in the opposite quadrant from the one containing (x, y).

Polar coordinates have many uses only one of which is the application to complex numbers.

Example 1. Find the rectangular coordinates of the point whose polar coordinates are $(3, \frac{1}{6}\pi)$. (See Fig. 13-4.)

Solution: From Equation (1) we have

$$x = r \cos \theta = 3 \cos \frac{1}{6}\pi = 3\left(\frac{1}{2}\sqrt{3}\right) = \frac{3}{2}\sqrt{3},$$

$$y = r \sin \theta = 3 \sin \frac{1}{6}\pi = 3\left(\frac{1}{2}\right) = \frac{3}{2}.$$

The required rectangular coordinates are $(\frac{3}{2}\sqrt{3}, \frac{3}{2})$.

Example 2. Find the polar coordinates of the point whose rectangular coordinates are $(2, -2\sqrt{3})$. (See Fig. 13-4.)

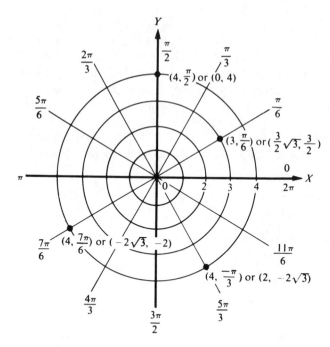

Figure 13-4

Solution: From Equation (2) we can take

$$r = \sqrt{2^2 + (-2\sqrt{3})^2} = \sqrt{4 + 12} = 4,$$

$$\theta = \arctan(-\sqrt{3}) = -\frac{1}{3}\pi.$$

The required coordinates can thus be taken as $(4, -\frac{1}{3}\pi)$. We could have taken $r = -4$ and $\theta = \frac{2}{3}\pi$. There are, in fact, infinitely many sets of polar coordinates for every point of the plane. However, it is usually convenient to take r positive and to give θ a numerically small correct value.

Exercises 13-5

In each of the Exercises 1–12 plot the point whose polar coordinates are given, and find its rectangular coordinates. Tables may be used when necessary.

1. $(4, \frac{1}{6}\pi)$.   2. $(2, \frac{1}{3}\pi)$.   3. $(-3, \pi)$.   4. $(-2, 2\pi)$.

5. $(8, -\frac{1}{2}\pi)$.   6. $(6, -\pi)$.   7. $(10, 200°)$.   8. $(10, 100°)$.

9. $(6, \frac{5}{2}\pi)$.   10. $(4, \frac{3}{4}\pi)$.   11. $(-2, \frac{5}{4}\pi)$.   12. $(6, -\frac{13}{6}\pi)$.

In each of the Exercises 13–24 the rectangular coordinates are given. Find one set of polar coordinates and plot the point. Tables may be used when necessary.

13. (2, 2).   **14.** $(-\sqrt{3}, 1)$.   **15.** (3, 0).   **16.** (0, 2).

17. (−5, 0).   **18.** (0, −2).   **19.** $(\sqrt{2}, -\sqrt{2})$.   **20.** (−1, −2).

21. (2, 0).   **22.** $(\frac{1}{2}, -\frac{1}{2}\sqrt{3})$.   **23.** (−2, −3).   **24.** (2, 1).

13-6 POLAR, OR TRIGONOMETRIC, FORM OF A COMPLEX NUMBER

A trigonometric representation of complex numbers is very useful in studying certain operations with these numbers.

Suppose that (r, θ) are the polar coordinates of the point $x + yi$ on the complex plane as shown in Figure 13-5. This number $(x + yi)$ determines both a point on the plane and also the length of a line from the origin to the point.

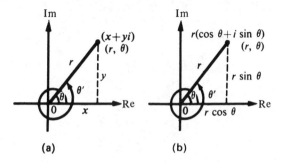

Figure 13-5   **(a)**   **(b)**

The positive (or zero) length, r, of this line, or *radius*, is called the **absolute value**, or **modulus**, of the complex number.

Absolute value or *modulus* of the complex number $(x + yi)$:

(1) $$|x + iy| = r = \sqrt{x^2 + y^2}.$$

The **amplitude** or **argument** (θ) of the complex number $(x + yi)$ is any angle whose terminal side passes through the point (x, y), where the angle is placed in standard position on the axes. It can be chosen as any one of infinitely many values that differ by multiples of 2π. Thus,

---

Amplitude or *argument* (θ) of the number $(x + yi)$:

(2) $$\theta = \arctan \frac{y}{x} \qquad x \neq 0.$$

In Formula (2) the angle θ must be chosen so as to put the point representing the complex number in the proper quadrant. This means that θ must satisfy the two equations $x = r \cos \theta$, $y = r \sin \theta$.

Since $x = r \cos \theta$ and $y = r \sin \theta$, we can write

$$x + iy = r \cos \theta + i(r \sin \theta),$$

or

Rectangular and **Polar form** of a complex number:

(3) $\qquad\qquad x + iy = r(\cos \theta + i \sin \theta).$

The left-hand member is the **rectangular form** and the right-hand member is the **polar**, or **trigonometric form**, of the complex number.

The expression $\cos \theta + i \sin \theta$ is sometimes abbreviated to the more compact symbol cis θ.

Example 1. Write the complex number $\sqrt{3} + i$ in polar form, and find its absolute value and amplitude.

Solution: Plot the point $(\sqrt{3}, 1)$ which represents the given number; draw the radius and sketch the reference triangle. (Fig. 13-6.)

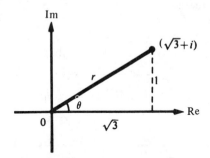

Figure 13-6

Find r and θ as follows:

$$r = \sqrt{x^2 + y^2} = \sqrt{3 + 1} = 2 \quad \text{(absolute value).}$$

The amplitude is $\theta = \arctan 1/\sqrt{3} = \frac{1}{6}\pi + 2n\pi$, where n is any integer. Applying Formula (3)

$$x + iy = r(\cos \theta + i \sin \theta),$$

and using $n = 0$, we get

$$\sqrt{3} + i = 2(\cos \tfrac{1}{6}\pi + i \sin \tfrac{1}{6}\pi).$$

We check the results as follows:

$$2(\cos \tfrac{1}{6}\pi + i \sin \tfrac{1}{6}\pi) = 2(\tfrac{1}{2}\sqrt{3} + i \cdot \tfrac{1}{2})$$
$$= \sqrt{3} + i.$$

Example 2. Write the complex number $4i$ in polar form and give several values of its argument.

Solution: The point that represents the complex number $4i$ is shown in Figure 13-2. Here we see that $4i$ is 4 units from the origin, making $r = 4$. The argument

$$\theta = \tfrac{1}{2}\pi \quad \text{or} \quad \tfrac{1}{2}\pi + 2n\pi.$$

Therefore,

$$4i = 4(\cos \tfrac{1}{2}\pi + i \sin \tfrac{1}{2}\pi).$$

A more general polar form for this complex number is

$$4i = [4 \cos (\tfrac{1}{2}\pi + 2n\pi) + i \sin (\tfrac{1}{2}\pi + 2n\pi)].$$

Example 3. The particular real and pure imaginary numbers 3, -5, and $-2i$ may be written in the polar form as follows:

$$3 = 3(\cos 0 + i \sin 0), \qquad -5 = 5(\cos \pi + i \sin \pi),$$
$$-2i = 2(\cos \tfrac{3}{2}\pi + i \sin \tfrac{3}{2}\pi).$$

Example 4. Express the complex number $8(\cos 120° + i \sin 120°)$ in the rectangular form $x + yi$.

Solution: Since $\cos 120° = -\tfrac{1}{2}$ and $\sin 120° = \tfrac{1}{2}\sqrt{3}$, we have

$$8(\cos 120° + i \sin 120°) = 8\left(-\frac{1}{2} + \frac{1}{2}\sqrt{3}i\right)$$
$$= -4 + 4\sqrt{3}i.$$

Example 5. Let $(2, \tfrac{5}{6}\pi)$ be the polar coordinates of a point on the complex plane. Plot this point, and then write the complex number represented by this point.

Solution: The polar coordinates of a point are understood to be indicated by (r, θ). Thus, for the point $(2, \tfrac{5}{6}\pi)$ it means that $r = 2$ and $\theta = \tfrac{5}{6}\pi$. Place $\theta = \tfrac{5}{6}\pi$ in standard position, and then locate a point on the terminal side of θ two units from the origin. (See Fig. 13-7.) In polar form this complex number is written

$$2\left(\cos \frac{5}{6}\pi + i \sin \frac{5}{6}\pi\right).$$

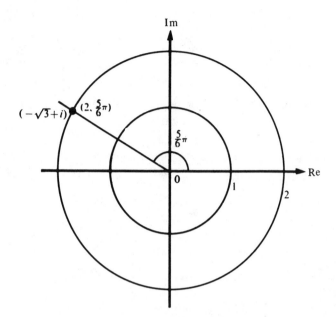

Figure 13-7

We can express this complex number in rectangular form as follows:

$$2\left(\cos\frac{5}{6}\pi + i\sin\frac{5}{6}\pi\right) = 2\left(-\frac{1}{2}\sqrt{3} + i\frac{1}{2}\right) = -\sqrt{3} + i.$$

Exercises 13-6

In each of the Exercises 1–16 represent the complex number graphically as a point and draw the radius; find the absolute value and one suitable value of the amplitude.

1. $2 + 2i$. 2. $2 - 2i$. 3. $-1 + \sqrt{3}i$.

4. $-1 + \sqrt{2}i$. 5. $1 - i$. 6. $1 + i$.

7. $-1 - \sqrt{2}i$. 8. $-3 - 3i$. 9. $3i$.

10. $-5i$. 11. -5. 12. 7.

13. 6. 14. -2. 15. $2 - 3i$.

16. $1 - 2i$.

In Exercises 17–24 represent graphically each of the complex numbers, and write the number in the rectangular form $x + yi$.

17. $3(\cos 60° + i \sin 60°)$. 18. $2(\cos \pi + i \sin \pi)$.

19. $4(\cos \frac{1}{4}\pi + i \sin \frac{1}{4}\pi)$. 20. $6(\cos \frac{5}{6}\pi + i \sin \frac{5}{6}\pi)$.

21. $\cos \pi + i \sin \pi$. 22. $4(\cos \frac{1}{2}\pi + i \sin \frac{1}{2}\pi)$.

23. $8(\cos 270° + i \sin 270°)$. 24. $2(\cos 4\pi + i \sin 4\pi)$.

In Exercises 25–30 write the complex number, in both polar and rectangular form, corresponding to the given set of polar coordinates (r, θ); draw the figure.

25. $(2, \frac{1}{6}\pi)$.   **26.** $(4, \frac{1}{4}\pi)$.   **27.** $(5, \pi)$.

28. $(3, \frac{3}{2}\pi)$.   **29.** $(4, \frac{3}{4}\pi)$.   **30.** $(4, 2\pi)$.

31. Prove that $|\cos \theta + i \sin \theta| = 1$.

32. Prove that the conjugate of $\cos \theta + i \sin \theta$ is $\cos (-\theta) + i \sin (-\theta)$.

13-7 MULTIPLICATION AND DIVISION OF COMPLEX NUMBERS IN POLAR FORM

Let z_1 and z_2 be two complex numbers, expressed in the polar form by

$$z_1 = r_1(\cos \theta_1 + i \sin \theta_1), \qquad z_2 = r_2(\cos \theta_2 + i \sin \theta_2).$$

Theorem I. *The product of two complex numbers is a complex number whose absolute value is the **product** of their absolute values and whose amplitude is the **sum** of the amplitudes of the two factors. Thus,*

(1) $$z_1 z_2 = r_1 r_2[\cos (\theta_1 + \theta_2) + i \sin (\theta_1 + \theta_2)].$$

Proof. Applying Operation (3) of Section 13-3, we have

$$z_1 z_2 = [r_1(\cos \theta_1 + i \sin \theta_1)][r_2(\cos \theta_2 + i \sin \theta_2)]$$

$$= (r_1 r_2)[(\cos \theta_1 \cos \theta_2 - \sin \theta_1 \sin \theta_2)$$

$$+ i(\sin \theta_1 \cos \theta_2 + \cos \theta_1 \sin \theta_2)].$$

But from Identity (15) of Section 8-2

$$\cos \theta_1 \cos \theta_2 - \sin \theta_1 \sin \theta_2 = \cos (\theta_1 + \theta_2),$$

and from Identity (17) of Section 8-3

$$\sin \theta_1 \cos \theta_2 + \cos \theta_1 \sin \theta_2 = \sin (\theta_1 + \theta_2).$$

Making these substitutions, we get

$$z_1 z_2 = (r_1 r_2)[\cos (\theta_1 + \theta_2) + i \sin (\theta_1 + \theta_2)].$$

Example 1. Multiply $\sqrt{2} - \sqrt{2}i$ and $1 + \sqrt{3}i$.

Solution: The absolute value of the first number is $\sqrt{2 + 2} = 2$, and its amplitude is arctan $(-1) = -45°$. The absolute value of the second number is $\sqrt{1 + 3} = 2$, and its amplitude is arctan $(\sqrt{3}) = 60°$. Therefore, the required product is

$$2[\cos (-45°) + i \sin (-45°)] \cdot 2[\cos 60° + i \sin 60°]$$
$$= 4(\cos 15° + i \sin 15°).$$

Theorem II. *If one complex number is divided by a second complex number, the quotient is a complex number whose absolute value is the quotient of the absolute value of the first divided by the absolute value of the second and whose amplitude is the amplitude of the first minus the amplitude of the second. Thus, if*

$$z_1 = r_1(\cos \theta_1 + i \sin \theta_1) \quad and \quad z_2 = r_2(\cos \theta_2 + i \sin \theta_2),$$

then

(2) $$\frac{z_1}{z_2} = \frac{r_1}{r_2}[\cos (\theta_1 - \theta_2) + i \sin (\theta_1 - \theta_2)].$$

Proof. Let $z_1 = r_1(\cos \theta_1 + i \sin \theta_1)$, $z_2 = r_2(\cos \theta_2 + i \sin \theta_2) \neq 0$ be any two given complex numbers. Then by multiplying both numerator and denominator of the fraction by the conjugate of the denominator, we get

$$\frac{z_1}{z_2} = \frac{r_1(\cos \theta_1 + i \sin \theta_1)}{r_2(\cos \theta_2 + i \sin \theta_2)} = \frac{r_1}{r_2} \frac{(\cos \theta_1 + i \sin \theta_1)(\cos \theta_2 - i \sin \theta_2)}{(\cos \theta_2 + i \sin \theta_2)(\cos \theta_2 - i \sin \theta_2)}$$

$$= \frac{r_1}{r_2} \frac{(\cos \theta_1 + i \sin \theta_1)(\cos \theta_2 - i \sin \theta_2)}{\cos^2 \theta_2 + \sin^2 \theta_2}.$$

Since $\sin (-\theta_2) = -\sin \theta_2$ and $\cos (-\theta_2) = \cos \theta_2$ [Identities (9) and (10) of Section 8-1], the factor $(\cos \theta_2 - i \sin \theta_2)$ can be written

$$[\cos (-\theta_2) + i \sin (-\theta_2)],$$

and since $\cos^2 \theta_2 + \sin^2 \theta_2 = 1$, the right-hand member of the above equation reduces to

$$\frac{r_1}{r_2}(\cos \theta_1 + i \sin \theta_1)[\cos (-\theta_2) + i \sin (-\theta_2)].$$

Then by applying Theorem I, we get

$$\frac{z_1}{z_2} = \frac{r_1}{r_2}[\cos (\theta_1 - \theta_2) + i \sin (\theta_1 - \theta_2)].$$

Example 2. Divide $6(\cos 170° + i \sin 170°)$ by $3(\cos 110° + i \sin 110°)$.

Solution: We apply Theorem II and get

$$[6(\cos 170° + i \sin 170°)] \div [3(\cos 110° + i \sin 110°)]$$

$$= \frac{6}{3}\cos (170° - 110°) + i \sin (170° - 110°)]$$

$$= 2(\cos 60° + i \sin 60°) = 2\left(\frac{1}{2} + i\frac{1}{2}\sqrt{3}\right)$$

$$= 1 + \sqrt{3}i.$$

13-8 POWERS OF COMPLEX NUMBERS: DE MOIVRE'S THEOREM

As a consequence of Theorem I, Section 13-7, when n is a positive integer

$$z_1 z_2 \cdots z_n = (r_1 r_2 \cdots r_n)[\cos(\theta_1 + \theta_2 + \cdots + \theta_n)$$
$$+ i \sin(\theta_1 + \theta_2 + \cdots + \theta_n)].$$

Therefore,

power of a complex number:

if $z = r(\cos\theta + i\sin\theta)$, and if n is a positive integer,

(1) $\qquad z^n = r^n(\cos n\theta + i \sin n\theta).$

When $r = 1$, this formula reduces to **De Moivre's theorem** for positive integral exponents,

De Moivre's theorem:

(2) $\qquad (\cos\theta + i\sin\theta)^n = \cos n\theta + i \sin n\theta.$

As a special case of Theorem II, Section 13-7, it follows that

$$\frac{1}{z} = \frac{\cos 0 + i \sin 0}{r(\cos\theta + i\sin\theta)} = \frac{1}{r}[\cos(0-\theta) + i\sin(0-\theta)]$$

$$= \frac{1}{r}(\cos\theta - i\sin\theta),$$

and, in applying Formula (1),

(3) $\qquad z^{-n} = \frac{1}{z^n} = \frac{1}{r^n}[\cos(-n\theta) + i\sin(-n\theta)].$

Therefore, both Formulas (1) and (2) are true when the exponent is any integer.

Example 1. Find $(1 + i)^{10}$.

Solution: We express $1 + i$ in polar form and then apply Formula (1).

$$(1 + i)^{10} = [\sqrt{2}(\cos 45° + i \sin 45°)]^{10}$$

$$= (\sqrt{2})^{10}(\cos 450° + i \sin 450°)$$

$$= 32(\cos 90° + i \sin 90°)$$

$$= 32(0 + i)$$

$$= 32i.$$

As an alternate solution, one could use the binomial theorem to expand $(1 + i)^{10}$, and then after simplifying and collecting terms we should have only the term $32i$ remaining. This latter solution would be long and certainly is not recommended, which indicates the importance of Formula (1) as applied to problems of this type.

Example 2. Find the value of $z = \dfrac{128}{(1 + i\sqrt{3})^6}$.

Solution: Let $z = \dfrac{128}{(1 + i\sqrt{3})^6} = 128(1 + i\sqrt{3})^{-6}$.

We express $1 + i\sqrt{3}$ in polar form.

$$1 + i\sqrt{3} = 2(\cos 60° + i \sin 60°).$$

Then, by applying Formula (2), we have

$$[2(\cos 60° + i \sin 60°)]^{-6} = 2^{-6}[\cos (-360°) + i \sin (-360°)]$$

$$= \frac{1}{64}(1 + 0).$$

Therefore,

$$z = \frac{128}{(1 + i\sqrt{3})^6} = 128\left(\frac{1}{64}\right) = 2.$$

Example 3. Derive identities for $\cos 3\theta$ and $\sin 3\theta$ by expanding

$$(\cos \theta + i \sin \theta)^3$$

by De Moivre's theorem and by the binomial theorem, and then equating the real parts and the imaginary parts.

Solution: By De Moivre's theorem

(4) $(\cos \theta + i \sin \theta)^3 = \cos 3\theta + i \sin 3\theta.$

We use the binomial theorem to expand the left-hand member of this equation; replace i^2 by -1, i^3 by $-i$, and arrange in the form $a + bi$.

$$\cos^3 \theta + 3(\cos \theta)^2(i \sin \theta) + 3(\cos \theta)(i \sin \theta)^2 + (i \sin \theta)^3$$

$$= \cos^3 \theta + (3 \cos^2 \theta \sin \theta)i - 3 \cos \theta \sin^2 \theta - (\sin^3 \theta)i$$

$$= (\cos^3 \theta - 3 \cos \theta \sin^2 \theta) + (3 \cos^2 \theta \sin \theta - \sin^3 \theta)i.$$

By replacing $\sin^2 \theta$ by $(1 - \cos^2 \theta)$ in the first term and $\cos^2 \theta$ by $(1 - \sin^2 \theta)$ in the second term, we reduce the expression as follows:

$$[\cos^3 \theta - 3 \cos \theta(1 - \cos^2 \theta)] + [3(1 - \sin^2 \theta) \sin \theta - \sin^3 \theta]i$$

$$= (4 \cos^3 \theta - 3 \cos \theta) + (3 \sin \theta - 4 \sin^3 \theta)i.$$

Replacing $(\cos \theta + i \sin \theta)^3$ of Equation (4) by this last expression, we have

$$(4 \cos^3 \theta - 3 \cos \theta) + (3 \sin \theta - 4 \sin^3 \theta)i = \cos 3\theta + (\sin 3\theta)i.$$

Applying II (definition of equality) of Section 13-2 we have the identities

$$\cos 3\theta = 4 \cos^3 \theta - 3 \cos \theta,$$

$$\sin 3\theta = 3 \sin \theta - 4 \sin^3 \theta.$$

Exercises 13-8

In Exercises 1–16 perform the indicated operations, giving the results in both polar and rectangular form; in the latter form, express the roots either exactly or to three decimal places.

1. $2(\cos 50° + i \sin 50°) \cdot 3(\cos 40° + i \sin 40°)$.

2. $4(\cos 100° + i \sin 100°) \cdot 5(\cos 80° + i \sin 80°)$.

3. $[8(\cos 305° + i \sin 305°)] \div [4(\cos 65° + i \sin 65°)]$.

4. $[10(\cos 293° + i \sin 293°)] \div [5(\cos 23° + i \sin 23°)]$.

5. $5(\cos 15° + i \sin 15°) \cdot 2(\cos 125° + i \sin 125°)$.

6. $5(\cos 50° + i \sin 50°) \cdot 2(\cos 110° + i \sin 110°)$.

7. $(\cos 15° + i \sin 15°)^{10}$.

8. $(\cos 75° + i \sin 75°)^{10}$.

9. $[2(\cos 12° + i \sin 12°)]^{-5}$.

10. $2(1 - i)^{-4}$.

11. $[3(-1 + i)]^4$.

12. $\dfrac{16}{(\sqrt{3} + i)^2}$.

13. $\dfrac{8}{(\sqrt{3} - i)^2}$.

14. $\dfrac{50}{(3 - 4i)^3}$.

15. $\dfrac{169}{(12 - 5i)^3}$.

16. $(1 + \sqrt{3}i)(1 - i)(-1 - i)$.

17. Derive identities for $\cos 2\theta$ and $\sin 2\theta$ by expanding $(\cos \theta + i \sin \theta)^2$ by De Moivre's theorem and by the binomial theorem, and then equating real and imaginary parts.

18. Derive identities for $\cos 4\theta$ and $\sin 4\theta$ by a method similar to that in Exercise 17.

13-9 ROOTS OF COMPLEX NUMBERS

The statement $\sqrt{9} = x$ implies $x^2 = 9$. The statement $\sqrt[n]{z} = w$ implies $w^n = z$. Thus, the problem of extracting the nth roots of a complex number z is one of solving the equation

(1) $w^n = z$

for w, when z and the positive integer n are given.

We express z in polar form,

$$z = r(\cos \theta + i \sin \theta)$$

and set

$$w = R(\cos \phi + i \sin \phi),$$

where R and ϕ are presently unknown. Then Equation (1) becomes

(2) $R^n(\cos n\phi + i \sin n\phi) = r(\cos \theta + i \sin \theta)$.

Since points that represent equal complex numbers coincide (Definition II, Section 13-2), it follows that their absolute values are equal and their amplitudes differ only by integral multiples of $360°$. From (2) we then obtain

(3) $R^n = r$ and $n\phi = \theta + k \cdot 360°$,

where k is an integer, or

(4) $R = \sqrt[n]{r}$ and $\phi = \dfrac{\theta}{n} + \dfrac{k\,360°}{n}$,

where $\sqrt[n]{r}$ denotes the principal nth root of the positive number r. Substituting these values of R and ϕ in $w = R(\cos \phi + i \sin \phi)$, we obtain nth roots of z. Write

(5) $$w_k = \sqrt[n]{r}\left[\cos\left(\frac{\theta}{n} + \frac{k \cdot 360°}{n}\right) + i \sin\left(\frac{\theta}{n} + \frac{k \cdot 360°}{n}\right)\right].$$

If we give k the n values $0, 1, 2, \ldots, n - 1$, we obtain n distinct complex numbers w_k, which are all nth roots of z. If we give k any other integral value, we find that we obtain one of the values w_k previously obtained.

If n is a positive integer and $z \neq 0$, a complex number $z = r(\cos \theta + i \sin \theta)$, real or imaginary, has n and only n distinct nth roots, which are given by the formula

(6) $$w_k = \sqrt[n]{r}\left[\cos\left(\frac{\theta}{n} + \frac{k \cdot 360°}{n}\right) + i \sin\left(\frac{\theta}{n} + \frac{k \cdot 360°}{n}\right)\right],$$

where k takes the values $0, 1, 2, \ldots, n - 1$.

Example 1. Find all the fourth roots of $-8 - 8\sqrt{3}i$, and plot points which represent these roots on the complex plane.

Solution: In polar form

$$-8 - 8\sqrt{3}i = 16(\cos 240° + i \sin 240°).$$

Therefore, $r = 16$ and $\theta = 240°$. Substituting these values for r and θ in Formula (6), we have

$$w_k = \sqrt[4]{16}\left[\cos\left(\frac{240°}{4} + \frac{k \cdot 360°}{4}\right) + i \sin\left(\frac{240°}{4} + \frac{k \cdot 360°}{4}\right)\right],$$

or $$w_k = 2\cos(60° + k90°) + i \sin(60° + k90°).$$

The values of k are $0, 1, 2,$ and 3, giving the following four roots:

$$w_0 = 2(\cos 60° + i \sin 60°) = 2\left(\frac{1}{2} + i\frac{1}{2}\sqrt{3}\right) = 1 + i\sqrt{3},$$

$$w_1 = 2(\cos 150° + i \sin 150°) = 2\left(-\frac{1}{2}\sqrt{3} + i\frac{1}{2}\right) = -\sqrt{3} + i,$$

$$w_2 = 2(\cos 240° + i \sin 240°) = 2\left[-\frac{1}{2} + i\left(-\frac{1}{2}\sqrt{3}\right)\right] = -1 - i\sqrt{3},$$

$$w_3 = 2(\cos 330° + i \sin 330°) = 2\left(\frac{1}{2}\sqrt{3} - i\frac{1}{2}\right) = \sqrt{3} - i.$$

Since the roots have equal amplitudes, they can be represented by points on a circle of radius 2 on the complex plane. (See Fig. 13-8.)

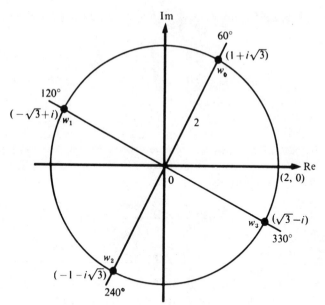

Figure 13-8

Example 2. Find all of the tenth roots of 1.

Solution: In polar form

$$1 = \cos 0° + i \sin 0°;$$

therefore, $r = 1$ and $\theta = 0°$. Substituting these values for r and θ in Formula (6), we have

$$w_k = \sqrt[10]{1}\left[\cos\left(\frac{0°}{10} + \frac{k \cdot 360°}{10}\right) + i \sin\left(\frac{0°}{10} + \frac{k \cdot 360°}{10}\right)\right]$$

or $$w_k = \cos k36° + i \sin k36°.$$

We assign the values $0, 1, 2, \ldots, 9$ for k, and obtain:

$$w_0 = \cos 0° + i \sin 0° = 1,$$

$$w_1 = \cos 36° + i \sin 36° = .8090 + .5878i,$$

$$w_2 = \cos 72° + i \sin 72° = .3090 + .9511i,$$

$$w_3 = \cos 108° + \sin 108° = -.3090 + .9511i,$$

$$w_4 = \cos 144° + i \sin 144° = -.8090 + .5878i,$$

$$w_5 = \cos 180° + i \sin 180° = -1,$$

$$w_6 = \cos 216° + i \sin 216° = -.8090 - .5878i,$$

$$w_7 = \cos 252° + i \sin 252° = -.3090 - .9511i,$$

$$w_8 = \cos 288° + i \sin 288° = .3090 - .9511i,$$

$$w_9 = \cos 324° + i \sin 324° = .8090 - .5878i.$$

We could have used the compact symbol cis θ (Section 13-6) to express the above roots as follows:

$$w_0 = \text{cis } 0°, \; w_1 = \text{cis } 36°, \; w_2 = \text{cis } 72°, \text{ etc.}$$

These roots are represented by points, spaced at equal intervals around a unit circle, on the complex plane. (See Fig. 13-9.)

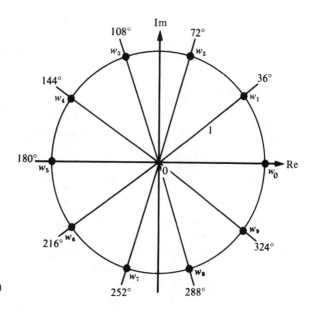

Figure 13-9

Exercises 13-9

In each of the Exercises 1–14 find all of the indicated roots in the rectangular form $a + bi$; and in this form express the roots either exactly or to three decimal places. Draw a figure and plot each of the required roots.

1. Cube roots of i.
2. Cube roots of $-i$.
3. Fourth roots of -1.
4. Sixth roots of 1.
5. Fifth roots of $4 + 4i$.
6. Fifth roots of $4 - 4i$.
7. Cube roots of 1.
8. Fourth roots of i.
9. Sixth roots of 64.
10. Square roots of $-1 + \sqrt{3}i$.
11. Fourth roots of $8 - 8\sqrt{3}i$.
12. Cube roots of $-4\sqrt{2} + 4\sqrt{2}i$.
13. Square roots of $-5 + 12i$.
14. Square roots of $4 - 3i$.

In each Exercise 15–20 solve the equation completely.

15. $x^5 + 1 = 0$.   16. $x^6 + 64 = 0$.

17. $x^3 + 27 = 0$.   18. $x^4 + 81 = 0$.

19. $x^4 - 2x^2 + 4 = 0$.   20. $x^4 + 4x^2 + 16 = 0$.

REVIEW EXERCISES

In Exercises 1–16 perform the indicated operations, simplify, and write the results in the form $a + bi$.

1. $(3 + 4i) + (2 - 5i)$.   2. $(4 + i) + (2 - 3i)$.

3. $(5 - 2i) - (3 - 4i)$.   4. $(-2 + 7i) - (3 + 5i)$.

5. $(4 + i)(4 - i)$.   6. $(3 - i)(3 + i)$.

7. $(5 + i) \div (5 - i)$.   8. $(7 - i) \div (7 + i)$.

9. $2(\cos 70° + i \sin 70°) \cdot 5(\cos 20° + i \sin 20°)$.

10. $3(\cos 110° + i \sin 110°) \cdot 4(\cos 70° + i \sin 70°)$.

11. $[6(\cos 146° + i \sin 146°)] \div [3(\cos 86° + i \sin 86°)]$.

12. $[8(\cos 104° + i \sin 104°)] \div [4(\cos 74° + i \sin 74°)]$.

13. $[2(\cos 15° + i \sin 15°)]^8$.

14. $[2(\cos 45° + i \sin 45°)]^4$.

15. $\dfrac{32}{(1 + i\sqrt{3})^4}$.   16. $\dfrac{64}{(\sqrt{3} - i)^5}$.

17. Find all the fifth roots (in polar form) of the imaginary unit i.

18. Find all the fourth roots (in polar form) of (-1).

19. Find the two square roots of $2 + 2\sqrt{3}i$. Express these roots in polar form and plot the points that represent these roots on the complex plane.

20. Find the three cube roots of $-4\sqrt{2} + 4\sqrt{2}i$. Express these roots in polar form and plot the points that represent these roots on the complex plane.

In Exercises 21 and 22 solve each equation for the real numbers x and y.

21. $(x + 2y) + (x - 3y)i = 8 - 2i$.

22. $(1 + xi)(y - 2i) = 7 + i$.

In Exercises 23 and 24 solve each equation for x.

23. $2x^2 + 6x + 13 = 0$.

24. $x^2 + 2x + 2 = 0$.

CHAPTER 13: DIAGNOSTIC TEST

The purpose of this test is to see how well you understand the work covered in Chapter 13. We recommend that you work this test before your instructor tests you on this chapter. Allow yourself approximately 50 minutes to do the test.

Solutions to the problems, together with section references, are given in the Answer Section at the end of this book. You should study the sections referred to for the problems you do incorrectly.

In Problems 1–9 perform the indicated operations, simplify, and write your answers in the form $a + bi$.

1. $(7 - 2i) + (3 + 5i)$.

2. $(9 - 4i) - (6 - 2i)$.

3. $(3 + 2i)(3 - 2i)$.

4. $(1 - 5i) \div (1 + 5i)$.

5. $(1 + i)^6$.

6. $12[\cos 80° + i \sin 80°] \div 3[\cos 20° + i \sin 20°]$.

7. $5[\cos 18° + i \sin 18°] \cdot 2[\cos 27° + i \sin 27°]$.

8. $\dfrac{\sqrt{3} - \sqrt{2}i}{\sqrt{3} + \sqrt{2}i}$.

9. $\dfrac{64}{(\sqrt{3} + i)^5}$.

10. Find the fourth roots (in polar form) of $-8 + 8\sqrt{3}i$, and plot points that represent these roots on the complex plane.

11. Solve the following equation for the real numbers x and y.

$$(x + i)(1 - yi) = 6 - 7i.$$

12. Solve the following equation for x.

$$2x^2 - 2x + 3 = 0.$$

Table 1:

Natural values and logarithms of the trigonometric functions

(to four decimal places)

0° — 9°

Degrees	Sine Value	Sine Log	Tangent Value	Tangent Log	Cotangent Value	Cotangent Log	Cosine Value	Cosine Log	
0° 00′	.0000	——	.0000	——	343.77	——	1.0000	0.0000	90° 00′
10	.0029	7.4637	.0029	7.4637	343.77	2.5363	1.0000	0.0000	50
20	.0058	7.7648	.0058	7.7648	171.89	2.2352	1.0000	0.0000	40
30	.0087	7.9408	.0087	7.9409	114.59	2.0591	1.0000	0.0000	30
40	.0116	8.0658	.0116	8.0658	85.940	1.9342	.9999	0.0000	20
50	.0145	8.1627	.0145	8.1627	68.750	1.8373	.9999	0.0000	10
1° 00′	.0175	8.2419	.0175	8.2419	57.290	1.7581	.9998	9.9999	89° 00′
10	.0204	8.3088	.0204	8.3089	49.104	1.6911	.9998	9.9999	50
20	.0233	8.3668	.0233	8.3669	42.964	1.6331	.9997	9.9999	40
30	.0262	8.4179	.0262	8.4181	38.188	1.5819	.9997	9.9999	30
40	.0291	8.4637	.0291	8.4638	34.368	1.5362	.9996	9.9998	20
50	.0320	8.5050	.0320	8.5053	31.242	1.4947	.9995	9.9998	10
2° 00′	.0349	8.5428	.0349	8.5431	28.636	1.4569	.9994	9.9997	88° 00′
10	.0378	8.5776	.0378	8.5779	26.432	1.4221	.9993	9.9997	50
20	.0407	8.6097	.0407	8.6101	24.542	1.3899	.9992	9.9996	40
30	.0436	8.6397	.0437	8.6401	22.904	1.3599	.9990	9.9996	30
40	.0465	8.6677	.0466	8.6682	21.470	1.3318	.9989	9.9995	20
50	.0494	8.6940	.0495	8.6945	20.206	1.3055	.9988	9.9995	10
3° 00′	.0523	8.7188	.0524	8.7194	19.081	1.2806	.9986	9.9994	87° 00′
10	.0552	8.7423	.0553	8.7429	18.075	1.2571	.9985	9.9993	50
20	.0581	8.7645	.0582	8.7652	17.169	1.2348	.9983	9.9993	40
30	.0610	8.7857	.0612	8.7865	16.350	1.2135	.9981	9.9992	30
40	.0640	8.8059	.0641	8.8067	15.605	1.1933	.9980	9.9991	20
50	.0669	8.8251	.0670	8.8261	14.924	1.1739	.9978	9.9990	10
4° 00′	.0698	8.8436	.0699	8.8446	14.301	1.1554	.9976	9.9989	86° 00′
10	.0727	8.8613	.0729	8.8624	13.727	1.1376	.9974	9.9989	50
20	.0756	8.8783	.0758	8.8795	13.197	1.1205	.9971	9.9988	40
30	.0785	8.8946	.0787	8.8960	12.706	1.1040	.9969	9.9987	30
40	.0814	8.9104	.0816	8.9118	12.251	1.0882	.9967	9.9986	20
50	.0843	8.9256	.0846	8.9272	11.826	1.0728	.9964	9.9985	10
5° 00′	.0872	8.9403	.0875	8.9420	11.430	1.0580	.9962	9.9983	85° 00′
10	.0901	8.9545	.0904	8.9563	11.059	1.0437	.9959	9.9982	50
20	.0929	8.9682	.0934	8.9701	10.712	1.0299	.9957	9.9981	40
30	.0958	8.9816	.0963	8.9836	10.385	1.0164	.9954	9.9980	30
40	.0987	8.9945	.0992	8.9966	10.078	1.0034	.9951	9.9979	20
50	.1016	9.0070	.1022	9.0093	9.7882	0.9907	.9948	9.9977	10
6° 00′	.1045	9.0192	.1051	9.0216	9.5144	0.9784	.9945	9.9976	84° 00′
10	.1074	9.0311	.1080	9.0336	9.2553	0.9664	.9942	9.9975	50
20	.1103	9.0426	.1110	9.0453	9.0098	0.9547	.9939	9.9973	40
30	.1132	9.0539	.1139	9.0567	8.7769	0.9433	.9936	9.9972	30
40	.1161	9.0648	.1169	9.0678	8.5555	0.9322	.9932	9.9971	20
50	.1190	9.0755	.1198	9.0786	8.3450	0.9214	.9929	9.9969	10
7° 00′	.1219	9.0859	.1228	9.0891	8.1443	0.9109	.9925	9.9968	83° 00′
10	.1248	9.0961	.1257	9.0995	7.9530	0.9005	.9922	9.9966	50
20	.1276	9.1060	.1287	9.1096	7.7704	0.8904	.9918	9.9964	40
30	.1305	9.1157	.1317	9.1194	7.5958	0.8806	.9914	9.9963	30
40	.1334	9.1252	.1346	9.1291	7.4287	0.8709	.9911	9.9961	20
50	.1363	9.1345	.1376	9.1385	7.2687	0.8615	.9907	9.9959	10
8° 00′	.1392	9.1436	.1405	9.1478	7.1154	0.8522	.9903	9.9958	82° 00′
10	.1421	9.1525	.1435	9.1569	6.9682	0.8431	.9899	9.9956	50
20	.1449	9.1612	.1465	9.1658	6.8269	0.8342	.9894	9.9954	40
30	.1478	9.1697	.1495	9.1745	6.6912	0.8255	.9890	9.9952	30
40	.1507	9.1781	.1524	9.1831	6.5606	0.8169	.9886	9.9950	20
50	.1536	9.1863	.1554	9.1915	6.4348	0.8085	.9881	9.9948	10
9° 00′	.1564	9.1943	.1584	9.1997	6.3138	0.8003	.9877	9.9946	81° 00′
	Value	Log	Value	Log	Value	Log	Value	Log	Degrees
	Cosine		Cotangent		Tangent		Sine		

81° — 90°

9° — 18°

Degrees	Sine		Tangent		Cotangent		Cosine		
	Value	Log	Value	Log	Value	Log	Value	Log	
9° 00′	.1564	9.1943	.1584	9.1997	6.3138	0.8003	.9877	9.9946	81° 00′
10	.1593	9.2022	.1614	9.2078	6.1970	0.7922	.9872	9.9944	50′
20	.1622	9.2100	.1644	9.2158	6.0844	0.7842	.9868	9.9942	40
30	.1650	9.2176	.1673	9.2236	5.9758	0.7764	.9863	9.9940	30
40	.1679	9.2251	.1703	9.2313	5.8708	0.7687	.9858	9.9938	20
50	.1708	9.2324	.1733	9.2389	5.7694	0.7611	.9853	9.9936	10
10° 00′	.1736	9.2397	.1763	9.2463	5.6713	0.7537	.9848	9.9934	80° 00′
10	.1765	9.2468	.1793	9.2536	5.5764	0.7464	.9843	9.9931	50
20	.1794	9.2538	.1823	9.2609	5.4845	0.7391	.9838	9.9929	40
30	.1822	9.2606	.1853	9.2680	5.3955	0.7320	.9833	9.9927	30
40	.1851	9.2674	.1883	9.2750	5.3093	0.7250	.9827	9.9924	20
50	.1880	9.2740	.1914	9.2819	5.2257	0.7181	.9822	9.9922	10
11° 00′	.1908	9.2806	.1944	9.2887	5.1446	0.7113	.9816	9.9919	79° 00′
10	.1937	9.2870	.1974	9.2953	5.0658	0.7047	.9811	9.9917	50
20	.1965	9.2934	.2004	9.3020	4.9894	0.6980	.9805	9.9914	40
30	.1994	9.2997	.2035	9.3085	4.9152	0.6915	.9799	9.9912	30
40	.2022	9.3058	.2065	9.3149	4.8430	0.6851	.9793	9.9909	20
50	.2051	9.3119	.2095	9.3212	4.7729	0.6788	.9787	9.9907	10
12° 00′	.2079	9.3179	.2126	9.3275	4.7046	0.6725	.9781	9.9904	78° 00′
10	.2108	9.3238	.2156	9.3336	4.6382	0.6664	.9775	9.9901	50
20	.2136	9.3296	.2186	9.3397	4.5736	0.6603	.9769	9.9899	40
30	.2164	9.3353	.2217	9.3458	4.5107	0.6542	.9763	9.9896	30
40	.2193	9.3410	.2247	9.3517	4.4494	0.6483	.9757	9.9893	20
50	.2221	9.3466	.2278	9.3576	4.3897	0.6424	.9750	9.9890	10
13° 00′	.2250	9.3521	.2309	9.3634	4.3315	0.6366	.9744	9.9887	77° 00′
10	.2278	9.3575	.2339	9.3691	4.2747	0.6309	.9737	9.9884	50
20	.2306	9.3629	.2370	9.3748	4.2193	0.6252	.9730	9.9881	40
30	.2334	9.3682	.2401	9.3804	4.1653	0.6196	.9724	9.9878	30
40	.2363	9.3734	.2432	9.3859	4.1126	0.6141	.9717	9.9875	20
50	.2391	9.3786	.2462	9.3914	4.0611	0.6086	.9710	9.9872	10
14° 00′	.2419	9.3837	.2493	9.3968	4.0108	0.6032	.9703	9.9869	76° 00′
10	.2447	9.3887	.2524	9.4021	3.9617	0.5979	.9696	9.9866	50
20	.2476	9.3937	.2555	9.4074	3.9136	0.5926	.9689	9.9863	40
30	.2504	9.3986	.2586	9.4127	3.8667	0.5873	.9681	9.9859	30
40	.2532	9.4035	.2617	9.4178	3.8208	0.5822	.9674	9.9856	20
50	.2560	9.4083	.2648	9.4230	3.7760	0.5770	.9667	9.9853	10
15° 00′	.2588	9.4130	.2679	9.4281	3.7321	0.5719	.9659	9.9849	75° 00′
10	.2616	9.4177	.2711	9.4331	3.6891	0.5669	.9652	9.9846	50
20	.2644	9.4223	.2742	9.4381	3.6470	0.5619	.9644	9.9843	40
30	.2672	9.4269	.2773	9.4430	3.6059	0.5570	.9636	9.9839	30
40	.2700	9.4314	.2805	9.4479	3.5656	0.5521	.9628	9.9836	20
50	.2728	9.4359	.2836	9.4527	3.5261	0.5473	.9621	9.9832	10
16° 00′	.2756	9.4403	.2867	9.4575	3.4874	0.5425	.9613	9.9828	74° 00′
10	.2784	9.4447	.2899	9.4622	3.4495	0.5378	.9605	9.9825	50
20	.2812	9.4491	.2931	9.4669	3.4124	0.5331	.9596	9.9821	40
30	.2840	9.4533	.2962	9.4716	3.3759	0.5284	.9588	9.9817	30
40	.2868	9.4576	.2994	9.4762	3.3402	0.5238	.9580	9.9814	20
50	.2896	9.4618	.3026	9.4808	3.3052	0.5192	.9572	9.9810	10
17° 00′	.2924	9.4659	.3057	9.4853	3.2709	0.5147	.9563	9.9806	73° 00′
10	.2952	9.4700	.3089	9.4898	3.2371	0.5102	.9555	9.9802	50
20	.2979	9.4741	.3121	9.4943	3.2041	0.5057	.9546	9.9798	40
30	.3007	9.4781	.3153	9.4987	3.1716	0.5013	.9537	9.9794	30
40	.3035	9.4821	.3185	9.5031	3.1397	0.4969	.9528	9.9790	20
50	.3062	9.4861	.3217	9.5075	3.1084	0.4925	.9520	9.9786	10
18° 00′	.3090	9.4900	.3249	9.5118	3.0777	0.4882	.9511	9.9782	72° 00′
	Value	Log	Value	Log	Value	Log	Value	Log	Degrees
	Cosine		Cotangent		Tangent		Sine		

72° — 81°

72°50 2952 ⟵ look for sin

EXAMPLE: LOOKING FOR   .2861 COS

72°40 2979 ⟶

Angles on right side column

18° — 27°

Degrees	Sine Value	Sine Log	Tangent Value	Tangent Log	Cotangent Value	Cotangent Log	Cosine Value	Cosine Log	Degrees
18° 00′	.3090	9.4900	.3249	9.5118	3.0777	0.4882	.9511	9.9782	**72° 00′**
10	.3118	9.4939	.3281	9.5161	3.0475	0.4839	.9502	9.9778	50
20	.3145	9.4977	.3314	9.5203	3.0178	0.4797	.9492	9.9774	40
30	.3173	9.5015	.3346	9.5245	2.9887	0.4755	.9483	9.9770	30
40	.3201	9.5052	.3378	9.5287	2.9600	0.4713	.9474	9.9765	20
50	.3228	9.5090	.3411	9.5329	2.9319	0.4671	.9465	9.9761	10
19° 00′	.3256	9.5126	.3443	9.5370	2.9042	0.4630	.9455	9.9757	**71° 00′**
10	.3283	9.5163	.3476	9.5411	2.8770	0.4589	.9446	9.9752	50
20	.3311	9.5199	.3508	9.5451	2.8502	0.4549	.9436	9.9748	40
30	.3338	9.5235	.3541	9.5491	2.8239	0.4509	.9426	9.9743	30
40	.3365	9.5270	.3574	9.5531	2.7980	0.4469	.9417	9.9739	20
50	.3393	9.5306	.3607	9.5571	2.7725	0.4429	.9407	9.9734	10
20° 00′	.3420	9.5341	.3640	9.5611	2.7475	0.4389	.9397	9.9730	**70° 00′**
10	.3448	9.5375	.3673	9.5650	2.7228	0.4350	.9387	9.9725	50
20	.3475	9.5409	.3706	9.5689	2.6985	0.4311	.9377	9.9721	40
30	.3502	9.5443	.3739	9.5727	2.6746	0.4273	.9367	9.9716	30
40	.3529	9.5477	.3772	9.5766	2.6511	0.4234	.9356	9.9711	20
50	.3557	9.5510	.3805	9.5804	2.6279	0.4196	.9346	9.9706	10
21° 00′	.3584	9.5543	.3839	9.5842	2.6051	0.4158	.9336	9.9702	**69° 00′**
10	.3611	9.5576	.3872	9.5879	2.5826	0.4121	.9325	9.9697	50
20	.3638	9.5609	.3906	9.5917	2.5605	0.4083	.9315	9.9692	40
30	.3665	9.5641	.3939	9.5954	2.5386	0.4046	.9304	9.9687	30
40	.3692	9.5673	.3973	9.5991	2.5172	0.4009	.9293	9.9682	20
50	.3719	9.5704	.4006	9.6028	2.4960	0.3972	.9283	9.9677	10
22° 00′	.3746	9.5736	.4040	9.6064	2.4751	0.3936	.9272	9.9672	**68° 00′**
10	.3773	9.5767	.4074	9.6100	2.4545	0.3900	.9261	9.9667	50
20	.3800	9.5798	.4108	9.6136	2.4342	0.3864	.9250	9.9661	40
30	.3827	9.5828	.4142	9.6172	2.4142	0.3828	.9239	9.9656	30
40	.3854	9.5859	.4176	9.6208	2.3945	0.3792	.9228	9.9651	20
50	.3881	9.5889	.4210	9.6243	2.3750	0.3757	.9216	9.9646	10
23° 00′	.3907	9.5919	.4245	9.6279	2.3559	0.3721	.9205	9.9640	**67° 00′**
10	.3934	9.5948	.4279	9.6314	2.3369	0.3686	.9194	9.9635	50
20	.3961	9.5978	.4314	9.6348	2.3183	0.3652	.9182	9.9629	40
30	.3987	9.6007	.4348	9.6383	2.2998	0.3617	.9171	9.9624	30
40	.4014	9.6036	.4383	9.6417	2.2817	0.3583	.9159	9.9618	20
50	.4041	9.6065	.4417	9.6452	2.2637	0.3548	.9147	9.9613	10
24° 00′	.4067	9.6093	.4452	9.6486	2.2460	0.3514	.9135	9.9607	**66° 00′**
10	.4094	9.6121	.4487	9.6520	2.2286	0.3480	.9124	9.9602	50
20	.4120	9.6149	.4522	9.6553	2.2113	0.3447	.9112	9.9596	40
30	.4147	9.6177	.4557	9.6587	2.1943	0.3413	.9100	9.9590	30
40	.4173	9.6205	.4592	9.6620	2.1775	0.3380	.9088	9.9584	20
50	.4200	9.6232	.4628	9.6654	2.1609	0.3346	.9075	9.9579	10
25° 00′	.4226	9.6259	.4663	9.6687	2.1445	0.3313	.9063	9.9573	**65° 00′**
10	.4253	9.6286	.4699	9.6720	2.1283	0.3280	.9051	9.9567	50
20	.4279	9.6313	.4734	9.6752	2.1123	0.3248	.9038	9.9561	40
30	.4305	9.6340	.4770	9.6785	2.0965	0.3215	.9026	9.9555	30
40	.4331	9.6366	.4806	9.6817	2.0809	0.3183	.9013	9.9549	20
50	.4358	9.6392	.4841	9.6850	2.0655	0.3150	.9001	9.9543	10
26° 00′	.4384	9.6418	.4877	9.6882	2.0503	0.3118	.8988	9.9537	**64° 00′**
10	.4410	9.6444	.4913	9.6914	2.0353	0.3086	.8975	9.9530	50
20	.4436	9.6470	.4950	9.6946	2.0204	0.3054	.8962	9.9524	40
30	.4462	9.6495	.4986	9.6977	2.0057	0.3023	.8949	9.9518	30
40	.4488	9.6521	.5022	9.7009	1.9912	0.2991	.8936	9.9512	20
50	.4514	9.6546	.5059	9.7040	1.9768	0.2960	.8923	9.9505	10
27° 00′	.4540	9.6570	.5095	9.7072	1.9626	0.2928	.8910	9.9499	**63° 00′**
	Value	Log	Value	Log	Value	Log	Value	Log	
	Cosine		Cotangent		Tangent		Sine		Degrees

63° — 72°

bottom read right ——> //

<— Top read left - 0

27° — 36°

Degrees	Sine		Tangent		Cotangent		Cosine		
	Value	Log	Value	Log	Value	Log	Value	Log	
27° 00′	.4540	9.6570	.5095	9.7072	1.9626	0.2928	.8910	9.9499	63° 00′
10	.4566	9.6595	.5132	9.7103	1.9486	0.2897	.8897	9.9492	50
20	.4592	9.6620	.5169	9.7134	1.9347	0.2866	.8884	9.9486	40
30	.4617	9.6644	.5206	9.7165	1.9210	0.2835	.8870	9.9479	30
40	.4643	9.6668	.5243	9.7196	1.9074	0.2804	.8857	9.9473	20
50	.4669	9.6692	.5280	9.7226	1.8940	0.2774	.8843	9.9466	10
28° 00′	.4695	9.6716	.5317	9.7257	1.8807	0.2743	.8829	9.9459	62° 00′
10	.4720	9.6740	.5354	9.7287	1.8676	0.2713	.8816	9.9453	50
20	.4746	9.6763	.5392	9.7317	1.8546	0.2683	.8802	9.9446	40
30	.4772	9.6787	.5430	9.7348	1.8418	0.2652	.8788	9.9439	30
40	.4797	9.6810	.5467	9.7378	1.8291	0.2622	.8774	9.9432	20
50	.4823	9.6833	.5505	9.7408	1.8165	0.2592	.8760	9.9425	10
29° 00′	.4848	9.6856	.5543	9.7438	1.8040	0.2562	.8746	9.9418	61° 00′
10	.4874	9.6878	.5581	9.7467	1.7917	0.2533	.8732	9.9411	50
20	.4899	9.6901	.5619	9.7497	1.7796	0.2503	.8718	9.9404	40
30	.4924	9.6923	.5658	9.7526	1.7675	0.2474	.8704	9.9397	30
40	.4950	9.6946	.5696	9.7556	1.7556	0.2444	.8689	9.9390	20
50	.4975	9.6968	.5735	9.7585	1.7437	0.2415	.8675	9.9383	10
30° 00′	.5000	9.6990	.5774	9.7614	1.7321	0.2386	.8660	9.9375	60° 00′
10	.5025	9.7012	.5812	9.7644	1.7205	0.2356	.8646	9.9368	50
20	.5050	9.7033	.5851	9.7673	1.7090	0.2327	.8631	9.9361	40
30	.5075	9.7055	.5890	9.7701	1.6977	0.2299	.8616	9.9353	30
40	.5100	9.7076	.5930	9.7730	1.6864	0.2270	.8601	9.9346	20
50	.5125	9.7097	.5969	9.7759	1.6753	0.2241	.8587	9.9338	10
31° 00′	.5150	9.7118	.6009	9.7788	1.6643	0.2212	.8572	9.9331	59° 00′
10	.5175	9.7139	.6048	9.7816	1.6534	0.2184	.8557	9.9323	50
20	.5200	9.7160	.6088	9.7845	1.6426	0.2155	.8542	9.9315	40
30	.5225	9.7181	.6128	9.7873	1.6319	0.2127	.8526	9.9308	30
40	.5250	9.7201	.6168	9.7902	1.6212	0.2098	.8511	9.9300	20
50	.5275	9.7222	.6208	9.7930	1.6107	0.2070	.8496	9.9292	10
32° 00′	.5299	9.7242	.6249	9.7958	1.6003	0.2042	.8480	9.9284	58° 00′
10	.5324	9.7262	.6289	9.7986	1.5900	0.2014	.8465	9.9276	50
20	.5348	9.7282	.6330	9.8014	1.5798	0.1986	.8450	9.9268	40
30	.5373	9.7302	.6371	9.8042	1.5697	0.1958	.8434	9.9260	30
40	.5398	9.7322	.6412	9.8070	1.5597	0.1930	.8418	9.9252	20
50	.5422	9.7342	.6453	9.8097	1.5497	0.1903	.8403	9.9244	10
33° 00′	.5446	9.7361	.6494	9.8125	1.5399	0.1875	.8387	9.9236	57° 00′
10	.5471	9.7380	.6536	9.8153	1.5301	0.1847	.8371	9.9228	50
20	.5495	9.7400	.6577	9.8180	1.5204	0.1820	.8355	9.9219	40
30	.5519	9.7419	.6619	9.8208	1.5108	0.1792	.8339	9.9211	30
40	.5544	9.7438	.6661	9.8235	1.5013	0.1765	.8323	9.9203	20
50	.5568	9.7457	.6703	9.8263	1.4919	0.1737	.8307	9.9194	10
34° 00′	.5592	9.7476	.6745	9.8290	1.4826	0.1710	.8290	9.9186	56° 00′
10	.5616	9.7494	.6787	9.8317	1.4733	0.1683	.8274	9.9177	50
20	.5640	9.7513	.6830	9.8344	1.4641	0.1656	.8258	9.9169	40
30	.5664	9.7531	.6873	9.8371	1.4550	0.1629	.8241	9.9160	30
40	.5688	9.7550	.6916	9.8398	1.4460	0.1602	.8225	9.9151	20
50	.5712	9.7568	.6959	9.8425	1.4370	0.1575	.8208	9.9142	10
35° 00′	.5736	9.7586	.7002	9.8452	1.4281	0.1548	.8192	9.9134	55° 00′
10	.5760	9.7604	.7046	9.8479	1.4193	0.1521	.8175	9.9125	50
20	.5783	9.7622	.7089	9.8506	1.4106	0.1494	.8158	9.9116	40
30	.5807	9.7640	.7133	9.8533	1.4019	0.1467	.8141	9.9107	30
40	.5831	9.7657	.7177	9.8559	1.3934	0.1441	.8124	9.9098	20
50	.5854	9.7675	.7221	9.8586	1.3848	0.1414	.8107	9.9089	10
36° 00′	.5878	9.7692	.7265	9.8613	1.3764	0.1387	.8090	9.9080	54° 00′
	Value	Log	Value	Log	Value	Log	Value	Log	
	Cosine		Cotangent		Tangent		Sine		Degrees

54° — 63°

Table 1

36° — 45°

Degrees	Sine		Tangent		Cotangent		Cosine		
	Value	Log	Value	Log	Value	Log	Value	Log	
36° 00′	.5878	9.7692	.7265	9.8613	1.3764	0.1387	.8090	9.9080	54° 00′
10	.5901	9.7710	.7310	9.8639	1.3680	0.1361	.8073	9.9070	50
20	.5925	9.7727	.7355	9.8666	1.3597	0.1334	.8056	9.9061	40
30	.5948	9.7744	.7400	9.8692	1.3514	0.1308	.8039	9.9052	30
40	.5972	9.7761	.7445	9.8718	1.3432	0.1282	.8021	9.9042	20
50	.5995	9.7778	.7490	9.8745	1.3351	0.1255	.8004	9.9033	10
37° 00′	.6018	9.7795	.7536	9.8771	1.3270	0.1229	.7986	9.9023	53° 00′
10	.6041	9.7811	.7581	9.8797	1.3190	0.1203	.7969	9.9014	50
20	.6065	9.7828	.7627	9.8824	1.3111	0.1176	.7951	9.9004	40
30	.6088	9.7844	.7673	9.8850	1.3032	0.1150	.7934	9.8995	30
40	.6111	9.7861	.7720	9.8876	1.2954	0.1124	.7916	9.8985	20
50	.6134	9.7877	.7766	9.8902	1.2876	0.1098	.7898	9.8975	10
38° 00′	.6157	9.7893	.7813	9.8928	1.2799	0.1072	.7880	9.8965	52° 00′
10	.6180	9.7910	.7860	9.8954	1.2723	0.1046	.7862	9.8955	50
20	.6202	9.7926	.7907	9.8980	1.2647	0.1020	.7844	9.8945	40
30	.6225	9.7941	.7954	9.9006	1.2572	0.0994	.7826	9.8935	30
40	.6248	9.7957	.8002	9.9032	1.2497	0.0968	.7808	9.8925	20
50	.6271	9.7973	.8050	9.9058	1.2423	0.0942	.7790	9.8915	10
39° 00′	.6293	9.7989	.8098	9.9084	1.2349	0.0916	.7771	9.8905	51° 00′
10	.6316	9.8004	.8146	9.9110	1.2276	0.0890	.7753	9.8895	50
20	.6338	9.8020	.8195	9.9135	1.2203	0.0865	.7735	9.8884	40
30	.6361	9.8035	.8243	9.9161	1.2131	0.0839	.7716	9.8874	30
40	.6383	9.8050	.8292	9.9187	1.2059	0.0813	.7698	9.8864	20
50	.6406	9.8066	.8342	9.9212	1.1988	0.0788	.7679	9.8853	10
40° 00′	.6428	9.8081	.8391	9.9238	1.1918	0.0762	.7660	9.8843	50° 00′
10	.6450	9.8096	.8441	9.9264	1.1847	0.0736	.7642	9.8832	50
20	.6472	9.8111	.8491	9.9289	1.1778	0.0711	.7623	9.8821	40
30	.6494	9.8125	.8541	9.9315	1.1708	0.0685	.7604	9.8810	30
40	.6517	9.8140	.8591	9.9341	1.1640	0.0659	.7585	9.8800	20
50	.6539	9.8155	.8642	9.9366	1.1571	0.0634	.7566	9.8789	10
41° 00′	.6561	9.8169	.8693	9.9392	1.1504	0.0608	.7547	9.8778	49° 00′
10	.6583	9.8184	.8744	9.9417	1.1436	0.0583	.7528	9.8767	50
20	.6604	9.8198	.8796	9.9443	1.1369	0.0557	.7509	9.8756	40
30	.6626	9.8213	.8847	9.9468	1.1303	0.0532	.7490	9.8745	30
40	.6648	9.8227	.8899	9.9494	1.1237	0.0506	.7470	9.8733	20
50	.6670	9.8241	.8952	9.9519	1.1171	0.0481	.7451	9.8722	10
42° 00′	.6691	9.8255	.9004	9.9544	1.1106	0.0456	.7431	9.8711	48° 00′
10	.6713	9.8269	.9057	9.9570	1.1041	0.0430	.7412	9.8699	50
20	.6734	9.8283	.9110	9.9595	1.0977	0.0405	.7392	9.8688	40
30	.6756	9.8297	.9163	9.9621	1.0913	0.0379	.7373	9.8676	30
40	.6777	9.8311	.9217	9.9646	1.0850	0.0354	.7353	9.8665	20
50	.6799	9.8324	.9271	9.9671	1.0786	0.0329	.7333	9.8653	10
43° 00′	.6820	9.8338	.9325	9.9697	1.0724	0.0303	.7314	9.8641	47° 00′
10	.6841	9.8351	.9380	9.9722	1.0661	0.0278	.7294	9.8629	50
20	.6862	9.8365	.9435	9.9747	1.0599	0.0253	.7274	9.8618	40
30	.6884	9.8378	.9490	9.9772	1.0538	0.0228	.7254	9.8606	30
40	.6905	9.8391	.9545	9.9798	1.0477	0.0202	.7234	9.8594	20
50	.6926	9.8405	.9601	9.9823	1.0416	0.0177	.7214	9.8582	10
44° 00′	.6947	9.8418	.9657	9.9848	1.0355	0.0152	.7193	9.8569	46° 00′
10	.6967	9.8431	.9713	9.9874	1.0295	0.0126	.7173	9.8557	50
20	.6988	9.8444	.9770	9.9899	1.0235	0.0101	.7153	9.8545	40
30	.7009	9.8457	.9827	9.9924	1.0176	0.0076	.7133	9.8532	30
40	.7030	9.8469	.9884	9.9949	1.0117	0.0051	.7112	9.8520	20
50	.7050	9.8482	.9942	9.9975	1.0058	0.0025	.7092	9.8507	10
45° 00′	.7071	9.8495	1.0000	0.0000	1.0000	0.0000	.7071	9.8495	45° 00′
	Value	Log	Value	Log	Value	Log	Value	Log	
	Cosine		Cotangent		Tangent		Sine		Degrees

45° — 54°

Table 2:
Common logarithms of numbers

(to four decimal places)

Four-place logarithms of numbers

n	0	1	2	3	4	5	6	7	8	9
10	0000	0043	0086	0128	0170	0212	0253	0294	0334	0374
11	0414	0453	0492	0531	0569	0607	0645	0682	0719	0755
12	0792	0828	0864	0899	0934	0969	1004	1038	1072	1106
13	1139	1173	1206	1239	1271	1303	1335	1367	1399	1430
14	1461	1492	1523	1553	1584	1614	1644	1673	1703	1732
15	1761	1790	1818	1847	1875	1903	1931	1959	1987	2014
16	2041	2068	2095	2122	2148	2175	2201	2227	2253	2279
17	2304	2330	2355	2380	2405	2430	2455	2480	2504	2529
18	2553	2577	2601	2625	2648	2672	2695	2718	2742	2765
19	2788	2810	2833	2856	2878	2900	2923	2945	2967	2989
20	3010	3032	3054	3075	3096	3118	3139	3160	3181	3201
21	3222	3243	3263	3284	3304	3324	3345	3365	3385	3404
22	3424	3444	3464	3483	3502	3522	3541	3560	3579	3598
23	3617	3636	3655	3674	3692	3711	3729	3747	3766	3784
24	3802	3820	3838	3856	3874	3892	3909	3927	3945	3962
25	3979	3997	4014	4031	4048	4065	4082	4099	4116	4133
26	4150	4166	4183	4200	4216	4232	4249	4265	4281	4298
27	4314	4330	4346	4362	4378	4393	4409	4425	4440	4456
28	4472	4487	4502	4518	4533	4548	4564	4579	4594	4609
29	4624	4639	4654	4669	4683	4698	4713	4728	4742	4757
30	4771	4786	4800	4814	4829	4843	4857	4871	4886	4900
31	4914	4928	4942	4955	4969	4983	4997	5011	5024	5038
32	5051	5065	5079	5092	5105	5119	5132	5145	5159	5172
33	5185	5198	5211	5224	5237	5250	5263	5276	5289	5302
34	5315	5328	5340	5353	5366	5378	5391	5403	5416	5428
35	5441	5453	5465	5478	5490	5502	5514	5527	5539	5551
36	5563	5575	5587	5599	5611	5623	5635	5647	5658	5670
37	5682	5694	5705	5717	5729	5740	5752	5763	5775	5786
38	5798	5809	5821	5832	5843	5855	5866	5877	5888	5899
39	5911	5922	5933	5944	5955	5966	5977	5988	5999	6010
40	6021	6031	6042	6053	6064	6075	6085	6096	6107	6117
41	6128	6138	6149	6160	6170	6180	6191	6201	6212	6222
42	6232	6243	6253	6263	6274	6284	6294	6304	6314	6325
43	6335	6345	6355	6365	6375	6385	6395	6405	6415	6425
44	6435	6444	6454	6464	6474	6484	6493	6503	6513	6522
45	6532	6542	6551	6561	6571	6580	6590	6599	6609	6618
46	6628	6637	6646	6656	6665	6675	6684	6693	6702	6712
47	6721	6730	6739	6749	6758	6767	6776	6785	6794	6803
48	6812	6821	6830	6839	6848	6857	6866	6875	6884	6893
49	6902	6911	6920	6928	6937	6946	6955	6964	6972	6981
50	6990	6998	7007	7016	7024	7033	7042	7050	7059	7067
51	7076	7084	7093	7101	7110	7118	7126	7135	7143	7152
52	7160	7168	7177	7185	7193	7202	7210	7218	7226	7235
53	7243	7251	7259	7267	7275	7284	7292	7300	7308	7316
54	7324	7332	7340	7348	7356	7364	7372	7380	7388	7396
n	0	1	2	3	4	5	6	7	8	9

Prop. Parts

	43	42	41	40
1	4.3	4.2	4.1	4.0
2	8.6	8.4	8.2	8.0
3	12.9	12.6	12.3	12.0
4	17.2	16.8	16.4	16.0
5	21.5	21.0	20.5	20.0
6	25.8	25.2	24.6	24.0
7	30.1	29.4	28.7	28.0
8	34.4	33.6	32.8	32.0
9	38.7	37.8	36.9	36.0

	39	38	37	36
1	3.9	3.8	3.7	3.6
2	7.8	7.6	7.4	7.2
3	11.7	11.4	11.1	10.8
4	15.6	15.2	14.8	14.4
5	19.5	19.0	18.5	18.0
6	23.4	22.8	22.2	21.6
7	27.3	26.6	25.9	25.2
8	31.2	30.4	29.6	28.8
9	35.1	34.2	33.3	32.4

	35	34	33	32
1	3.5	3.4	3.3	3.2
2	7.0	6.8	6.6	6.4
3	10.5	10.2	9.9	9.6
4	14.0	13.6	13.2	12.8
5	17.5	17.0	16.5	16.0
6	21.0	20.4	19.8	19.2
7	24.5	23.8	23.1	22.4
8	28.0	27.2	26.4	25.6
9	31.5	30.6	29.7	28.8

	31	30	29	28
1	3.1	3.0	2.9	2.8
2	6.2	6.0	5.8	5.6
3	9.3	9.0	8.7	8.4
4	12.4	12.0	11.6	11.2
5	15.5	15.0	14.5	14.0
6	18.6	18.0	17.4	16.8
7	21.7	21.0	20.3	19.6
8	24.8	24.0	23.2	22.4
9	27.9	27.0	26.1	25.2

	27	26	25	24
1	2.7	2.6	2.5	2.4
2	5.4	5.2	5.0	4.8
3	8.1	7.8	7.5	7.2
4	10.8	10.4	10.0	9.6
5	13.5	13.0	12.5	12.0
6	16.2	15.6	15.0	14.4
7	18.9	18.2	17.5	16.8
8	21.6	20.8	20.0	19.2
9	24.3	23.4	22.5	21.6

Prop. Parts

	23	22	21	20
1	2.3	2.2	2.1	2.0
2	4.6	4.4	4.2	4.0
3	6.9	6.6	6.3	6.0
4	9.2	8.8	8.4	8.0
5	11.5	11.0	10.5	10.0
6	13.8	13.2	12.6	12.0
7	16.1	15.4	14.7	14.0
8	18.4	17.6	16.8	16.0
9	20.7	19.8	18.9	18.0

	19	18	17	16
1	1.9	1.8	1.7	1.6
2	3.8	3.6	3.4	3.2
3	5.7	5.4	5.1	4.8
4	7.6	7.2	6.8	6.4
5	9.5	9.0	8.5	8.0
6	11.4	10.8	10.2	9.6
7	13.3	12.6	11.9	11.2
8	15.2	14.4	13.6	12.8
9	17.1	16.2	15.3	14.4

	15	14	13	12
1	1.5	1.4	1.3	1.2
2	3.0	2.8	2.6	2.4
3	4.5	4.2	3.9	3.6
4	6.0	5.6	5.2	4.8
5	7.5	7.0	6.5	6.0
6	9.0	8.4	7.8	7.2
7	10.5	9.8	9.1	8.4
8	12.0	11.2	10.4	9.6
9	13.5	12.6	11.7	10.8

	11	10	9	8
1	1.1	1.0	0.9	0.8
2	2.2	2.0	1.8	1.6
3	3.3	3.0	2.7	2.4
4	4.4	4.0	3.6	3.2
5	5.5	5.0	4.5	4.0
6	6.6	6.0	5.4	4.8
7	7.7	7.0	6.3	5.6
8	8.8	8.0	7.2	6.4
9	9.9	9.0	8.1	7.2

	7	6	5	4
1	0.7	0.6	0.5	0.4
2	1.4	1.2	1.0	0.8
3	2.1	1.8	1.5	1.2
4	2.8	2.4	2.0	1.6
5	3.5	3.0	2.5	2.0
6	4.2	3.6	3.0	2.4
7	4.9	4.2	3.5	2.8
8	5.6	4.8	4.0	3.2
9	6.3	5.4	4.5	3.6

n	0	1	2	3	4	5	6	7	8	9
55	7404	7412	7419	7427	7435	7443	7451	7459	7466	7474
56	7482	7490	7497	7505	7513	7520	7528	7536	7543	7551
57	7559	7566	7574	7582	7589	7597	7604	7612	7619	7627
58	7634	7642	7649	7657	7664	7672	7679	7686	7694	7701
59	7709	7716	7723	7731	7738	7745	7752	7760	7767	7774
60	7782	7789	7796	7803	7810	7818	7825	7832	7839	7846
61	7853	7860	7868	7875	7882	7889	7896	7903	7910	7917
62	7924	7931	7938	7945	7952	7959	7966	7973	7980	7987
63	7993	8000	8007	8014	8021	8028	8035	8041	8048	8055
64	8062	8069	8075	8082	8089	8096	8102	8109	8116	8122
65	8129	8136	8142	8149	8156	8162	8169	8176	8182	8189
66	8195	8202	8209	8215	8222	8228	8235	8241	8248	8254
67	8261	8267	8274	8280	8287	8293	8299	8306	8312	8319
68	8325	8331	8338	8344	8351	8357	8363	8370	8376	8382
69	8388	8395	8401	8407	8414	8420	8426	8432	8439	8445
70	8451	8457	8463	8470	8476	8482	8488	8494	8500	8506
71	8513	8519	8525	8531	8537	8543	8549	8555	8561	8567
72	8573	8579	8585	8591	8597	8603	8609	8615	8621	8627
73	8633	8639	8645	8651	8657	8663	8669	8675	8681	8686
74	8692	8698	8704	8710	8716	8722	8727	8733	8739	8745
75	8751	8756	8762	8768	8774	8779	8785	8791	8797	8802
76	8808	8814	8820	8825	8831	8837	8842	8848	8854	8859
77	8865	8871	8876	8882	8887	8893	8899	8904	8910	8915
78	8921	8927	8932	8938	8943	8949	8954	8960	8965	8971
79	8976	8982	8987	8993	8998	9004	9009	9015	9020	9025
80	9031	9036	9042	9047	9053	9058	9063	9069	9074	9079
81	9085	9090	9096	9101	9106	9112	9117	9122	9128	9133
82	9138	9143	9149	9154	9159	9165	9170	9175	9180	9186
83	9191	9196	9201	9206	9212	9217	9222	9227	9232	9238
84	9243	9248	9253	9258	9263	9269	9274	9279	9284	9289
85	9294	9299	9304	9309	9315	9320	9325	9330	9335	9340
86	9345	9350	9355	9360	9365	9370	9375	9380	9385	9390
87	9395	9400	9405	9410	9415	9420	9425	9430	9435	9440
88	9445	9450	9455	9460	9465	9469	9474	9479	9484	9489
89	9494	9499	9504	9509	9513	9518	9523	9528	9533	9538
90	9542	9547	9552	9557	9562	9566	9571	9576	9581	9586
91	9590	9595	9600	9605	9609	9614	9619	9624	9628	9633
92	9638	9643	9647	9652	9657	9661	9666	9671	9675	9680
93	9685	9689	9694	9699	9703	9708	9713	9717	9722	9727
94	9731	9736	9741	9745	9750	9754	9759	9763	9768	9773
95	9777	9782	9786	9791	9795	9800	9805	9809	9814	9818
96	9823	9827	9832	9836	9841	9845	9850	9854	9859	9863
97	9868	9872	9877	9881	9886	9890	9894	9899	9903	9908
98	9912	9917	9921	9926	9930	9934	9939	9943	9948	9952
99	9956	9961	9965	9969	9974	9978	9983	9987	9991	9996

Prop. Parts

n	0	1	2	3	4	5	6	7	8	9

Table 3:
Trigonometric Functions: Radian and Degree Measures

Angle Rad.	Degrees	Sin	Cos	Tan
.00	0° 00.0′	.00000	1.0000	.00000
.01	0° 34.4′	.01000	.99995	.01000
.02	1° 08.8′	.02000	.99980	.02000
.03	1° 43.1′	.03000	.99955	.03001
.04	2° 17.5′	.03999	.99920	.04002
.05	2° 51.9′	.04998	.99875	.05004
.06	3° 26.3′	.05996	.99820	.06007
.07	4° 00.6′	.06994	.99755	.07011
.08	4° 35.0′	.07991	.99680	.08017
.09	5° 09.4′	.08988	.99595	.09024
.10	5° 43.8′	.09983	.99500	.10033
.11	6° 18.2′	.10978	.99396	.11045
.12	6° 52.5′	.11971	.99281	.12058
.13	7° 26.9′	.12963	.99156	.13074
.14	8° 01.3′	.13954	.99022	.14092
.15	8° 35.7′	.14944	.98877	.15114
.16	9° 10.0′	.15932	.98723	.16138
.17	9° 44.4′	.16918	.98558	.17166
.18	10° 18.8′	.17903	.98384	.18197
.19	10° 53.2′	.18886	.98200	.19232
.20	11° 27.5′	.19867	.98007	.20271
.21	12° 01.9′	.20846	.97803	.21314
.22	12° 36.3′	.21823	.97590	.22362
.23	13° 10.7′	.22798	.97367	.23414
.24	13° 45.1′	.23770	.97134	.24472
.25	14° 19.4′	.24740	.96891	.25534
.26	14° 53.8′	.25708	.96639	.26602
.27	15° 28.2′	.26673	.96377	.27676
.28	16° 02.6′	.27636	.96106	.28755
.29	16° 36.9′	.28595	.95824	.29841
.30	17° 11.3′	.29552	.95534	.30934
.31	17° 45.7′	.30506	.95233	.32033
.32	18° 20.1′	.31457	.94924	.33139
.33	18° 54.5′	.32404	.94604	.34252
.34	19° 28.8′	.33349	.94275	.35374
.35	20° 03.2′	.34290	.93937	.36503
.36	20° 37.6′	.35227	.93590	.37640
.37	21° 12.0′	.36162	.93233	.38786
.38	21° 46.3′	.37092	.92866	.39941
.39	22° 20.7′	.38019	.92491	.41106
.40	22° 55.1′	.38942	.92106	.42279
.41	23° 29.5′	.39861	.91712	.43463
.42	24° 03.9′	.40776	.91309	.44657
.43	24° 38.2′	.41687	.90897	.45862
.44	25° 12.6′	.42594	.90475	.47078
.45	25° 47.0′	.43497	.90045	.48305
.46	26° 21.4′	.44395	.89605	.49545
.47	26° 55.7′	.45289	.89157	.50795
.48	27° 30.1′	.46178	.88699	.52061
.49	28° 04.5′	.47063	.88233	.53339
.50	28° 38.9′	.47943	.87758	.54630
.51	29° 13.3′	.48818	.87274	.55936
.52	29° 47.6′	.49688	.86782	.57256
.53	30° 22.0′	.50553	.86281	.58592
.54	30° 56.4′	.51414	.85771	.59943
.55	31° 30.8′	.52269	.85252	.61311
.56	32° 05.1′	.53119	.84726	.62695
.57	32° 39.5′	.53963	.84190	.64097
.58	33° 13.9′	.54802	.83646	.65517
.59	33° 48.3′	.55636	.83094	.66956
.60	34° 22.6′	.56464	.82534	.68414

Angle Rad.	Degrees	Sin	Cos	Tan
.60	34° 22.6′	.56464	.82534	.68414
.61	34° 57.0′	.57287	.81965	.69892
.62	35° 31.4′	.58104	.81388	.71391
.63	36° 05.8′	.58914	.80803	.72911
.64	36° 40.2′	.59720	.80210	.74454
.65	37° 14.5′	.60519	.79608	.76020
.66	37° 48.9′	.61312	.78999	.77610
.67	38° 23.3′	.62099	.78382	.79225
.68	38° 57.7′	.62879	.77757	.80866
.69	39° 32.0′	.63654	.77125	.82533
.70	40° 06.4′	.64422	.76484	.84229
.71	40° 40.8′	.65183	.75836	.85953
.72	41° 15.2′	.65938	.75181	.87707
.73	41° 49.6′	.66687	.74517	.89492
.74	42° 23.9′	.67429	.73847	.91309
.75	42° 58.3′	.68164	.73169	.93160
.76	43° 32.7′	.68892	.72484	.95055
.77	44° 07.1′	.69614	.71791	.96967
.78	44° 41.4′	.70328	.71091	.98926
.79	45° 15.8′	.71035	.70385	1.0092
.80	45° 50.2′	.71736	.69671	1.0296
.81	46° 24.6′	.72429	.68950	1.0505
.82	46° 59.0′	.73115	.68222	1.0717
.83	47° 33.3′	.73793	.67488	1.0934
.84	48° 07.7′	.74464	.66746	1.1156
.85	48° 42.1′	.75128	65998	1.1383
.86	49° 16.5′	.75784	.65244	1.1616
.87	49° 50.8′	.76433	.64483	1.1853
.88	50° 25.2′	.77074	.63715	1.2097
.89	50° 59.6′	.77707	.62941	1.2346
.90	51° 34.0′	.78333	.62161	1.2602
.91	52° 08.3′	.78950	.61375	1.2864
.92	52° 42.7′	.79560	.60582	1.3133
.93	53° 17.1′	.80162	.59783	1.3409
.94	53° 51.5′	.80756	.58979	1.3692
.95	54° 25.9′	.81342	.58168	1.3984
.96	55° 00.2′	.81919	.57352	1.4284
.97	55° 34.6′	.82489	.56530	1.4592
.98	56° 09.0′	.83050	.55702	1.4910
.99	56° 43.4′	.83603	.54869	1.5237
1.00	57° 17.7′	.84147	.54030	1.5574
1.01	57° 52.1′	.84683	.53186	1.5922
1.02	58° 26.5′	.85211	.52337	1.6281
1.03	59° 00.9′	.85730	.51482	1.6652
1.04	59° 35.3′	.86240	.50622	1.7036
1.05	60° 09.6′	.86742	.49757	1.7433
1.06	60° 44.0′	.87236	.48887	1.7844
1.07	61° 18.4′	.87720	.48012	1.8270
1.08	61° 52.8′	.88196	.47133	1.8712
1.09	62° 27.1′	.88663	.46249	1.9171
1.10	63° 01.5′	.89121	.45360	1.9648
1.11	63° 35.9′	.89570	.44466	2.0143
1.12	64° 10.3′	.90010	.43568	2.0660
1.13	64° 44.7′	.90441	.42666	2.1198
1.14	65° 19.0′	.90863	.41759	2.1759
1.15	65° 53.4′	.91276	.40849	2.2345
1.16	66° 27.8′	.91680	.39934	2.2958
1.17	67° 02.2′	.92075	.39015	2.3600
1.18	67° 36.5′	.92461	.38092	2.4273
1.19	68° 10.9′	.92837	.37166	2.4979
1.20	68° 45.3′	.93204	.36236	2.5722

Angle		Sin	Cos	Tan	Angle		Sin	Cos	Tan
Rad.	Degrees				Rad.	Degrees			
1.20	68° 45.3′	.93204	.36236	2.5722	**1.40**	80° 12.8′	.98545	.16997	5.7979
1.21	69° 19.7′	.93562	.35302	2.6503	1.41	80° 47.2′	.98710	.16010	6.1654
1.22	69° 54.1′	.93910	.34365	2.7328	1.42	81° 21.6′	.98865	.15023	6.5811
1.23	70° 28.4′	.94249	.33424	2.8198	1.43	81° 56.0′	.99010	.14033	7.0555
1.24	71° 02.8′	.94578	.32480	2.9119	1.44	82° 30.4′	.99146	.13042	7.6018
1.25	71° 37.2′	.94898	.31532	3.0096	**1.45**	83° 04.7′	.99271	.12050	8.2381
1.26	72° 11.6′	.95209	.30582	3.1133	1.46	83° 39.1′	.99387	.11057	8.9886
1.27	72° 45.9′	.95510	.29628	3.2236	1.47	84° 13.5′	.99492	.10063	9.8874
1.28	73° 20.3′	.95802	.28672	3.3413	1.48	84° 47.9′	.99588	.09067	10.983
1.29	73° 54.7′	.96084	.27712	3.4672	1.49	85° 22.2′	.99674	.08071	12.350
1.30	74° 29.1′	.96356	.26750	3.6021	**1.50**	85° 56.6′	.99749	.07074	14.101
1.31	75° 03.4′	.96618	.25785	3.7470	1.51	86° 31.0′	.99815	.06076	16.428
1.32	75° 37.8′	.96872	.24818	3.9033	1.52	87° 05.4′	.99871	.05077	19.670
1.33	76° 12.2′	.97115	.23848	4.0723	1.53	87° 39.8′	.99917	.04079	24.498
1.34	76° 46.6′	.97348	.22875	4.2556	1.54	88° 14.1′	.99953	.03079	32.461
1.35	77° 21.0′	.97572	.21901	4.4552	**1.55**	88° 48.5′	.99978	.02079	48.078
1.36	77° 55.3′	.97786	.20924	4.6734	1.56	89° 22.9′	.99994	.01080	92.621
1.37	78° 29.7′	.97991	.19945	4.9131	1.57	89° 57.3′	1.0000	.00080	1255.8
1.38	79° 04.1′	.98185	.18964	5.1774	1.58	90° 31.6′	.99996	−.00920	−108.65
1.39	79° 38.5′	.98370	.17981	5.4707	1.59	91° 06.0′	.99982	−.01920	−52.067
1.40	80° 12.8′	.98545	.16997	5.7979	**1.60**	91° 40.4′	.99957	−.02920	−34.233

Table 4:
Degrees and Minutes in Radians

Degrees in Radians

0°	0.00000	15°	0.26180	30°	0.52360	45°	0.78540	60°	1.04720	75°	1.30900
1	0.01745	16	0.27925	31	0.54105	46	0.80285	61	1.06465	76	1.32645
2	0.03491	17	0.29671	32	0.55851	47	0.82030	62	1.08210	77	1.34390
3	0.05236	18	0.31416	33	0.57596	48	0.83776	63	1.09956	78	1.36136
4	0.06981	19	0.33161	34	0.59341	49	0.85521	64	1.11701	79	1.37881
5	0.08727	20	0.34907	35	0.61087	50	0.87266	65	1.13446	80	1.39626
6	0.10472	21	0.36652	36	0.62832	51	0.89012	66	1.15192	81	1.41372
7	0.12217	22	0.38397	37	0.64577	52	0.90757	67	1.16937	82	1.43117
8	0.13963	23	0.40143	38	0.66323	53	0.92502	68	1.18682	83	1.44862
9	0.15708	24	0.41888	39	0.68068	54	0.94248	69	1.20428	84	1.46608
10	0.17453	25	0.43633	40	0.69813	55	0.95993	70	1.22173	85	1.48353
11	0.19199	26	0.45379	41	0.71558	56	0.97738	71	1.23918	86	1.50098
12	0.20944	27	0.47124	42	0.73304	57	0.99484	72	1.25664	87	1.51844
13	0.22689	28	0.48869	43	0.75049	58	1.01229	73	1.27409	88	1.53589
14	0.24435	29	0.50615	44	0.76794	59	1.02974	74	1.29154	89	1.55334
15	0.26180	30	0.52360	45	0.78540	60	1.04720	75	1.30900	90	1.57080

Minutes in Radians

0′	0.00000	10′	0.00291	20′	0.00582	30′	0.00873	40′	0.01164	50′	0.01454
1	0.00029	11	0.00320	21	0.00611	31	0.00902	41	0.01193	51	0.01484
2	0.00058	12	0.00349	22	0.00640	32	0.00931	42	0.01222	52	0.01513
3	0.00087	13	0.00378	23	0.00669	33	0.00960	43	0.01251	53	0.01542
4	0.00116	14	0.00407	24	0.00698	34	0.00989	44	0.01280	54	0.01571
5	0.00145	15	0.00436	25	0.00727	35	0.01018	45	0.01309	55	0.01600
6	0.00174	16	0.00465	26	0.00756	36	0.01047	46	0.01338	56	0.01629
7	0.00204	17	0.00495	27	0.00785	37	0.01076	47	0.01367	57	0.01658
8	0.00233	18	0.00524	28	0.00814	38	0.01105	48	0.01396	58	0.01687
9	0.00262	19	0.00553	29	0.00844	39	0.01134	49	0.01425	59	0.01716
10	0.00291	20	0.00582	30	0.00873	40	0.01164	50	0.01454	60	0.01745

1 second ≈ 0.0000048481 radians. Therefore to change seconds to radians, multiply by this number.

Table 5:

*Squares,
Square roots,
Reciprocals*

Table 5

n	n^2	$\sqrt{n}$	$\sqrt{10n}$	$1/n$	n	n^2	$\sqrt{n}$	$\sqrt{10n}$	$1/n$
1.00	1.0000	1.00000	3.16228	1.00000	**1.50**	2.2500	1.22474	3.87298	.666667
1.01	1.0201	1.00499	3.17805	.990099	1.51	2.2801	1.22882	3.88587	.662252
1.02	1.0404	1.00995	3.19374	.980392	1.52	2.3104	1.23288	3.89872	.657895
1.03	1.0609	1.01489	3.20936	.970874	1.53	2.3409	1.23693	3.91152	.653595
1.04	1.0816	1.01980	3.22490	.961538	1.54	2.3716	1.24097	3.92428	.649351
1.05	1.1025	1.02470	3.24037	.952381	1.55	2.4025	1.24499	3.93700	.645161
1.06	1.1236	1.02956	3.25576	.943396	1.56	2.4336	1.24900	3.94968	.641026
1.07	1.1449	1.03441	3.27109	.934579	1.57	2.4649	1.25300	3.96232	.636943
1.08	1.1664	1.03923	3.28634	.925926	1.58	2.4964	1.25698	3.97492	.632911
1.09	1.1881	1.04403	3.30151	.917431	1.59	2.5281	1.26095	3.98748	.628931
1.10	1.2100	1.04881	3.31662	.909091	**1.60**	2.5600	1.26491	4.00000	.625000
1.11	1.2321	1.05357	3.33167	.900901	1.61	2.5921	1.26886	4.01248	.621118
1.12	1.2544	1.05830	3.34664	.892857	1.62	2.6244	1.27279	4.02492	.617284
1.13	1.2769	1.06301	3.36155	.884956	1.63	2.6569	1.27671	4.03733	.613497
1.14	1.2996	1.06771	3.37639	.877193	1.64	2.6896	1.28062	4.04969	.609756
1.15	1.3225	1.07238	3.39116	.869565	1.65	2.7225	1.28452	4.06202	.606061
1.16	1.3456	1.07703	3.40588	.862069	1.66	2.7556	1.28841	4.07431	.602410
1.17	1.3689	1.08167	3.42053	.854701	1.67	2.7889	1.29228	4.08656	.598802
1.18	1.3924	1.08628	3.43511	.847458	1.68	2.8224	1.29615	4.09878	.595238
1.19	1.4161	1.09087	3.44964	.840336	1.69	2.8561	1.30000	4.11096	.591716
1.20	1.4400	1.09545	3.46410	.833333	**1.70**	2.8900	1.30384	4.12311	.588235
1.21	1.4641	1.10000	3.47851	.826446	1.71	2.9241	1.30767	4.13521	.584795
1.22	1.4884	1.10454	3.49285	.819672	1.72	2.9584	1.31149	4.14729	.581395
1.23	1.5129	1.10905	3.50714	.813008	1.73	2.9929	1.31529	4.15933	.578035
1.24	1.5376	1.11355	3.52136	.806452	1.74	3.0276	1.31909	4.17133	.574713
1.25	1.5625	1.11803	3.53553	.800000	1.75	3.0625	1.32288	4.18330	.571429
1.26	1.5876	1.12250	3.54965	.793651	1.76	3.0976	1.32665	4.19524	.568182
1.27	1.6129	1.12694	3.56371	.787402	1.77	3.1329	1.33041	4.20714	.564972
1.28	1.6384	1.13137	3.57771	.781250	1.78	3.1684	1.33417	4.21900	.561798
1.29	1.6641	1.13578	3.59166	.775194	1.79	3.2041	1.33791	4.23084	.558659
1.30	1.6900	1.14018	3.60555	.769231	**1.80**	3.2400	1.34164	4.24264	.555556
1.31	1.7161	1.14455	3.61939	.763359	1.81	3.2761	1.34536	4.25441	.552486
1.32	1.7424	1.14891	3.63318	.757576	1.82	3.3124	1.34907	4.26615	.549451
1.33	1.7689	1.15326	3.64692	.751880	1.83	3.3489	1.35277	4.27785	.546448
1.34	1.7956	1.15758	3.66060	.746269	1.84	3.3856	1.35647	4.28952	.543478
1.35	1.8225	1.16190	3.67423	.740741	1.85	3.4225	1.36015	4.30116	.540541
1.36	1.8496	1.16619	3.68782	.735294	1.86	3.4596	1.36382	4.31277	.537634
1.37	1.8769	1.17047	3.70135	.729927	1.87	3.4969	1.36748	4.32435	.534759
1.38	1.9044	1.17473	3.71484	.724638	1.88	3.5344	1.37113	4.33590	.531915
1.39	1.9321	1.17898	3.72827	.719424	1.89	3.5721	1.37477	4.34741	.529101
1.40	1.9600	1.18322	3.74166	.714286	**1.90**	3.6100	1.37840	4.35890	.526316
1.41	1.9881	1.18743	3.75500	.709220	1.91	3.6481	1.38203	4.37035	.523560
1.42	2.0164	1.19164	3.76829	.704225	1.92	3.6864	1.38564	4.38178	.520833
1.43	2.0449	1.19583	3.78153	.699301	1.93	3.7249	1.38924	4.39318	.518135
1.44	2.0736	1.20000	3.79473	.694444	1.94	3.7636	1.39284	4.40454	.515464
1.45	2.1025	1.20416	3.80789	.689655	1.95	3.8025	1.39642	4.41588	.512821
1.46	2.1316	1.20830	3.82099	.684932	1.96	3.8416	1.40000	4.42719	.510204
1.47	2.1609	1.21244	3.83406	.680272	1.97	3.8809	1.40357	4.43847	.507614
1.48	2.1904	1.21655	3.84708	.675676	1.98	3.9204	1.40712	4.44972	.505051
1.49	2.2201	1.22066	3.86005	.671141	1.99	3.9601	1.41067	4.46094	.502513
1.50	2.2500	1.22474	3.87298	.666667	**2.00**	4.0000	1.41421	4.47214	.500000
n	n^2	$\sqrt{n}$	$\sqrt{10n}$	$1/n$	n	n^2	$\sqrt{n}$	$\sqrt{10n}$	$1/n$

n	n^2	$\sqrt{n}$	$\sqrt{10n}$	$1/n$	n	n^2	$\sqrt{n}$	$\sqrt{10n}$	$1/n$
2.00	4.0000	1.41421	4.47214	.500000	**2.50**	6.2500	1.58114	5.00000	.400000
2.01	4.0401	1.41774	4.48330	.497512	2.51	6.3001	1.58430	5.00999	.398406
2.02	4.0804	1.42127	4.49444	.495050	2.52	6.3504	1.58745	5.01996	.396825
2.03	4.1209	1.42478	4.50555	.492611	2.53	6.4009	1.59060	5.02991	.395257
2.04	4.1616	1.42829	4.51664	.490196	2.54	6.4516	1.59374	5.03984	.393701
2.05	4.2025	1.43178	4.52769	.487805	2.55	6.5025	1.59687	5.04975	.392157
2.06	4.2436	1.43527	4.53872	.485437	2.56	6.5536	1.60000	5.05964	.390625
2.07	4.2849	1.43875	4.54973	.483092	2.57	6.6049	1.60312	5.06952	.389105
2.08	4.3264	1.44222	4.56070	.480769	2.58	6.6564	1.60624	5.07937	.387597
2.09	4.3681	1.44568	4.57165	.478469	2.59	6.7081	1.60935	5.08920	.386100
2.10	4.4100	1.44914	4.58258	.476190	**2.60**	6.7600	1.61245	5.09902	.384615
2.11	4.4521	1.45258	4.59347	.473934	2.61	6.8121	1.61555	5.10882	.383142
2.12	4.4944	1.45602	4.60435	.471698	2.62	6.8644	1.61864	5.11859	.381679
2.13	4.5369	1.45945	4.61519	.469434	2.63	6.9169	1.62173	5.12835	.380228
2.14	4.5796	1.46287	4.62601	.467290	2.64	6.9696	1.62481	5.13809	.378788
2.15	4.6225	1.46629	4.63681	.465116	2.65	7.0225	1.62788	5.14782	.377358
2.16	4.6656	1.46969	4.64758	.462963	2.66	7.0756	1.63095	5.15752	.375940
2.17	4.7089	1.47309	4.65833	.460829	2.67	7.1289	1.63401	5.16720	.374532
2.18	4.7524	1.47648	4.66905	.458716	2.68	7.1824	1.63707	5.17687	.373134
2.19	4.7961	1.47986	4.67974	.456621	2.69	7.2361	1.64012	5.18652	.371747
2.20	4.8400	1.48324	4.69042	.454545	**2.70**	7.2900	1.64317	5.19615	.370370
2.21	4.8841	1.48661	4.70106	.452489	2.71	7.3441	1.64621	5.20577	.369004
2.22	4.9284	1.48997	4.71169	.450450	2.72	7.3984	1.64924	5.21536	.367647
2.23	4.9729	1.49332	4.72229	.448430	2.73	7.4529	1.65227	5.22494	.366300
2.24	5.0176	1.49666	4.73286	.446429	2.74	7.5076	1.65529	5.23450	.364964
2.25	5.0625	1.50000	4.74342	.444444	2.75	7.5625	1.65831	5.24404	.363636
2.26	5.1076	1.50333	4.75395	.442478	2.76	7.6176	1.66132	5.25357	.362319
2.27	5.1529	1.50665	4.76445	.440529	2.77	7.6729	1.66433	5.26308	.361011
2.28	5.1984	1.50997	4.77493	.438596	2.78	7.7284	1.66733	5.27257	.359712
2.29	5.2441	1.51327	4.78539	.436681	2.79	7.7841	1.67033	5.28205	.358423
2.30	5.2900	1.51658	4.79583	.434783	**2.80**	7.8400	1.67332	5.29150	.357143
2.31	5.3361	1.51987	4.80625	.432900	2.81	7.8961	1.67631	5.30094	.355872
2.32	5.3824	1.52315	4.81664	.431034	2.82	7.9524	1.67929	5.31037	.354610
2.33	5.4289	1.52643	4.82701	.429185	2.83	8.0089	1.68226	5.31977	.353357
2.34	5.4756	1.52971	4.83735	.427350	2.84	8.0656	1.68523	5.32917	.352113
2.35	5.5225	1.53297	4.84768	.425532	2.85	8.1225	1.68819	5.33854	.350877
2.36	5.5696	1.53623	4.85798	.423729	2.86	8.1796	1.69115	5.34790	.349650
2.37	5.6169	1.53948	4.86826	.421941	2.87	8.2369	1.69411	5.35724	.348432
2.38	5.6644	1.54272	4.87852	.420168	2.88	8.2944	1.69706	5.36656	.347222
2.39	5.7121	1.54596	4.88876	.418410	2.89	8.3521	1.70000	5.37587	.346021
2.40	5.7600	1.54919	4.89898	.416667	**2.90**	8.4100	1.70294	5.38516	.344828
2.41	5.8081	1.55242	4.90918	.414938	2.91	8.4681	1.70587	5.39444	.343643
2.42	5.8564	1.55563	4.91935	.413223	2.92	8.5264	1.70880	5.40370	.342466
2.43	5.9049	1.55885	4.92950	.411523	2.93	8.5849	1.71172	5.41295	.341297
2.44	5.9536	1.56205	4.93964	.409836	2.94	8.6436	1.71464	5.42218	3.40136
2.45	6.0025	1.56525	4.94975	.408163	2.95	8.7025	1.71756	5.43139	.338983
2.46	6.0516	1.56844	4.95984	.406504	2.96	8.7616	1.72047	5.44059	.337838
2.47	6.1009	1.57162	4.96991	.404858	2.97	8.8209	1.72337	5.44977	.336700
2.48	6.1504	1.57480	4.97996	.403226	2.98	8.8804	1.72627	5.45894	.335570
2.49	6.2001	1.57797	4.98999	.401606	2.99	8.9401	1.72916	5.46809	.334448
2.50	6.2500	1.58114	5.00000	.400000	**3.00**	9.0000	1.73205	5.47723	.333333
n	n^2	$\sqrt{n}$	$\sqrt{10n}$	$1/n$	n	n^2	$\sqrt{n}$	$\sqrt{10n}$	$1/n$

n	n^2	$\sqrt{n}$	$\sqrt{10n}$	$1/n$	n	n^2	$\sqrt{n}$	$\sqrt{10n}$	$1/n$
3.00	9.0000	1.73205	5.47723	.333333	**3.50**	12.2500	1.87083	5.91608	.285714
3.01	9.0601	1.73494	5.48635	.332226	3.51	12.3201	1.87350	5.92453	.284900
3.02	9.1204	1.73781	5.49545	.331126	3.52	12.3904	1.87617	5.93296	.284091
3.03	9.1809	1.74069	5.50454	.330033	3.53	12.4609	1.87883	5.94138	.283286
3.04	9.2416	1.74356	5.51362	.328947	3.54	12.5316	1.88149	5.94979	.282486
3.05	9.3025	1.74642	5.52268	.327869	3.55	12.6025	1.88414	5.95819	.281690
3.06	9.3636	1.74929	5.53173	.326797	3.56	12.6736	1.88680	5.96657	.280899
3.07	9.4249	1.75214	5.54076	.325733	3.57	12.7449	1.88944	5.97495	.280112
3.08	9.4864	1.75499	5.54977	.324675	3.58	12.8164	1.89209	5.98331	.279330
3.09	9.5481	1.75784	5.55878	.323625	3.59	12.8881	1.89473	5.99166	.278552
3.10	9.6100	1.76068	5.56776	.322581	**3.60**	12.9600	1.89737	6.00000	.277778
3.11	9.6721	1.76352	5.57674	.321543	3.61	13.0321	1.90000	6.00833	.277008
3.12	9.7344	1.76635	5.58570	.320513	3.62	13.1044	1.90263	6.01664	.276243
3.13	9.7969	1.76918	5.59464	.319489	3.63	13.1769	1.90526	6.02495	.275482
3.14	9.8596	1.77200	5.60357	.318471	3.64	13.2496	1.90788	6.03324	.274725
3.15	9.9225	1.77482	5.61249	.317460	3.65	13.3225	1.91050	6.04152	.273973
3.16	9.9856	1.77764	5.62139	.316456	3.66	13.3956	1.91311	6.04979	.273224
3.17	10.0489	1.78045	5.63028	.315457	3.67	13.4689	1.91572	6.05805	.272480
3.18	10.1124	1.78326	5.63915	.314465	3.68	13.5424	1.91833	6.06630	.271739
3.19	10.1761	1.78606	5.64801	.313480	3.69	13.6161	1.92094	6.07454	.271003
3.20	10.2400	1.78885	5.65685	.312500	**3.70**	13.6900	1.92354	6.08276	.270270
3.21	10.3041	1.79165	5.66569	.311526	3.71	13.7641	1.92614	6.09098	.269542
3.22	10.3684	1.79444	5.67450	.310559	3.72	13.8384	1.92873	6.09918	.268817
3.23	10.4329	1.79722	5.68331	.309598	3.73	13.9129	1.93132	6.10737	.268097
3.24	10.4976	1.80000	5.69210	.308642	3.74	13.9876	1.93391	6.11555	.267380
3.25	10.5625	1.80278	5.70088	.307692	3.75	14.0625	1.93649	6.12372	.266667
3.26	10.6276	1.80555	5.70964	.306748	3.76	14.1376	1.93907	6.13188	.265957
3.27	10.6929	1.80831	5.71839	.305810	3.77	14.2129	1.94165	6.14003	.265252
3.28	10.7584	1.81108	5.72713	.304878	3.78	14.2884	1.94422	6.14817	.264550
3.29	10.8241	1.81384	5.73585	.303951	3.79	14.3641	1.94679	6.15630	.263852
3.30	10.8900	1.81659	5.74456	.303030	**3.80**	14.4400	1.94936	6.16441	.263158
3.31	10.9561	1.81934	5.75326	.302115	3.81	14.5161	1.95192	6.17252	.262467
3.32	11.0224	1.82209	5.76194	.301205	3.82	14.5924	1.95448	6.18061	.261780
3.33	11.0889	1.82483	5.77062	.300300	3.83	14.6689	1.95704	6.18870	.261097
3.34	11.1556	1.82757	5.77927	.299401	3.84	14.7456	1.95959	6.19677	.260417
3.35	11.2225	1.83030	5.78792	.298507	3.85	14.8225	1.96214	6.20484	.259740
3.36	11.2896	1.83303	5.79655	.297619	3.86	14.8996	1.96469	6.21289	.259067
3.37	11.3569	1.83576	5.80517	.296736	3.87	14.9769	1.96723	6.22093	.258398
3.38	11.4244	1.83848	5.81378	.295858	3.88	15.0544	1.96977	6.22896	.257732
3.39	11.4921	1.84120	5.82237	.294985	3.89	15.1321	1.97231	6.23699	.257069
3.40	11.5600	1.84391	5.83095	.294118	**3.90**	15.2100	1.97484	6.24500	.256410
3.41	11.6281	1.84662	5.83952	.293255	3.91	15.2881	1.97737	6.25300	.255754
3.42	11.6964	1.84932	5.84808	.292398	3.92	15.3664	1.97990	6.26099	.255102
3.43	11.7649	1.85203	5.85662	.291545	3.93	15.4449	1.98242	6.26897	.254453
3.44	11.8336	1.85472	5.86515	.290698	3.94	15.5236	1.98494	6.27694	.253807
3.45	11.9025	1.85742	5.87367	.289855	3.95	15.6025	1.98746	6.28490	.253165
3.46	11.9716	1.86011	5.88218	.289017	3.96	15.6816	1.98997	6.29285	.252525
3.47	12.0409	1.86279	5.89067	.288184	3.97	15.7609	1.99249	6.30079	.251889
3.48	12.1104	1.86548	5.89915	.287356	3.98	15.8404	1.99499	6.30872	.251256
3.49	12.1801	1.86815	5.90762	.286533	3.99	15.9201	1.99750	6.31664	.250627
3.50	12.2500	1.87083	5.91608	.285714	**4.00**	16.0000	2.00000	6.32456	.250000
n	n^2	$\sqrt{n}$	$\sqrt{10n}$	$1/n$	n	n^2	$\sqrt{n}$	$\sqrt{10n}$	$1/n$

n	n^2	$\sqrt{n}$	$\sqrt{10n}$	$1/n$	n	n^2	$\sqrt{n}$	$\sqrt{10n}$	$1/n$
4.00	16.0000	2.00000	6.32456	.250000	**4.50**	20.2500	2.12132	6.70820	.222222
4.01	16.0801	2.00250	6.33246	.249377	4.51	20.3401	2.12368	6.71565	.221729
4.02	16.1604	2.00499	6.34035	.248756	4.52	20.4304	2.12603	6.72309	.221239
4.03	16.2409	2.00749	6.34823	.248139	4.53	20.5209	2.12838	6.73053	.220751
4.04	16.3216	2.00998	6.35610	.247525	4.54	20.6116	2.13073	6.73795	.220264
4.05	16.4025	2.01246	6.36396	.246914	4.55	20.7025	2.13307	6.74537	.219780
4.06	16.4836	2.01494	6.37181	.246305	4.56	20.7936	2.13542	6.75278	.219298
4.07	16.5649	2.01742	6.37966	.245700	4.57	20.8849	2.13776	6.76018	.218818
4.08	16.6464	2.01990	6.38749	.245098	4.58	20.9764	2.14009	6.76757	.218341
4.09	16.7281	2.02237	6.39531	.244499	4.59	21.0681	2.14243	6.77495	.217865
4.10	16.8100	2.02485	6.40312	.243902	**4.60**	21.1600	2.14476	6.78233	.217391
4.11	16.8921	2.02731	6.41093	.243309	4.61	21.2521	2.14709	6.78970	.216920
4.12	16.9744	2.02978	6.41872	.242718	4.62	21.3444	2.14942	6.79706	.216450
4.13	17.0569	2.03224	6.42651	.242131	4.63	21.4369	2.15174	6.80441	.215983
4.14	17.1396	2.03470	6.43428	.241546	4.64	21.5296	2.15407	6.81175	.215517
4.15	17.2225	2.03715	6.44205	.240964	4.65	21.6225	2.15639	6.81909	.215054
4.16	17.3056	2.03961	6.44981	.240385	4.66	21.7156	2.15870	6.82642	.214592
4.17	17.3889	2.04206	6.45755	.239808	4.67	21.8089	2.16102	6.83374	.214133
4.18	17.4724	2.04450	6.46529	.239234	4.68	21.9024	2.16333	6.84105	.213675
4.19	17.5561	2.04695	6.47302	.238663	4.69	21.9961	2.16564	6.84836	.213220
4.20	17.6400	2.04939	6.48074	.238095	**4.70**	22.0900	2.16795	6.85565	.212766
4.21	17.7241	2.05183	6.48845	.237530	4.71	22.1841	2.17025	6.86294	.212314
4.22	17.8084	2.05426	6.49615	.236967	4.72	22.2784	2.17256	6.87023	.211864
4.23	17.8929	2.05670	6.50384	.236407	4.73	22.3729	2.17486	6.87750	.211416
4.24	17.9776	2.05913	6.51153	.235849	4.74	22.4676	2.17715	6.88477	.210970
4.25	18.0625	2.06155	6.51920	.235294	4.75	22.5625	2.17945	6.89202	.210526
4.26	18.1476	2.06398	6.52687	.234742	4.76	22.6576	2.18174	6.89928	.210084
4.27	18.2329	2.06640	6.53452	.234192	4.77	22.7529	2.18403	6.90652	.209644
4.28	18.3184	2.06882	6.54217	.233645	4.78	22.8484	2.18632	6.91375	.209205
4.29	18.4041	2.07123	6.54981	.233100	4.79	22.9441	2.18861	6.92098	.208768
4.30	18.4900	2.07364	6.55744	.232558	**4.80**	23.0400	2.19089	6.92820	.208333
4.31	18.5761	2.07605	6.56506	.232019	4.81	23.1361	2.19317	6.93542	.207900
4.32	18.6624	2.07846	6.57267	.231481	4.82	23.2324	2.19545	6.94262	.207469
4.33	18.7489	2.08087	6.58027	.230947	4.83	23.3289	2.19773	6.94982	.207039
4.34	18.8356	2.08327	6.58787	.230415	4.84	23.4256	2.20000	6.95701	.206612
4.35	18.9225	2.08567	6.59545	.229885	4.85	23.5225	2.20227	6.96419	.206186
4.36	19.0096	2.08806	6.60303	.229358	4.86	23.6196	2.20454	6.97137	.205761
4.37	19.0969	2.09045	6.61060	.228833	4.87	23.7169	2.20681	6.97854	.205339
4.38	19.1844	2.09284	6.61816	.228311	4.88	23.8144	2.20907	6.98570	.204918
4.39	19.2721	2.09523	6.62571	.227790	4.89	23.9121	2.21133	6.99285	.204499
4.40	19.3600	2.09762	6.63325	.227273	**4.90**	24.0100	2.21359	7.00000	.204082
4.41	19.4481	2.10000	6.64078	.226757	4.91	24.1081	2.21585	7.00714	.203666
4.42	19.5364	2.10238	6.64831	.226244	4.92	24.2064	2.21811	7.01427	.203252
4.43	19.6249	2.10476	6.65582	.225734	4.93	24.3049	2.22036	7.02140	.202840
4.44	19.7136	2.10713	6.66333	.225225	4.94	24.4036	2.22261	7.02851	.202429
4.45	19.8025	2.10950	6.67083	.224719	4.95	24.5025	2.22486	7.03562	.202020
4.46	19.8916	2.11187	6.67832	.224215	4.96	24.6016	2.22711	7.04273	.201613
4.47	19.9809	2.11424	6.68581	.223714	4.97	24.7009	2.22935	7.04982	.201207
4.48	20.0704	2.11660	6.69328	.223214	4.98	24.8004	2.23159	7.05691	.200803
4.49	20.1601	2.11896	6.70075	.222717	4.99	24.9001	2.23383	7.06399	.200401
4.50	20.2500	2.12132	6.70820	.222222	**5.00**	25.0000	2.23607	7.07107	.200000
n	n^2	$\sqrt{n}$	$\sqrt{10n}$	$1/n$	n	n^2	$\sqrt{n}$	$\sqrt{10n}$	$1/n$

Table 5

n	n^2	$\sqrt{n}$	$\sqrt{10n}$	$1/n$	n	n^2	$\sqrt{n}$	$\sqrt{10n}$	$1/n$
5.00	25.0000	2.23607	7.07107	.200000	**5.50**	30.2500	2.34521	7.41620	.181818
5.01	25.1001	2.23830	7.07814	.199601	5.51	30.3601	2.34734	7.42294	.181488
5.02	25.2004	2.24054	7.08520	.199203	5.52	30.4704	2.34947	7.42967	.181159
5.03	25.3009	2.24277	7.09225	.198807	5.53	30.5809	2.35160	7.43640	.180832
5.04	25.4016	2.24499	7.09930	.198413	5.54	30.6916	2.35372	7.44312	.180505
5.05	25.5025	2.24722	7.10634	.198020	5.55	30.8025	2.35584	7.44983	.180180
5.06	25.6036	2.24944	7.11337	.197628	5.56	30.9136	2.35797	7.45654	.179856
5.07	25.7049	2.25167	7.12039	.197239	5.57	31.0249	2.36008	7.46324	.179533
5.08	25.8064	2.25389	7.12741	.196850	5.58	31.1364	2.36220	7.46994	.179211
5.09	25.9081	2.25610	7.13442	.196464	5.59	31.2481	2.36432	7.47663	.178891
5.10	26.0100	2.25832	7.14143	.196078	**5.60**	31.3600	2.36643	7.48331	.178571
5.11	26.1121	2.26053	7.14843	.195695	5.61	31.4721	2.36854	7.48999	.178253
5.12	26.2144	2.26274	7.15542	.195312	5.62	31.5844	2.37065	7.49667	.177936
5.13	26.3169	2.26495	7.16240	.194932	5.63	31.6969	2.37276	7.50333	.177620
5.14	26.4196	2.26716	7.16938	.194553	5.64	31.8096	2.37487	7.50999	.177305
5.15	26.5225	2.26936	7.17635	.194175	5.65	31.9225	2.37697	7.51665	.176991
5.16	26.6256	2.27156	7.18331	.193798	5.66	32.0356	2.37908	7.52330	.176678
5.17	26.7289	2.27376	7.19027	.193424	5.67	32.1489	2.38118	7.52994	.176367
5.18	26.8324	2.27596	7.19722	.193050	5.68	32.2624	2.38328	7.53658	.176056
5.19	26.9361	2.27816	7.20417	.192678	5.69	32.3761	2.38537	7.54321	.175747
5.20	27.0400	2.28035	7.21110	.192308	**5.70**	32.4900	2.38747	7.54983	.175439
5.21	27.1441	2.28254	7.21803	.191939	5.71	32.6041	2.38956	7.55645	.175131
5.22	27.2484	2.28473	7.22496	.191571	5.72	32.7184	2.39165	7.56307	.174825
5.23	27.3529	2.28692	7.23187	.191205	5.73	32.8329	2.39374	7.56968	.174520
5.24	27.4576	2.28910	7.23878	.190840	5.74	32.9476	2.39583	7.57628	.174216
5.25	27.5625	2.29129	7.24569	.190476	5.75	33.0625	2.39792	7.58288	.173913
5.26	27.6676	2.29347	7.25259	.190114	5.76	33.1776	2.40000	7.58947	.173611
5.27	27.7729	2.29565	7.25948	.189753	5.77	33.2929	2.40208	7.59605	.173310
5.28	27.8784	2.29783	7.26636	.189394	5.78	33.4084	2.40416	7.60263	.173010
5.29	27.9841	2.30000	7.27324	.189036	5.79	33.5241	2.40624	7.60920	.172712
5.30	28.0900	2.30217	7.28011	.188679	**5.80**	33.6400	2.40832	7.61577	.172414
5.31	28.1961	2.30434	7.28697	.188324	5.81	33.7561	2.41039	7.62234	.172117
5.32	28.3024	2.30651	7.29383	.187970	5.82	33.8724	2.41247	7.62889	.171821
5.33	28.4089	2.30868	7.30068	.187617	5.83	33.9889	2.41454	7.63544	.171527
5.34	28.5156	2.31084	7.30753	.187266	5.84	34.1056	2.41661	7.64199	.171233
5.35	28.6225	2.31301	7.31437	.186916	5.85	34.2225	2.41868	7.64853	.170940
5.36	28.7296	2.31517	7.32120	.186567	5.86	34.3396	2.42074	7.65506	.170649
5.37	28.8369	2.31733	7.32803	.186220	5.87	34.4569	2.42281	7.66159	.170358
5.38	28.9444	2.31948	7.33485	.185874	5.88	34.5744	2.42487	7.66812	.170068
5.39	29.0521	2.32164	7.34166	.185529	5.89	34.6921	2.42693	7.67463	.169779
5.40	29.1600	2.32379	7.34847	.185185	**5.90**	34.8100	2.42899	7.68115	.169492
5.41	29.2681	2.32594	7.35527	.184843	5.91	34.9281	2.43105	7.68765	.169205
5.42	29.3764	2.32809	7.36206	.184502	5.92	35.0464	2.43311	7.69415	.168919
5.43	29.4849	2.33024	7.36885	.184162	5.93	35.1649	2.43516	7.70065	.168634
5.44	29.5936	2.33238	7.37564	.183824	5.94	35.2836	2.43721	7.70714	.168350
5.45	29.7025	2.33452	7.38241	.183486	5.95	35.4025	2.43926	7.71362	.168067
5.46	29.8116	2.33666	7.38918	.183150	5.96	35.5216	2.44131	7.72010	.167785
5.47	29.9209	2.33880	7.39594	.182815	5.97	35.6409	2.44336	7.72658	.167504
5.48	30.0304	2.34094	7.40270	.182482	5.98	35.7604	2.44540	7.73305	.167224
5.49	30.1401	2.34307	7.40945	.182149	5.99	35.8801	2.44745	7.73951	.166945
5.50	30.2500	2.34521	7.41620	.181818	**6.00**	36.0000	2.44949	7.74597	.166667
n	n^2	$\sqrt{n}$	$\sqrt{10n}$	$1/n$	n	n^2	$\sqrt{n}$	$\sqrt{10n}$	$1/n$

n	n^2	$\sqrt{n}$	$\sqrt{10n}$	$1/n$	n	n^2	$\sqrt{n}$	$\sqrt{10n}$	$1/n$
6.00	36.0000	2.44949	7.74597	.166667	**6.50**	42.2500	2.54951	8.06226	.153846
6.01	36.1201	2.45153	7.75242	.166389	6.51	42.3801	2.55147	8.06846	.153610
6.02	36.2404	2.45357	7.75887	.166113	6.52	42.5104	2.55343	8.07465	.153374
6.03	36.3609	2.45561	7.76531	.165837	6.53	42.6409	2.55539	8.08084	.153139
6.04	36.4816	2.45764	7.77174	.165563	6.54	42.7716	2.55734	8.08703	.152905
6.05	36.6025	2.45967	7.77817	.165289	6.55	42.9025	2.55930	8.09321	.152672
6.06	36.7236	2.46171	7.78460	.165017	6.56	43.0336	2.56125	8.09938	.152439
6.07	36.8449	2.46374	7.79102	.164745	6.57	43.1649	2.56320	8.10555	.152207
6.08	36.9664	2.46577	7.79744	.164474	6.58	43.2964	2.56515	8.11172	.151976
6.09	37.0881	2.46779	7.80385	.164204	6 59	43.4281	2.56710	8.11788	.151745
6.10	37.2100	2.46982	7.81025	.163934	**6.60**	43.5600	2.56905	8.12404	.151515
6.11	37.3321	2.47184	7.81665	.163666	6.61	43.6921	2.57099	8.13019	.151286
6.12	37.4544	2.47386	7.82304	.163399	6.62	43.8244	2.57294	8.13634	.151057
6.13	37.5769	2.47588	7.82943	.163132	6.63	43.9569	2.57488	8.14248	.150830
6.14	37.6996	2.47790	7.83582	.162866	6.64	44.0896	2.57682	8.14862	.150602
6.15	37.8225	2.47992	7.84219	.162602	6.65	44.2225	2.57876	8.15475	.150376
6.16	37.9456	2.48193	7.84857	.162338	6.66	44.3556	2.58070	8.16088	.150150
6.17	38.0689	2.48395	7.85493	.162075	6.67	44.4889	2.58263	8.16701	.149925
6.18	38.1924	2.48596	7.86130	.161812	6.68	44.6224	2.58457	8.17313	.149701
6.19	38.3161	2.48797	7.86766	.161551	6.69	44.7561	2.58650	8.17924	.149477
6.20	38.4400	2.48998	7.87401	.161290	**6.70**	44.8900	2.58844	8.18535	.149254
6.21	38.5641	2.49199	7.88036	.161031	6.71	45.0241	2.59037	8.19146	.149031
6.22	38.6884	2.49399	7.88670	.160772	6.72	45.1584	2.59230	8.19756	.148810
6.23	38.8129	2.49600	7.89303	.160514	6.73	45.2929	2.59422	8.20366	.148588
6.24	38.9376	2.49800	7.89937	.160256	6.74	45.4276	2.59615	8.20975	.148368
6.25	39.0625	2.50000	7.90569	.160000	6.75	45.5625	2.59808	8.21584	.148148
6.26	39.1876	2.50200	7.91202	.159744	6.76	45.6976	2.60000	8.22192	.147929
6.27	39.3129	2.50400	7.91833	.159490	6.77	45.8329	2.60192	8.22800	.147710
6.28	39.4384	2.50599	7.92465	.159236	6.78	45.9684	2.60384	8.23408	.147493
6.29	39.5641	2.50799	7.93095	.158983	6.79	46.1041	2.60576	8.24015	.147275
6.30	39.6900	2.50998	7.93725	.158730	**6.80**	46.2400	2.60768	8.24621	.147059
6.31	39.8161	2.51197	7.94355	.158479	6.81	46.3761	2.60960	8.25227	.146843
6.32	39.9424	2.51396	7.94984	.158228	6.82	46.5124	2.61151	8.25833	.146628
6.33	40.0689	2.51595	7.95613	.157978	6.83	46.6489	2.61343	8.26438	.146413
6.34	40.1956	2.51794	7.96241	.157729	6.84	46.7856	2.61534	8.27043	.146199
6.35	40.3225	2.51992	7.96869	.157480	6.85	46.9225	2.61725	8.27647	.145985
6 36	40.4496	2.52190	7.97496	.157233	6.86	47.0596	2.61916	8.28251	.145773
6.37	40.5769	2.52389	7.98123	.156986	6.87	47.1969	2.62107	8.28855	.145560
6.38	40.7044	2.52587	7.98749	.156740	6.88	47.3344	2.62298	8.29458	.145349
6.39	40.8321	2.52784	7.99375	.156495	6.89	47.4721	2.62488	8.30060	.145138
6.40	40.9600	2.52982	8.00000	.156250	**6.90**	47.6100	2.62679	8.30662	.144928
6.41	41.0881	2.53180	8.00625	.156006	6.91	47.7481	2.62869	8.31264	.144718
6.42	41.2164	2.53377	8.01249	.155763	6.92	47.8864	2.63059	8.31865	.144509
6.43	41.3449	2.53574	8.01873	.155521	6.93	48.0249	2.63249	8.32466	.144300
6.44	41.4736	2.53772	8.02496	.155280	6.94	48.1636	2.63439	8.33067	.144092
6.45	41.6025	2.53969	8.03119	.155039	6.95	48.3025	2.63629	8.33667	.143885
6.46	41.7316	2.54165	8.03741	.154799	6.96	48.4416	2.63818	8.34266	.143678
6.47	41.8609	2.54362	8.04363	.154560	6.97	48.5809	2.64008	8.34865	.143472
6.48	41.9904	2.54558	8.04984	.154321	6.98	48.7204	2.64197	8.35464	.143266
6.49	42.1201	2.54755	8.05605	.154083	6.99	48.8601	2.64386	8.36062	.143062
6.50	42.2500	2.54951	8.06226	.153846	**7.00**	49.0000	2.64575	8.36660	.142857
n	n^2	$\sqrt{n}$	$\sqrt{10n}$	$1/n$	n	n^2	$\sqrt{n}$	$\sqrt{10n}$	$1/n$

Table 5

n	n²	√n	√10n	1/n	n	n²	√n	√10n	1/n
7.00	49.0000	2.64575	8.36660	.142857	7.50	56.2500	2.73861	8.66025	.133333
7.01	49.1401	2.64764	8.37257	.142653	7.51	56.4001	2.74044	8.66603	.133156
7.02	49.2804	2.64953	8.37854	.142450	7.52	56.5504	2.74226	8.67179	.132979
7.03	49.4209	2.65141	8.38451	.142248	7.53	56.7009	2.74408	8.67756	.132802
7.04	49.5616	2.65330	8.39047	.142045	7.54	56.8516	2.74591	8.68332	.132626
7.05	49.7025	2.65518	8.39643	.141844	7.55	57.0025	2.74773	8.68907	.132450
7.06	49.8436	2.65707	8.40238	.141643	7.56	57.1536	2.74955	8.69483	.132275
7.07	49.9849	2.65895	8.40833	.141443	7.57	57.3049	2.75136	8.70057	.132100
7.08	50.1264	2.66083	8.41427	.141243	7.58	57.4564	2.75318	8.70632	.131926
7.09	50.2681	2.66271	8.42021	.141044	7.59	57.6081	2.75500	8.71206	.131752
7.10	50.4100	2.66458	8.42615	.140845	7.60	57.7600	2.75681	8.71780	.131579
7.11	50.5521	2.66646	8.43208	.140647	7.61	57.9121	2.75862	8.72353	.131406
7.12	50.6944	2.66833	8.43801	.140449	7.62	58.0644	2.76043	8.72926	.131234
7.13	50.8369	2.67021	8.44393	.140252	7.63	58.2169	2.76225	8.73499	.131062
7.14	50.9796	2.67208	8.44985	.140056	7.64	58.3696	2.76405	8.74071	.130890
7.15	51.1225	2.67395	8.45577	.139860	7.65	58.5225	2.76586	8.74643	.130719
7.16	51.2656	2.67582	8.46168	.139665	7.66	58.6756	2.76767	8.75214	.130548
7.17	51.4089	2.67769	8.46759	.139470	7.67	58.8289	2.76948	8.75785	.130378
7.18	51.5524	2.67955	8.47349	.139276	7.68	58.9824	2.77128	8.76356	.130208
7.19	51.6961	2.68142	8.47939	.139082	7.69	59.1361	2.77308	8.76926	.130039
7.20	51.8400	2.68328	8.48528	.138889	7.70	59.2900	2.77489	8.77496	.129870
7.21	51.9841	2.68514	8.49117	.138696	7.71	59.4441	2.77669	8.78066	.129702
7.22	52.1284	2.68701	8.49706	.138504	7.72	59.5984	2.77849	8.78635	.129534
7.23	52.2729	2.68887	8.50294	.138313	7.73	59.7529	2.78029	8.79204	.129366
7.24	52.4176	2.69072	8.50882	.138122	7.74	59.9076	2.78209	8.79773	.129199
7.25	52.5625	2.69258	8.51469	.137931	7.75	60.0625	2.78388	8.80341	.129032
7.26	52.7076	2.69444	8.52056	.137741	7.76	60.2176	2.78568	8.80909	.128866
7.27	52.8529	2.69629	8.52643	.137552	7.77	60.3729	2.78747	8.81476	.128700
7.28	52.9984	2.69815	8.53229	.137363	7.78	60.5284	2.78927	8.82043	.128535
7.29	53.1441	2.70000	8.53815	.137174	7.79	60.6841	2.79106	8.82610	.128370
7.30	53.2900	2.70185	8.54400	.136986	7.80	60.8400	2.79285	8.83176	.128205
7.31	53.4361	2.70370	8.54985	.136799	7.81	60.9961	2.79464	8.83742	.128041
7.32	53.5824	2.70555	8.55570	.136612	7.82	61.1524	2.79643	8.84308	.127877
7.33	53.7289	2.70740	8.56154	.136426	7.83	61.3089	2.79821	8.84873	.127714
7.34	53.8756	2.70924	8.56738	.136240	7.84	61.4656	2.80000	8.85438	.127551
7.35	54.0225	2.71109	8.57321	.136054	7.85	61.6225	2.80179	8.86002	.127389
7.36	54.1696	2.71293	8.57904	.135870	7.86	61.7796	2.80357	8.86566	.127226
7.37	54.3169	2.71477	8.58487	.135685	7.87	61.9369	2.80535	8.87130	.127065
7.38	54.4644	2.71662	8.59069	.135501	7.88	62.0944	2.80713	8.87694	.126904
7.39	54.6121	2.71846	8.59651	.135318	7.89	62.2521	2.80891	8.88257	.126743
7.40	54.7600	2.72029	8.60233	.135135	7.90	62.4100	2.81069	8.88819	.126582
7.41	54.9081	2.72213	8.60814	.134953	7.91	62.5681	2.81247	8.89382	.126422
7.42	55.0564	2.72397	8.61394	.134771	7.92	62.7264	2.81425	8.89944	.126263
7.43	55.2049	2.72580	8.61974	.134590	7.93	62.8849	2.81603	8.90505	.126103
7.44	55.3536	2.72764	8.62554	.134409	7.94	63.0436	2.81780	8.91067	.125945
7.45	55.5025	2.72947	8.63134	.134228	7.95	63.2025	2.81957	8.91628	.125786
7.46	55.6516	2.73130	8.63713	.134048	7.96	63.3616	2.82135	8.92188	.125628
7.47	55.8009	2.73313	8.64292	.133869	7.97	63.5209	2.82312	8.92749	.125471
7.48	55.9504	2.73496	8.64870	.133690	7.98	63.6804	2.82489	8.93308	.125313
7.49	56.1001	2.73679	8.65448	.133511	7.99	63.8401	2.82666	8.93868	.125156
7.50	56.2500	2.73861	8.66025	.133333	8.00	64.0000	2.82843	8.94427	.125000
n	n²	√n	√10n	1/n	n	n²	√n	√10n	1/n

n	n^2	$\sqrt{n}$	$\sqrt{10n}$	$1/n$	n	n^2	$\sqrt{n}$	$\sqrt{10n}$	$1/n$
8.00	64.0000	2.82843	8.94427	.125000	**8.50**	72.2500	2.91548	9.21954	.117647
8.01	64.1601	2.83019	8.94986	.124844	8.51	72.4201	2.91719	9.22497	.117509
8.02	64.3204	2.83196	8.95545	.124688	8.52	72.5904	2.91890	9.23038	.117371
8.03	64.4809	2.83373	8.96103	.124533	8.53	72.7609	2.92062	9.23580	.117233
8.04	64.6416	2.83549	8.96660	.124378	8.54	72.9316	2.92233	9.24121	.117096
8.05	64.8025	2.83725	8.97218	.124224	8.55	73.1025	2.92404	9.24662	.116959
8.06	64.9636	2.83901	8.97775	.124069	8.56	73.2736	2.92575	9.25203	.116822
8.07	65.1249	2.84077	8.98332	.123916	8.57	73.4449	2.92746	9.25743	.116686
8.08	65.2864	2.84253	8.98888	.123762	8.58	73.6164	2.92916	9.26283	.116550
8.09	65.4481	2.84429	8.99444	.123609	8.59	73.7881	2.93087	9.26823	.116414
8.10	65.6100	2.84605	9.00000	.123457	**8.60**	73.9600	2.93258	9.27362	.116279
8.11	65.7721	2.84781	9.00555	.123305	8.61	74.1321	2.93428	9.27901	.116144
8.12	65.9344	2.84956	9.01110	.123153	8.62	74.3044	2.93598	9.28440	.116009
8.13	66.0969	2.85132	9.01665	.123001	8.63	74.4769	2.93769	9.28978	.115875
8.14	66.2596	2.85307	9.02219	.122850	8.64	74.6496	2.93939	9.29516	.115741
8.15	66.4225	2.85482	9.02774	.122699	8.65	74.8225	2.94109	9.30054	.115607
8.16	66.5856	2.85657	9.03327	.122549	8.66	74.9956	2.94279	9.30591	.115473
8.17	66.7489	2.85832	9.03881	.122399	8.67	75.1689	2.94449	9.31128	.115340
8.18	66.9124	2.86007	9.04434	.122249	8.68	75.3424	2.94618	9.31665	.115207
8.19	67.0761	2.86182	9.04986	.122100	8.69	75.5161	2.94788	9.32202	.115075
8.20	67.2400	2.86356	9.05539	.121951	**8.70**	75.6900	2.94958	9.32738	.114943
8.21	67.4041	2.86531	9.06091	.121803	8.71	75.8641	2.95127	9.33274	.114811
8.22	67.5684	2.86705	9.06642	.121655	8.72	76.0384	2.95296	9.33809	.114679
8.23	67.7329	2.86880	9.07193	.121507	8.73	76.2129	2.95466	9.34345	.114548
8.24	67.8976	2.87054	9.07744	.121359	8.74	76.3876	2.95635	9.34880	.114416
8.25	68.0625	2.87228	9.08295	.121212	8.75	76.5625	2.95804	9.35414	.114286
8.26	68.2276	2.87402	9.08845	.121065	8.76	76.7376	2.95973	9.35949	.114155
8.27	68.3929	2.87576	9.09395	.120919	8.77	76.9129	2.96142	9.36483	.114025
8.28	68.5584	2.87750	9.09945	.120773	8.78	77.0884	2.96311	9.37017	.113895
8.29	68.7241	2.87924	9.10494	.120627	8.79	77.2641	2.96479	9.37550	.113766
8.30	68.8900	2.88097	9,11043	.120482	**8.80**	77.4400	2.96648	9.38083	.113636
8.31	69.0561	2.88271	9.11592	.120337	8.81	77.6161	2.96816	9.38616	.113507
8.32	69.2224	2.88444	9.12140	.120192	8.82	77.7924	2.96985	9.39149	.113379
8.33	69.3889	2.88617	9.12688	.120048	8.83	77.9689	2.97153	9.39681	.113250
8.34	69.5556	2.88791	9.13236	.119904	8.84	78.1456	2.97321	9.40213	.113122
8.35	69.7225	2.88964	9.13783	.119760	8.85	78.3225	2.97489	9.40744	.112994
8.36	69.8896	2.89137	9.14330	.119617	8.86	78.4996	2.97658	9.41276	.112867
8.37	70.0569	2.89310	9.14877	.119474	8.87	78.6769	2.97825	9.41807	.112740
8.38	70.2244	2.89482	9.15423	.119332	8.88	78.8544	2.97993	9.42338	.112613
8.39	70.3921	2.89655	9.15969	.119190	8.89	79.0321	2.98161	9.42868	.112486
8.40	70.5600	2.89828	9.16515	.119048	**8.90**	79.2100	2.98329	9.43398	.112360
8.41	70.7281	2.90000	9.17061	.118906	8.91	79.3881	2.98496	9.43928	.112233
8.42	70.8964	2.90172	9.17606	.118765	8.92	79.5664	2.98664	9.44458	.112108
8.43	71.0649	2.90345	9.18150	.118624	8.93	79.7449	2.98831	9.44987	.111982
8.44	71.2336	2.90517	9.18695	.118483	8.94	79.9236	2.98998	9.45516	.111857
8.45	71.4025	2.90689	9.19239	.118343	8.95	80.1025	2.99166	9.46044	.111732
8.46	71.5716	2.90861	9.19783	.118203	8.96	80.2816	2.99333	9.46573	.111607
8.47	71.7409	2.91033	9.20326	.118064	8.97	80.4609	2.99500	9.47101	.111483
8.48	71.9104	2.91204	9.20869	.117925	8.98	80.6404	2.99666	9.47629	.111359
8.49	72.0801	2.91376	9.21412	.117786	8.99	80.8201	2.99833	9.48156	.111235
8.50	72.2500	2.91548	9.21954	.117647	**9.00**	81.0000	3.00000	9.48683	.111111
n	n^2	$\sqrt{n}$	$\sqrt{10n}$	$1/n$	n	n^2	$\sqrt{n}$	$\sqrt{10n}$	$1/n$

n	n^2	$\sqrt{n}$	$\sqrt{10n}$	$1/n$
9.00	81.0000	3.00000	9.48683	.111111
9.01	81.1801	3.00167	9.49210	.110988
9.02	81.3604	3.00333	9.49737	.110865
9.03	81.5409	3.00500	9.50263	.110742
9.04	81.7216	3.00666	9.50789	.110619
9.05	81.9025	3.00832	9.51315	.110497
9.06	82.0836	3.00998	9.51840	.110375
9.07	82.2649	3.01164	9.52365	.110254
9.08	82.4464	3.01330	9.52890	.110132
9.09	82.6281	3.01496	9.53415	.110011
9.10	82.8100	3.01662	9.53939	.109890
9.11	82.9921	3.01828	9.54463	.109769
9.12	83.1744	3.01993	9.54987	.109649
9.13	83.3569	3.02159	9.55510	.109529
9.14	83.5396	3.02324	9.56033	.109409
9.15	83.7225	3.02490	9.56556	.109290
9.16	83.9056	3.02655	9.57079	.109170
9.17	84.0889	3.02820	9.57601	.109051
9.18	84.2724	3.02985	9.58123	.108932
9.19	84.4561	3.03150	9.58645	.108814
9.20	84.6400	3.03315	9.59166	.108696
9.21	84.8241	3.03480	9.59687	.108578
9.22	85.0084	3.03645	9.60208	.108460
9.23	85.1929	3.03809	9.60729	.108342
9.24	85.3776	3.03974	9.61249	.108225
9.25	85.5625	3.04138	9.61769	.108108
9.26	85.7476	3.04302	9.62289	.107991
9.27	85.9329	3.04467	9.62808	.107875
9.28	86.1184	3.04631	9.63328	.107759
9.29	86.3041	3.04795	9.63846	.107643
9.30	86.4900	3.04959	9.64365	.107527
9.31	86.6761	3.05123	9.64883	.107411
9.32	86.8624	3.05287	9.65401	.107296
9.33	87.0489	3.05450	9.65919	.107181
9.34	87.2356	3.05614	9.66437	.107066
9.35	87.4225	3.05778	9.66954	.106952
9.36	87.6096	3.05941	9.67471	.106838
9.37	87.7969	3.06105	9.67988	.106724
9.38	87.9844	3.06268	9.68504	.106610
9.39	88.1721	3.06431	9.69020	.106496
9.40	88.3600	3.06594	9.69536	.106383
9.41	88.5481	3.06757	9.70052	.106270
9.42	88.7364	3.06920	9.70567	.106157
9.43	88.9249	3.07083	9.71082	.106045
9.44	89.1136	3.07246	9.71597	.105932
9.45	89.3025	3.07409	9.72111	.105820
9.46	89.4916	3.07571	9.72625	.105708
9.47	89.6809	3.07734	9.73139	.105597
9.48	89.8704	3.07896	9.73653	.105485
9.49	90.0601	3.08058	9.74166	.105374
9.50	90.2500	3.08221	9.74679	.105263
n	n^2	$\sqrt{n}$	$\sqrt{10n}$	$1/n$

n	n^2	$\sqrt{n}$	$\sqrt{10n}$	$1/n$
9.50	90.2500	3.08221	9.74679	.105263
9.51	90.4401	3.08383	9.75192	.105152
9.52	90.6304	3.08545	9.75705	.105042
9.53	90.8209	3.08707	9.76217	.104932
9.54	91.0116	3.08869	9.76729	.104822
9.55	91.2025	3.09031	9.77241	.104712
9.56	91.3936	3.09192	9.77753	.104603
9.57	91.5849	3.09354	9.78264	.104493
9.58	91.7764	3.09516	9.78775	.104384
9.59	91.9681	3.09677	9.79285	.104275
9.60	92.1600	3.09839	9.79796	.104167
9.61	92.3521	3.10000	9.80306	.104058
9.62	92.5444	3.10161	9.80816	.103950
9.63	92.7369	3.10322	9.81326	.103842
9.64	92.9296	3.10483	9.81835	.103734
9.65	93.1225	3.10644	9.82344	.103627
9.66	93.3156	3.10805	9.82853	.103520
9 67	93.5089	3.10966	9.83362	.103413
9.68	93.7024	3.11127	9.83870	.103306
9.69	93.8961	3.11288	9.84378	.103199
9.70	94.0900	3.11448	9.84886	.103093
9.71	94.2841	3.11609	9.85393	.102987
9.72	94.4784	3.11769	9.85901	.102881
9.73	94.6729	3.11929	9.86408	.102775
9.74	94.8676	3.12090	9.86914	.102669
9.75	95.0625	3.12250	9.87421	.102564
9.76	95.2576	3.12410	9.87927	.102459
9.77	95.4529	3.12570	9.88433	.102354
9.78	95.6484	3.12730	9.88939	.102249
9.79	95.8441	3.12890	9.89444	.102145
9.80	96.0400	3.13050	9.89949	.102041
9.81	96.2361	3.13209	9.90454	.101937
9.82	96.4324	3.13369	9.90959	.101833
9.83	96.6289	3.13528	9.91464	.101729
9.84	96.8256	3.13688	9.91968	.101626
9.85	97.0225	3.13847	9.92472	.101523
9.86	97.2196	3.14006	9.92975	.101420
9.87	97.4169	3.14166	9.93479	.101317
9.88	97.6144	3.14325	9.93982	.101215
9.89	97.8121	3.14484	9.94485	.101112
9.90	98.0100	3.14643	9.94987	.101010
9.91	98.2081	3.14802	9.95490	.100908
9.92	98.4064	3.14960	9.95992	.100806
9.93	98.6049	3.15119	9.96494	.100705
9.94	98.8036	3.15278	9.96995	.100604
9.95	99.0025	3.15436	9.97497	.100503
9.96	99.2016	3.15595	9.97998	.100402
9.97	99.4009	3.15753	9.98499	.100301
9.98	99.6004	3.15911	9.98999	.100200
9.99	99.8001	3.16070	9.99500	.100100
10.00	100.000	3.16228	10.0000	.100000
n	n^2	$\sqrt{n}$	$\sqrt{10n}$	$1/n$

Answers to Odd-numbered Exercises

Exercises 1–8   page 13

1.	2.	**3.**	3.	**5.**	4.
7.	2.	**9.**	1.	**11.**	2.
13.	80°.	**15.**	30°.	**17.**	85°.
19.	85°.	**21.**	40°.	**23.**	20°.
25.	30°.	**27.**	80°.	**29.**	0°.
31.	90°.	**33.**	0°.	**35.**	10°.
37.	10° 15′.	**39.**	40° 10′.	**41.**	39° 35′.
43.	45° 10.3′.	**45.**	59° 49.3′.	**47.**	28° 44.6′.
49.	5.	**51.**	13.	**53.**	5.
55.	5.	**57.**	11.40 18.	**59.**	8.24621.
61.	13.91119.				

Exercises 1–11   page 17

1.	$2.5\sqrt{3}, 2.5$	**3.**	36 and 48.
5.	$y = 4.$	**7.**	$x = -12.$
9.	$y = -\sqrt{3}.$	**11.**	$x = -1.$

Chapter 1: Review Exercises   page 18

1. 50.	**3.** 11.	**5.** 4.	**7.** Q III.
9. Q II.	**11.** Q I.	**13.** Q IV.	**15.** 10°.
17. 60°.	**19.** 90°.	**21.** 13.	**23.** 13.
25. 5.	**27.** 6.70,820.	**29.** $4\sqrt{3}$.	**31.** 7.2 inches.

Chapter 1: Diagnostic Test   page 18

Following each problem number is the number of the textbook section (in parentheses) in which that kind of problem is discussed.

1. (1-2) (a) 27, (b) 5, (c) 18.
2. (1-6) Yes.   **3.** (1-6) Yes.
4. (1-8) (a) 10°, (b) 30°, (c) 90°, (d) 30°, (e) 0°, (f) 30°, (g) 49° 44.4′.
5. (1-6) (a) Q III, (b) Q IV, (c) Q II, (d) Q I.
6. (1-6) 5.   **7.** (1-7) 5.   **8.** (1-10) $3\sqrt{2}$.
9. (1-11) −4.   **10.** (1-9) $2\sqrt{3}$.

Exercises 2-7   page 27

1. $\frac{1}{2}(1 + \sqrt{3})$.	**3.** −2.	**5.** $\frac{1}{2}(\sqrt{2} - 1)$.
7. 1.	**9.** $2\frac{3}{4}$.	**11.** $-\frac{1}{2}$.
13. $\frac{15}{16}$.	**15.** $1\frac{1}{2}$.	**17.** 0.
19. $6\frac{1}{2}$.	**21.** $\frac{4}{5}$.	**23.** $\frac{4}{5}$.
25. $-\frac{1}{2}\sqrt{2}$.	**27.** 0.	

Exercises 2-9   page 31

1. .1736.	**3.** .3420.	**5.** −.6428.
7. −.9848.	**9.** −.1736.	**11.** −.9848.
13. .3529.	**15.** .9822.	**17.** −.9971.
19. .6428.	**21.** .7071.	**23.** 23° 00′.
25. 43° 40′.	**27.** 79° 20′.	**29.** 170° 00′.
31. 123° 50′.	**33.** 213° 00′.	**35.** 244° 10′.
37. 334° 00′.	**39.** 297° 10′.	

Exercises 2–13   page 37

1. 4.	**3.** 4.	**5.** 5.
7. 12.2.	**9.** 4.0.	**11.** 2000.
13. 9.3×10^7.	**15.** 3.147×10^1.	**17.** 1.5×10^{-2}.
19. 4.00×10^3.	**21.** 9.30×10^7.	**23.** .00150 or 1.50×10^{-3}.
25. 2.812.	**27.** 1936.	**29.** 10° 50′.
31. 56° 00′.	**33.** 14 feet.	

Exercises 2–14   page 40

1. $a = 1.02$, $c = 1.17$, $C = 65° 00′$.   **3.** $b = 1.78$, $c = 3.19$, $B = 31° 30′$.
5. 86 feet.   **7.** 263 feet.

Exercises 2–15   page 45

1. $B = 36° 40'$, $C = 102° 20'$, $c = 16.4$.
3. No triangle.
5. $B_1 = 58° 30'$, $C_1 = 78° 30'$, $c_1 = 11.5$.
   $B_2 = 121° 30'$, $C_2 = 15° 30'$, $c_2 = 3.13$.
7. Right triangle. $A = 90°$, $C = 61° 50'$, $b = 4.72$, $c = 8.82$.
9. $A_1 = 81° 50'$, $B_1 = 35° 10'$, $b_1 = 5.82$.
   $A_2 = 98° 10'$, $B_2 = 18° 50'$, $b_2 = 3.26$.

Exercises 2–16   page 48

1. $90°$.
5. $60°$, $120°$.
9. No solution.
13. $0°$, $180°$, $270°$.
17. $30°$, $150°$, $270°$.
21. $0°$, $90°$, $180°$, $210°$, $330°$.
25. $0°$, $90°$, $135°$, $180°$, $270°$, $315°$.

3. $0°$, $180°$.
7. $225°$, $315°$.
11. $210°$, $330°$.
15. $30°$, $150°$, $210°$, $330°$.
19. $270°$.
23. $45°$, $225°$.
27. $15°$, $75°$, $135°$, $195°$, $255°$, $315°$.

Exercises 2–17   page 52

1.

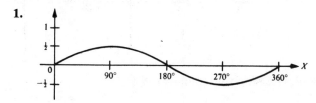

3.

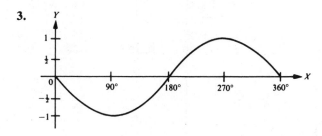

5.

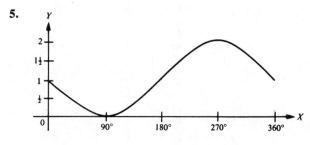

7.

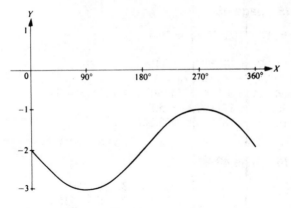

9.

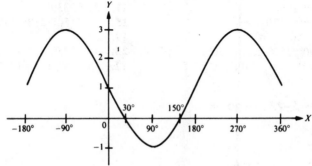

Chapter 2: Review Exercises   page 52

1. 0.	**3.** 1.	**5.** 0.
7. $\frac{3}{5}$.	**9.** .2644.	**11.** .7642.
13. 43° 10′.	**15.** 353° 50′.	**17.** 3.
19. 4.	**21.** .368.	**23.** 69.9 feet.
25. 210°, 330°	**27.** 0°, 30°, 150°, 180°.	

29.

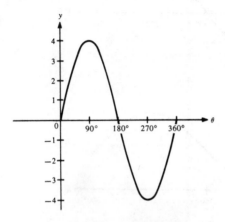

Chapter 2: Diagnostic Test   page 53

Following each problem number is the number of the textbook section (in parentheses) in which that kind of problem is discussed.

1. (2-7) $\frac{12}{13}$.

2. (2-7) (a) $-\dfrac{\sqrt{3}}{2}$, (b) -1, (c) $\frac{1}{2}$.

3. (2-9) (a) .8355, (b) $-.7451$, (c) $-.6670$.

4. (2-9) (a) 135° 50′, (b) 217° 10′, (c) 355° 30′.

5. (2-11) (a) 70.50, (b) .0362, (c) 190.000.

6. (2-10) (a) 3.5×10^3, (b) 6.15×10^{-2}, (c) 8.56×10^1.

7. (2-13) 6.18.

8. (2-14) 99.6.

9. (2-15) None.

10. (2-16) 0°, 180°.

11. (2-17).

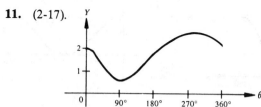

Exercises 3-5   page 61

1. 0.

3. 1.

5. 1.

7. 1.

9. $\sqrt{3} - 1$.

11. $\sqrt{3}$.

13. 2.

15. $\frac{1}{2}(4 + \sqrt{3})$.

17. $-\frac{1}{2}\sqrt{3}$.

19. $6\frac{1}{2}$.

21. $\sin \theta = \frac{3}{5}, \cos \theta = \frac{4}{5}$.

23. $\sin \theta = -\frac{4}{5}, \cos \theta = \frac{3}{5}$.

25. $\sin \theta = \frac{1}{2}\sqrt{3}, \cos \theta = -\frac{1}{2}$.

27. .9848.

29. $-.9397$.

31. $-.7660$.

33. .1736.

35. .9848.

37. $-.1736$.

39. $-.1880$.

41. $-.0843$.

43. .7660.

45. 18° 30′.

47. 69° 10′.

49. 88° 30′.

Exercises 3-6   page 65

1. .9254.

3. .7589.

5. .2580.

7. $-.1086$.

9. .9650.

11. .0724.

13. 19° 13′.

15. 19° 28′.

17. 54° 11′.

19. 135° 57′.

21. 155° 47′.

23. 171° 12′.

25. 0°.

27. 90°, 270°.

29. 30°, 330°.

31. 150°, 210°.

33. 120°, 240°.

35. 90°, 180°, 270°.

37. 60°, 120°, 240°, 300°.

39. 60°, 180°, 300°.

41. 0°.

43. 0°, 90°, 120°, 240°, 270°.

45. 45°, 135°, 225°, 315°.

47. 45°, 90°, 135°, 225°, 270°, 315°.

49. 30°, 90°, 150°, 210°, 270°, 330°.

51. 0°.

53. 30°, 150°, 180°.

Exercises 3-8   page 71

1. $B = 65° 50'$, $a = 8.19$, $b = 18.2$.   **3.** $B = 22° 17'$, $a = 9.253$, $b = 3.792$.
5. 2.76.   **7.** 6.68.
9. $A = 28°$, $B = 41°$, $C = 111°$.   **11.** 4250 feet.
15. 868 miles per hour.

Exercises 3-10   page 76

1. Max. 3, min. 1.   **3.** Max. 2, min. 0.
5. Max. 4, min. 2.   **7.** Max. 2, min. 1.

9. 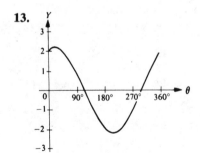   **11.**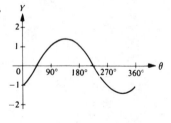

13. 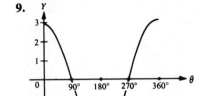   **15.**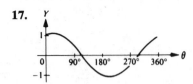

17.

Chapter 3: Review Exercises   page 76

1. $-\frac{1}{2}\sqrt{2}$.   **3.** $1\frac{1}{2}$.   **5.** 1.
7. (a) $\frac{3}{5}$, (b) $-\frac{4}{5}$.   **9.** $-.7707$.   **11.** .9828.
13. $152° 10'$.   **15.** $60°, 90°, 270°, 300°$.
17. $0°, 90°, 180°, 270°$.   **19.** 102 feet.
21. 0.

Chapter 3: Diagnostic Test   page 77

Following each problem number is the number of the textbook section (in parentheses) in which that kind of problem is discussed.

1. (3-2, 3-4, 3-5)   (a) $\frac{1}{2}$, (b) $-\frac{1}{2}\sqrt{3}$, (c) -1, (d) 0, (e) -1, (f) 0, (g) 1, (h) 3.

2. (3-2)   (a) $-\frac{1}{2}$, (b) $\dfrac{\sqrt{3}}{2}$.

3. (3-5)   (a) .9959, (b) .2034, (c) .9189.

4. (3-5, 3-6)   (a) 62° 10′, (b) 125° 40′.

5. (3-6)   90°, 120°, 240°, 270°.

6. (3-6)   0°, 180°.

7. (3-6)   30°, 150°, 210°, 330°.

8. (3-11)   4.

9. (3-9)   26.9.

10. (3-10, 3-11)

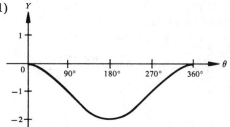

Exercises 4-3   page 83

1. $\frac{1}{3}$.

3. $\frac{1}{3}(4 + 3\sqrt{3})$.

5. $1 - \sqrt{3}$.

7. $1 + \sqrt{2}$.

9. -1.

11. 4.

13. $\frac{1}{4}(2 - \sqrt{2})$.

15. .700.

17. .8192.

19. 11.43.

21. -1.4281.

23. .8391.

25. -1.1778.

27. -1.1918.

29. .5581.

31. III.

33. II.

35. IV.

37. $1\frac{11}{20}$.

39. 20.

41. $1\frac{7}{25}$.

Exercises 4-6   page 87

1. 2.65.

3. 1.10×10^2.

5. 36° 44′.

7. 5° 57′.

9. 3° 50′.

11. 120 feet.

13. 241 feet.

15. 29 miles, *N* 31° *W*.

17. 4910 feet.

19. 365 feet.

21. 45°, 225°.

23. 135°, 315°.

25. 30°, 210°.	**27.** 0°, 180°.
29. 60°, 120°, 240°, 300°.	**31.** 0°, 135°, 180°, 315°.
33. 90°.	**35.** 0°.
37. 135°, 315°.	**39.** 0°, $67\frac{1}{2}°$, 90°, $157\frac{1}{2}°$, 180°, $247\frac{1}{2}°$, 270°,
41. 35° 00′, 215° 00′.	$337\frac{1}{2}°$.
45. 160° 50′, 340° 50′.	**43.** 80° 00′, 260° 00′.

Exercises 4-7   page 94

1. (a) 4, (b) 14, (c) 20, (d) 25, (e) 28.
3. 58 pounds; 29° with 40-pound force and 41° with 30-pound force.
5. Horizontal, 32 pounds; vertical, 45 pounds.
7. 283 pounds, 100 pounds.
9. Horizontal, 57 pounds; vertical, 40 pounds.
11. 173 pounds.
13. Speed, 270 miles per hour; course 10°:
15. 94 miles per hour.

Exercises 4-8   page 96

1.
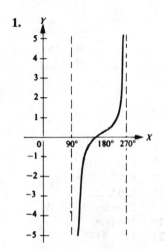

3.
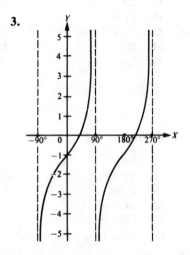

5.

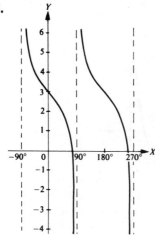

Chapter 4: Review Exercises   page 96

1. −1.   **3.** 2.   **5.** .7865.   **7.** .7212.

9. 127° 45′.   **11.** III.   **13.** $\frac{1}{5}$.   **15.** 7.31.

17. 0°, 45°, 180°, 225°.   **19.** 30°, 150°, 210°, 330°.

21.

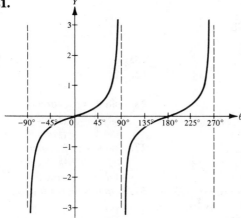

Chapter 4: Diagnostic Test   page 98

Following each problem number is the number of the textbook section (in parentheses) in which that kind of problem is discussed.

1. (4-3) $-\frac{4}{3}$.   **2.** (4-3) (a) 1, (b) $-\sqrt{3}$, (c) 0.

3. (4-3) (a) IV, (b) III.   **4.** (4-3) 7.

5. (4-3) (a) 0°, 60°, 180°, 240°; (b) 45°, 225°.

6. (4-8)

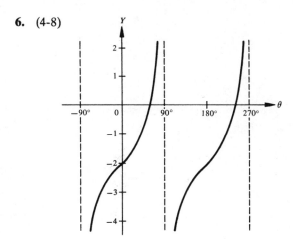

7. (4-3) −.7454.   **8.** (4-3) 35° 45′.   **9.** (4-6) 54° 15′.   **10.** (4-7) 26.5.

Exercises 5-3   page 105

1. 5.   **3.** 2.   **5.** 0.   **7.** 1.

9. 1.   **11.** 2.   **13.** $2\frac{2}{3}$.   **15.** $-\sqrt{3}$.

17. $4\frac{2}{3}$.   **19.** −1.

21. $\sin\theta = \frac{3}{5}$, $\cot\theta = \frac{4}{3}$,   **23.** $\sin\theta = \frac{5}{13}$, $\cot\theta = -\frac{12}{5}$,

      $\cos\theta = \frac{4}{5}$, $\sec\theta = \frac{5}{4}$, $\cos\theta = -\frac{12}{13}$, $\sec\theta = -\frac{13}{12}$,

      $\tan\theta = \frac{3}{4}$, $\csc\theta = \frac{5}{3}$. $\tan\theta = -\frac{5}{12}$, $\csc\theta = \frac{13}{5}$.

25. $\sin\theta = -\frac{3}{5}$, $\cot\theta = -\frac{4}{3}$,   **27.** $\sin\theta = -\frac{5}{13}$, $\cot\theta = \frac{12}{5}$,

      $\cos\theta = \frac{4}{5}$, $\sec\theta = \frac{5}{4}$, $\cos\theta = -\frac{12}{13}$, $\sec\theta = -\frac{13}{12}$,

      $\tan\theta = -\frac{3}{4}$, $\csc\theta = -\frac{5}{3}$. $\tan\theta = \frac{5}{12}$, $\csc\theta = -\frac{13}{5}$.

29. $\sin\theta = \frac{2}{13}\sqrt{13}$, $\cot\theta = -\frac{3}{2}$,   **31.** $\sin\theta = 0$, $\cot\theta$ undefined,

      $\cos\theta = -\frac{3}{13}\sqrt{13}$, $\sec\theta = -\frac{1}{3}\sqrt{13}$, $\cos\theta = -1$, $\sec\theta = -1$,

      $\tan\theta = -\frac{2}{3}$, $\csc\theta = \frac{1}{2}\sqrt{13}$. $\tan\theta = 0$, $\csc\theta$ undefined.

33. III.   **35.** IV.

37. II.   **39.** Impossible.

41. Possible.   **43.** Impossible.

45. Possible.   **47.** Possible.

49. Close to 0°.   **51.** Close to 90°.

53. Close to 0°.   **55.** Close to 0°.

57. 74° 50′.   **59.** 20°.

61. 80°.

Exercises 5-7   page 110

1.

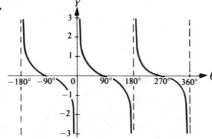

3.

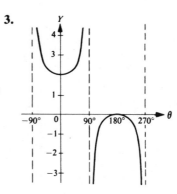

5.

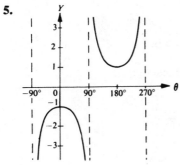

7.

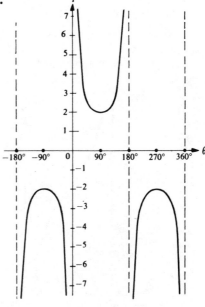

9. 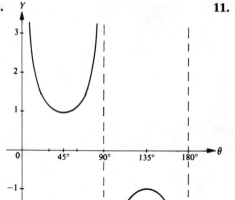   **11.**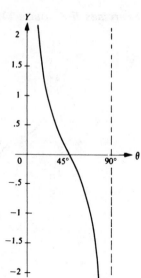

Chapter 5: Review Exercises   page 111

1. (a) -1, (b) 2, (c) 2, (d) $-\sqrt{2}$.

3. (a) 1, (b) 1.

5. (a) $\frac{5}{3}$, (b) $-\frac{5}{4}$, (c) $-\frac{3}{4}$.

7. IV.   **9.** I.   **11.** Possible.   **13.** Impossible.

15. $6°$.   **17.** $45°$, $225°$.

19. $\cos 86° = \sin 4°$ Cofunctions of complementary angles.

$\dfrac{1}{\csc 4°} = \sin 4°$ Reciprocal functions.

Therefore, $\cos 86° = \dfrac{1}{\csc 4°}$, both equal to $\sin 4°$.

21.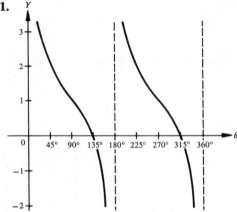

Chapter 5: Diagnostic Test page 111

Following each problem number is the number of the textbook section (in parentheses) in which that kind of problem is discussed.

1. (5-1, 5-2) (a) -2, (b) 1, (c) 2, (d) -2.
2. (5-1, 5-2) (a) $\frac{1}{3}$, (b) $\frac{4}{3}$, (c) 3, (d) $\frac{4}{3}$.
3. (5-1) (a) $\frac{4}{3}$, (b) $-\frac{5}{4}$, (c) $-\frac{5}{3}$.
4. (5-2) IV.
5. (5-1) (a) Impossible, (b) possible, (c) possible, (d) impossible.
6. (5-3) 40°.
7. (5-3) 30°, 150°, 270°.

8.

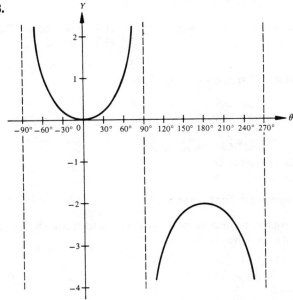

Exercises 6-1 page 118

1. 60°.	**3.** 90°.	**5.** 150°.
7. 360°.	**9.** 210°.	**11.** 300°.
13. 450°.	**15.** 65° 53.4′.	**17.** 116° 18.6′.
19. 43° 25.8′.	**21.** $\frac{1}{6}\pi$.	**23.** $\frac{1}{4}\pi$.
25. $\frac{5}{6}\pi$.	**27.** $\frac{5}{4}\pi$.	**29.** $\frac{3}{2}\pi$.
31. $\frac{11}{6}\pi$.	**33.** .3142.	**35.** 1.3134.
37. .2609.	**39.** 2.1859.	**41.** $5\frac{1}{2}$.
43. Undefined.	**45.** $-\frac{9}{2}$.	**47.** 8.
49. 0.	**51.** .99492.	**53.** .45862.
55. $-.17154$.	**57.** .26866.	

Exercises 6-2   page 121

1. .67.   3. 2.40 inches.
5. 6064 feet when Table 4 is used. (More accurate data would give 6080 feet.)
7. 865,800 miles.
9. (a) 78.2 radians per second, (b) 747 revolutions per minute.
11. 3870 miles.
13. 791 statute miles, 687 nautical miles.
15. 5.73 inches.
17. 69.6 miles per hour.
19. 88.4 inches.

Exercises 6-3   page 125

1. 87 square inches, 38 square inches.   3. 200 square inches, 238 square inches.
5. 61 square inches.   7. 8808 gallons.

Chapter 6: Review Exercises   page 126

1. 135°.   3. 60°.   5. 450°.   7. 71° 37.2′.
9. $5\pi/4$.   11. .88532.   13. $2\frac{1}{2}$. 15. .3090.
17. −5.4707.   19. 1.5.
21. 2208 nautical miles, 2544 statute miles.
23. 770 revolutions per minute.

Chapter 6: Diagnostic Test   page 127

Following each problem number is the number of the textbook section (in parentheses)
in which that kind of problem is discussed.

1. (6-1)   (a) 270°, (b) 18°, (c) 720°, (d) 85° 56.6′.
2. (6-1)   (a) $3\pi/4$, (b) π, (c) $5\pi/3$.
3. (6-1)   2.19360.
4. (6-1)   (a) $\frac{1}{2}$, (b) $-\frac{1}{2}$, (c) −1.
5. (6-1)   (a) .92837, (b) −.22875, (c) .14092.
6. (6-1)   1.3 radians.   7. (6-2) ≈4.71 inches.
8. (6-2) ≈62.18 square inches.   9. (6-2) ≈314 square inches.
10. (6-3) ≈20.7 square inches.

Exercises 7-1   page 133

1. $\sin \theta = \frac{4}{5}$, $\tan \theta = -\frac{4}{3}$, $\cot \theta = -\frac{3}{4}$, $\sec \theta = -\frac{5}{3}$, $\csc \theta = \frac{5}{4}$.
3. $\sin \theta = \frac{4}{5}$, $\cos \theta = \frac{3}{5}$, $\cot \theta = \frac{3}{4}$, $\sec \theta = \frac{5}{3}$, $\csc \theta = \frac{5}{4}$.
5. $\sin \theta = -\frac{12}{13}$, $\cos \theta = -\frac{5}{13}$, $\tan \theta = \frac{12}{5}$, $\sec \theta = -\frac{13}{5}$, $\csc \theta = -\frac{13}{12}$.
7. 1.   9. $\sec x \csc x$.   11. $\cos x$.
13. $\sec^2 \theta$.   15. $\sec \theta$.   17. $4 \sin \theta \cos \theta$.
19. $-2 \sec^2 \theta$.   21. $\csc \theta$.   23. $\cot^2 x$.

Exercises 7-4 page 143

1. 60°, 240°.	**3.** 60°, 300°.
5. 60°, 180°, 300°.	**7.** 60°, 120°, 240°, 300°.
9. 0°.	**11.** 90°, 270°.
13. 270°.	**15.** 45°, 135°.
17. 227° 04', 312° 56'.	**19.** 120°, 180°.

Chapter 7: Review Exercises page 143

1. (a) $\sin^2 \theta + \cos^2 \theta = 1$
$$(\tfrac{3}{5})^2 + \cos^2 \theta = 1$$
$$\cos^2 \theta = 1 - \tfrac{9}{25} = \tfrac{16}{25}$$
$$\cos \theta = \pm\tfrac{4}{5}.$$
But θ is in Q II.
Therefore, $\cos \theta = -\tfrac{4}{5}$.

(b) $\sec \theta = \dfrac{1}{\cos \theta} = -\tfrac{5}{4}$.

(c) $\tan \theta = \dfrac{\sin \theta}{\cos \theta} = \dfrac{3/5}{-4/5} = -\tfrac{3}{4}$.

(d) $\cot \theta = \dfrac{1}{\tan \theta} = -\tfrac{4}{3}$.

(e) $\csc \theta = \dfrac{1}{\sin \theta} = \tfrac{5}{3}$.

3. $\sec \theta \tan \theta \cos^2 \theta \csc \theta = \dfrac{1}{\cancel{\cos \theta}} \cdot \dfrac{\sin \theta}{\cancel{\cos \theta}} \cdot \dfrac{\cancel{\cos^2 \theta}}{1} \cdot \dfrac{1}{\cancel{\sin \theta}} = 1.$

5. $(\sin \theta + \cos \theta)^2 - 2 \sin \theta \cos \theta$
$= \sin^2 \theta + \cancel{2 \sin \theta \cos \theta} + \cos^2 \theta - \cancel{2 \sin \theta \cos \theta}$
$= \sin^2 \theta + \cos^2 \theta = 1.$

7. $\dfrac{1}{1 - \cos \theta} + \dfrac{1}{1 + \cos \theta}$

$= \dfrac{1 + \cos \theta + 1 - \cos \theta}{(1 - \cos \theta)(1 + \cos \theta)}$

$= \dfrac{2}{1 - \cos^2 \theta} = \dfrac{2}{\sin^2 \theta} = 2\left(\dfrac{1}{\sin^2 \theta}\right) = 2 \csc^2.$

9. $\dfrac{1 + \tan^2 \theta}{\sin^2 + \cos^2 \theta} = \dfrac{\sec^2}{1} = \sec^2.$

11. $\dfrac{(\sin x + \cos x)^2}{\sin x} - \csc x$

$= \dfrac{\sin^2 x + \cos^2 x + 2\sin x \cos x}{\sin x} - \dfrac{1}{\sin x}$

$= \dfrac{1 + 2\sin x \cos x - 1}{\sin x}$

$= \dfrac{2\,\cancel{\sin x}\cos x}{\cancel{\sin x}} = 2\cos x.$

13. $(\sec x + 2)(\tan x - 1) = 0$

$\qquad\qquad\qquad \sec x + 2 = 0 \qquad\quad \tan x - 1 = 0$

$\qquad\qquad\qquad\quad \sec x = -2 \qquad\qquad \tan x = 1$

$\qquad\qquad\qquad\qquad x = 120°, 240° \qquad x = 45°, 225°.$

Therefore, $x = 45°, 120°, 225°, 240°.$

15. $2\sin^2 x = 1$

$\sin x = \pm\sqrt{\tfrac{1}{2}} = \pm\dfrac{1}{\sqrt{2}}.$

Therefore, $x = 45°, 135°, 225°, 315°.$

17. $\tan^2 x - \cos^2 x = \sin^2 x$

$\tan^2 x = \sin^2 x + \cos^2 x$

$\tan^2 x = 1$

$\tan x = \pm 1.$

Therefore, $x = 45°, 135°, 225°, 315°.$

19. $\cos x - \sqrt{1 - \cos^2 x} = 0$

$\cos x = \sqrt{1 - \cos^2 x}$ (In this exercise squaring both sides

$\cos^2 x = 1 - \cos^2 x$ is one way of working the problem, but

$2\cos^2 x = 1$ it does give some extraneous roots.)

$\cos x = \pm\dfrac{1}{\sqrt{2}}.$ (cos x must be positive to satisfy the original equation.)

Therefore, $x = 45°, 315°.$

21. We transform the terms of the left side of the identity to an expression over a common denominator and then reduce it to $\tan\theta$ (the expression on the right side of the identity).

$\dfrac{1 - 2\cos^2\theta}{\sin\theta\cos\theta} + \cot\theta = \tan\theta$

$= \dfrac{1 - 2\cos^2\theta}{\sin\theta\cos\theta} + \dfrac{\cos\theta}{\sin\theta}\cdot\dfrac{\boxed{\cos\theta}}{\boxed{\cos\theta}}$ (getting both fractions over the same denominator)

$= \dfrac{1 - 2\cos^2\theta}{\sin\theta\cos\theta} + \dfrac{\cos^2\theta}{\sin\theta\cos\theta}$

$= \dfrac{1 - \cos^2\theta}{\sin\theta\cos\theta} = \dfrac{\sin^2\theta}{\sin\theta\cos\theta} = \dfrac{\sin\theta}{\cos\theta} = \tan\theta.$

Chapter 7: Diagnostic Test   page 144

Following each problem number is the number of the textbook section (in parentheses) in which that kind of problem is discussed.

1. (7-1) (a) $\cot\theta = \dfrac{1}{\tan\theta} = \dfrac{1}{-3/4} = -\frac{4}{3}$,

(b) $\csc^2\theta = 1 + \cot^2\theta$
$$= 1 + (-\tfrac{4}{3})^2 = 1 + \tfrac{16}{9} = \tfrac{25}{9}$$
$\csc\theta = \pm\sqrt{\tfrac{25}{9}} = \pm\tfrac{5}{3}$. In Q IV $\csc\theta$ is negative.
Therefore, $\csc\theta = -\tfrac{5}{3}$.

(c) $\sin\theta = \dfrac{1}{\csc\theta} = \dfrac{1}{-5/3} = -\tfrac{3}{5}$.

2. (7-1) $\csc\theta \tan\theta \sec\theta \cot^2\theta$
$$= \frac{1}{\cancel{\sin\theta}} \cdot \frac{\cancel{\sin\theta}}{\cancel{\cos\theta}} \cdot \frac{1}{\cancel{\cos\theta}} \cdot \frac{\cancel{\cos^2\theta}}{\sin^2\theta} = \frac{1}{\sin^2\theta} = \csc^2\theta.$$

3. (7-1) $(1 + \cot^2\alpha)\sin^2\alpha + \tan^2\alpha$
$$= (\csc^2\alpha)\sin^2\alpha + \tan^2\alpha$$
$$= 1 \qquad\quad + \tan^2\alpha = \sec^2\alpha.$$

4. (7-1) $(\sin^2 x + \cos^2 x)(\sin^2 x - \cos^2 x) + \cos^2 x$
$$= \quad 1 \qquad (\sin^2 x - \cos^2 x) + \cos^2 x$$
$$= \qquad\quad \sin^2 x - \cos^2 x + \cos^2 x = \sin^2 x.$$

5. (7-1) $\dfrac{\sec\theta}{\tan\theta + \cot\theta} = \dfrac{\dfrac{1}{\cos\theta}}{\dfrac{\sin\theta}{\cos\theta} + \dfrac{\cos\theta}{\sin\theta}} \cdot \boxed{\dfrac{\sin\theta\cos\theta}{\dfrac{1}{\sin\theta\cos\theta}}{1}} = \dfrac{\sin\theta}{\sin^2\theta + \cos^2\theta} = \sin\theta.$

6. (7-1) $\dfrac{\cos\theta - \sin\theta}{\cos\theta} + \tan\theta$
$$= \frac{\cos\theta}{\cos\theta} - \frac{\sin\theta}{\cos\theta} + \tan\theta$$
$$= 1 - \tan\theta + \tan\theta = 1.$$

7. (7-4) $\sec^2 x = 4$
$\sec x = \pm 2$ (both ± 2 satisfy the original equation).
Therefore, $x = 60°, 120°, 240°, 300°$.

8. (7-4) $(\sin x + \cos x)^2 = 1$
$$\sin^2 x + 2\sin x \cos x + \cos^2 x = 1$$
$$\sin^2 x + \cos^2 x + 2\sin x \cos x = 1$$
$$1 \qquad\quad + 2\sin x \cos x = 1$$
$$2\sin x \cos x = 0.$$

Therefore, $\sin x = 0$ or $\cos x = 0$
$$x = 0°, 180° \qquad x = 90°, 270°.$$

Thus, $x = 0°, 90°, 180°, 270°$.

9. (7-4) $\cos^2 x = 3 \sin^2 x$
$$1 - \sin^2 x = 3 \sin^2 x$$
$$1 = 4 \sin^2 x$$
$$\sin x = \pm\tfrac{1}{2}.$$
Therefore, $x = 30°, 150°, 210°, 330°$.

10. (7-3) We reduce the left side of the identity to that given on the right side.

$$\frac{\sin \theta}{1 + \cos \theta} + \cot \theta = \csc \theta$$

$$\frac{\sin \theta}{1 + \cos \theta} + \frac{\cos \theta}{\sin \theta}$$

$$= \frac{\sin^2 \theta + \cos^2 \theta + \cos \theta}{(1 + \cos \theta)(\sin \theta)} = \frac{\cancel{1 + \cos \theta}}{(\cancel{1 + \cos \theta}) \sin \theta} = \frac{1}{\sin \theta} = \csc \theta.$$

Exercises 8-2 page 152

3. $\frac{1}{4}\sqrt{2}(1 - \sqrt{3})$.
11. -1.
15. $\frac{1}{2}$.

5. (a) $-\frac{56}{65}$, (b) $-\frac{16}{65}$.
13. $\cos \theta$.

Exercises 8-4 page 155

1. $-1, 0$, undefined, $A + B = 270°$; $\frac{7}{25}, -\frac{24}{25}, -\frac{7}{24}, A - B$ in Q II.
3. $-\frac{1}{2}\sqrt{3}, -\frac{1}{2}, \sqrt{3}$.
21. $-\tan x$.
25. $\cos 6x$.

19. $\cos 9x$.
23. $\sin 2x$.
27. $\sin \theta$.

Exercises 8-6 page 162

7. $\sin 105° = \frac{1}{2}\sqrt{2 + \sqrt{3}}$, $\cos 105° = -\frac{1}{2}\sqrt{2 - \sqrt{3}}$, $\tan 105° = -2 - \sqrt{3}$.
9. $\sin 80°$.
13. $\tan 20°$.
17. $3 \sin 6\alpha$.
21. $-\frac{120}{169}, \frac{119}{169}, -\frac{120}{119}, \frac{1}{26}\sqrt{26}, -\frac{5}{26}\sqrt{26}, -\frac{1}{5}$, IV, II.

11. $\cos 100°$.
15. $\cos 35°$.
19. $\pm 5 \tan \frac{5}{2}\beta$.

Exercises 8-8 page 166

1. $\sin 60° + \sin 10°$.
5. $\frac{3}{4}(\cos \beta - \cos 3\beta)$.
9. $2 \cos 4\theta \cos \theta$
31. $\frac{1}{6}\pi, \frac{5}{6}\pi, \frac{3}{2}\pi$.
35. $\frac{1}{4}\pi, \frac{1}{3}\pi, \frac{5}{4}\pi, \frac{3}{4}\pi, \frac{5}{3}\pi, \frac{7}{4}\pi$.
37. $0, \frac{1}{4}\pi, \pi, \frac{5}{4}\pi$.
41. $\frac{1}{4}\pi, \frac{1}{2}\pi, \frac{3}{4}\pi, \frac{5}{4}\pi, \frac{3}{2}\pi, \frac{7}{4}\pi$.
45. $0, \pi, \frac{7}{6}\pi, \frac{11}{6}\pi$.
47. $0, \frac{1}{8}\pi, \frac{3}{8}\pi, \frac{5}{8}\pi, \frac{7}{8}\pi, \frac{9}{8}\pi, \frac{11}{8}\pi, \frac{13}{8}\pi, \frac{15}{8}\pi$.
49. $0, \frac{1}{3}\pi, \frac{2}{3}\pi, \pi, \frac{4}{3}\pi, \frac{5}{3}\pi$.

3. $3(\cos 8\theta + \cos 2\theta)$.
7. $2 \sin 35° \cos 5°$.
11. $2 \cos \frac{5}{2}\alpha \sin \alpha$.
33. π.

39. $0, \frac{1}{3}\pi, \pi, \frac{5}{3}\pi$.
43. $\frac{1}{6}\pi, \frac{1}{2}\pi, \frac{5}{6}\pi, \frac{3}{2}\pi$.

Exercises 8-9 page 172

1. $13 \cos (\theta - 67° 23')$.   **3.** $13 \cos (\theta - 292° 37')$.
5. $13 \cos (\theta - 202° 37')$.   **7.** $13 \cos (\theta - 157° 23')$.
9. $\sqrt{13} \cos (\theta - 123° 41')$.   **11.** Maximum $\sqrt{5}$ when
13. $22° 37'$. $\theta = 296° 34' + n (360°)$.

15.
$$a \cos \theta + b \sin \theta = \sqrt{a^2 + b^2} \left[(\cos \theta) \frac{a}{\sqrt{a^2 + b^2}} + (\sin \theta) \frac{b}{\sqrt{a^2 + b^2}} \right]$$
$$= \sqrt{a^2 + b^2} [\cos \theta \sin \beta + \sin \theta \cos \beta]$$
$$= \sqrt{a^2 + b^2} [\sin \theta \cos \beta + \cos \theta \sin \beta]$$
$$= \sqrt{a^2 + b^2} \sin (\theta + \beta)$$

where $k = \sqrt{a^2 + b^2}$, $\sin \beta = \dfrac{a}{k}$, $\cos \beta = \dfrac{b}{k}$, $\tan \beta = \dfrac{a}{b}$.

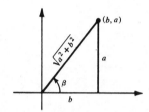

Chapter 8: Review Exercises page 172

1. (a) $\sin (A + B) = \sin A \cos B + \cos A \sin B$
$$= \tfrac{4}{5}(-\tfrac{12}{13}) + \tfrac{3}{5}(-\tfrac{5}{13})$$
$$= -\tfrac{48}{65} - \tfrac{15}{65} = -\tfrac{63}{65},$$

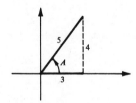

(b) $\cos (A - B) = \cos A \cos B + \sin A \sin B$
$$= \tfrac{3}{5}(-\tfrac{12}{13}) + \tfrac{4}{5}(-\tfrac{5}{13})$$
$$= -\tfrac{36}{65} - \tfrac{20}{65} = -\tfrac{56}{65}.$$

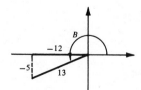

3. $\tan (A + B) = \dfrac{\tan A + \tan B}{1 - \tan A \tan B}$

$= \dfrac{\tan 60° + \tan 45°}{1 - \tan 60° \tan 45°}$

$= \dfrac{\sqrt{3} + 1}{1 - \sqrt{3}} \cdot \dfrac{1 + \sqrt{3}}{1 + \sqrt{3}} = \dfrac{3 + 2\sqrt{3} + 1}{1 - 3} = \dfrac{4 + 2\sqrt{3}}{-2} = -2 - \sqrt{3}.$

5. $\tan (\pi - \theta) = \dfrac{\tan \pi - \tan \theta}{1 - \tan \pi \tan \theta}$

$= \dfrac{0 - \tan \theta}{1 - 0} = -\tan \theta.$

7. $\sin (\pi + \theta) = \sin \pi \cos \theta + \cos \pi \sin \theta$
   $= 0 \cdot \cos \theta + (-1) \sin \theta$
   $= 0 - \sin \theta = -\sin \theta.$

9. $\dfrac{\sin 2x}{\sin x} - \dfrac{\cos 2x}{\cos x} = \dfrac{\sin 2x \cos x - \cos 2x \sin x}{\sin x \cos x}$

$= \dfrac{\sin (2x - x)}{\sin x \cos x} = \dfrac{\sin x}{\sin x \cos x} = \dfrac{1}{\cos x} = \sec x.$

11. $2 \sin \theta \cos \theta = \sin 2\theta.$
   $2 \sin 25° \cos 25° = \sin 2(25°) = \sin 50°.$

13. $\dfrac{2 \tan \theta}{1 - \tan^2 \theta} = \tan 2\theta.$

$\dfrac{2 \tan 20°}{1 - \tan^2 20°} = \tan 2(20°) = \tan 40°.$

15. $\sin (210° + x) + \sin (210° - x)$
   $= \sin 210° \cos x + \cos 210° \sin x + \sin 210° \cos x - \cos 210° \sin x$
   $= 2 \sin 210° \cos x$
   $= 2(-\tfrac{1}{2}) \cos x = -\cos x.$

17. $\sin (-x) + \cos (-x) + \sin x + \cos x$
   $= -\sin x + \cos x + \sin x + \cos x = 2 \cos x.$

19. $\sin 2x + \cos x = 0$
   $2 \sin x \cos x + \cos x = 0$
   $\cos x (2 \sin x + 1) = 0$

   $\cos x = 0 \qquad\qquad$ or $\quad 2 \sin x + 1 = 0$
   $x = \pi/2, 3\pi/2 \qquad\qquad\qquad \sin x = -\tfrac{1}{2}$
   $\qquad\qquad\qquad\qquad\qquad\qquad x = 7\pi/6, 11\pi/6.$

   Therefore, $x = \pi/2, 7\pi/6, 3\pi/2, 11\pi/6.$

21. $\dfrac{2 \tan \theta}{1 - \tan^2 \theta} = \tan 2\theta$

$\dfrac{2 \tan \theta}{1 - \tan^2 \theta} = \frac{4}{3}$ [clear fractions by multiplying through by $3(1 - \tan^2 \theta)$].

$6 \tan \theta = 4 - 4 \tan^2 \theta$

$4 \tan^2 \theta + 6 \tan \theta - 4 = 0$

$2[2 \tan^2 \theta + 3 \tan \theta - 2] = 0$

$2(2 \tan \theta - 1)(\tan \theta + 2) = 0$

$2 \tan \theta - 1 = 0 \quad \text{or} \quad \tan \theta + 2 = 0$

$\tan \theta = \frac{1}{2}$ $\quad$ $\begin{array}{|l|}\hline \tan \theta = -2. \\ \text{Not true because } 0° < \theta < 90°. \\ \hline \end{array}$

Therefore, $\tan \theta = \frac{1}{2}$.

Chapter 8: Diagnostic Test page 174

Following each problem number is the number of the textbook section (in parentheses) in which that kind of problem is discussed.

1. (a) (8-5) $\sin 2\theta = 2 \sin \theta \cos \theta$

$\qquad = 2(\frac{4}{5})(\frac{3}{5}) = \frac{24}{25}$,

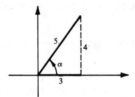

(b) (8-6) $\cos \frac{1}{2}\theta = \sqrt{\dfrac{1 + \cos \theta}{2}}$

$\qquad\qquad = \sqrt{\dfrac{1 + \frac{3}{5}}{2}} = \dfrac{2}{\sqrt{5}} = \dfrac{2\sqrt{5}}{5}$,

(c) (8-5) $\tan 2\theta = \dfrac{2 \tan \theta}{1 - \tan^2 \theta}$

$\qquad\qquad = \dfrac{2(\frac{4}{3})}{1 - (\frac{4}{3})^2} = \dfrac{\frac{8}{3}}{-\frac{7}{9}} = -\frac{24}{7}$.

2. (a) (8-3)

$\sin (A - B) = \sin A \cos B - \cos A \sin B$

$\qquad = \frac{4}{5}(\frac{3}{5}) - (-\frac{3}{5})(\frac{4}{5})$

$\qquad = \frac{12}{25} + \frac{12}{25} = \frac{24}{25}$,

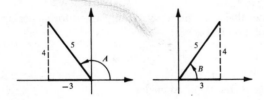

(b) (8-2)

$$\cos (A + B) = \cos A \cos B - \sin A \sin B$$
$$= (-\tfrac{3}{5})(\tfrac{3}{5}) - (\tfrac{4}{5})(\tfrac{4}{5})$$
$$= -\tfrac{9}{25} - \tfrac{16}{25} = -1 \quad \text{(this means } A + B = 180°}$$
$$\text{or coterminal with } 180°),$$

(c) (8-4)

$$\tan (A - B) = \frac{\tan A - \tan B}{1 + \tan A \tan B}$$

$$= \frac{-\tfrac{4}{3} - \tfrac{4}{3}}{1 + (-\tfrac{4}{3})(\tfrac{4}{3})} = \frac{-\tfrac{8}{3}}{+\tfrac{7}{9}} = -\tfrac{24}{7}.$$

3. (a) (8-4) $\tan (\pi + \theta) = \dfrac{\tan \pi + \tan \theta}{1 - \tan \pi \tan \theta} = \dfrac{0 + \tan \theta}{1 - 0} = \tan \theta,$

   (b) (8-2) $\cos(\pi/2 - \theta) = \cos \pi/2 \cos \theta + \sin \pi/2 \sin \theta$
   $$= 0(\cos \theta) + 1(\sin \theta) = \sin \theta.$$

4. (a) (8-5) $\dfrac{2 \tan 40°}{1 - \tan^2 40°} = \tan 2(40°) = \tan 80°,$

   (b) (8-5) $2 \sin 3x \cos 3x = \sin 2(3x) = \sin 6x,$

   (c) (8-6) $\sqrt{\dfrac{1 - \cos 4x}{1 + \cos 4x}} = \tan \tfrac{1}{2}(4x) = \tan 2x,$

   (d) (8-1) $\tan x + \tan (-x) + \cos (-x)$
   $$= \tan x - \tan x + \cos x = \cos x.$$

5. [(8-8)   Identity 31]
   $$\sin x + \sin y = 2 \sin \tfrac{1}{2}(x + y) \cos \tfrac{1}{2}(x - y)$$
   $$\sin 7\theta + \sin \theta = 2 \sin \tfrac{1}{2}(7\theta + \theta) \cos \tfrac{1}{2}(7\theta - \theta)$$
   $$= 2 \sin 4\theta \cos 3\theta.$$

6. (8-5) $\sin 2x + \sin x = 0$
   $$2 \sin x \cos x + \sin x = 0$$
   $$\sin x(2 \cos x + 1) = 0$$

   $\sin x = 0$ $2 \cos x + 1 = 0$
   $\quad x = 0, \pi$ $\cos x = -\tfrac{1}{2}$
   $$x = 2\pi/3, 4\pi/3.$$
   Therefore, $x = 0, 2\pi/3, \pi, 4\pi/3.$

7. (8-8) $\tan \tfrac{1}{2}x = \sqrt{3}$
   $$\tfrac{1}{2}x = \pi/3, 4\pi/3$$
   $$x = 2\pi/3, 8\pi/3,$$
   but $8\pi/3$ is greater than 2π, and we were only asked for angles up to 2π. Therefore, the solution is $2\pi/3$.

8. (8-8) $\cos 2x = 1$
   $$2x = 0, 2\pi, 4\pi, \text{ etc.}$$
   Therefore, $x = 0, \pi$ (the required angles).

9. [(8-8)   Identities 32 and 33]

$$\frac{\sin 5\theta - \sin 3\theta}{\cos 5\theta + \cos 3\theta} = \frac{2 \cos \frac{1}{2}(5\theta + 3\theta) \sin \frac{1}{2}(5\theta - 3\theta)}{2 \cos \frac{1}{2}(5\theta + 3\theta) \cos \frac{1}{2}(5\theta - 3\theta)}$$

$$= \frac{2 \cancel{\cos 4\theta} \sin \theta}{2 \cancel{\cos 4\theta} \cos \theta} = \tan \theta.$$

10. [(8-9)   Identity 35]

$-3 \cos \theta + 4 \sin \theta = c \cos (\theta - \alpha)$

$a = -3, b = 4,$

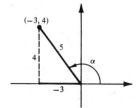

$c = \sqrt{a^2 + b^2}$ $\sin \alpha = \dfrac{b}{c} = \dfrac{4}{5}, \cos \alpha = \dfrac{a}{c} = \dfrac{-3}{5}$

$= \sqrt{(-3)^2 + (4)^2} = 5.$ $\tan a = -\frac{4}{3}$

$\alpha \approx 126° 52'.$

$\overline{\text{Since } \sin \alpha > 0 \text{ and } \cos \alpha < 0, \alpha \text{ is in Q II.}}$

Therefore, $-3 \cos \theta + 4 \sin \theta \approx 5 \cos (\theta - 126° 52')$.

Exercises 9-2   page 181

1. $\pi, 3.$ **3.** $3\pi, 5.$ **5.** $12, 2.$

7. $2\pi, 4.$ **9.** $10, 5.$

11.
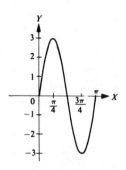

13.
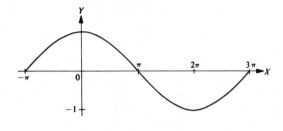

15.

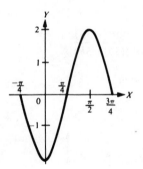

17.

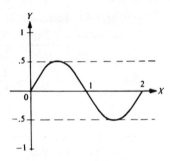

19.

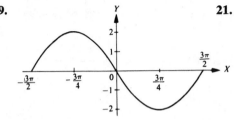

21.

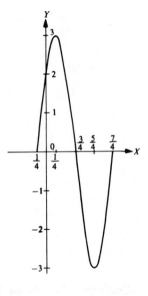

23. $y = \sqrt{10} \cos (x - 71° 34')$.

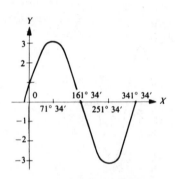

25. $y = 5 \cos (x - 143° \, 08')$.

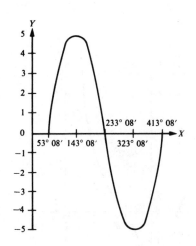

Exercises 9-5   page 188

1.

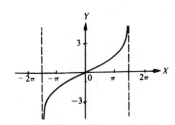

3.

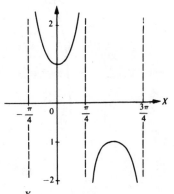

5.

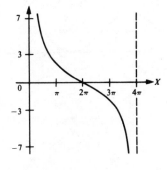

7.

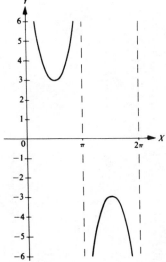

9.

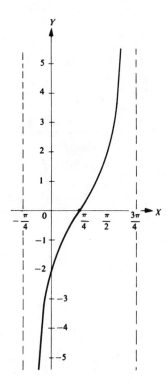

11.

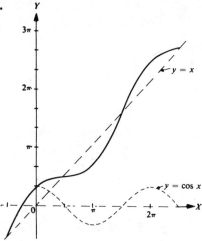

13.

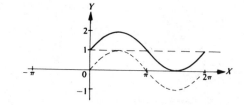

15.

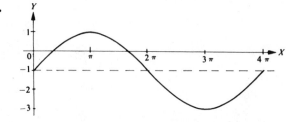

17.

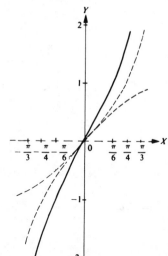

19.

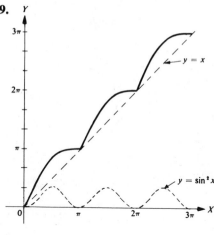

Chapter 9: Review Exercises   page 188

1. Amplitude 3, period $= 2\pi/2 = \pi$.

3. Amplitude infinite, period $\pi/3$.

5. Amplitude $\frac{1}{2}$,
   period $2\pi/2 = \pi$,
   x-intercepts: $x = n(\pi/2)$
   $\qquad\qquad = 0, \pi/2, \pi$, etc.

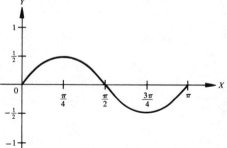

7. Amplitude 3, period $2\dfrac{\pi}{\frac{1}{2}\pi} = 4$,

   x-intercepts: $\frac{1}{2}\pi x = n\pi$
   $\qquad\qquad x = 2n$
   $\qquad\qquad\quad = 0, 2, 4$, etc.

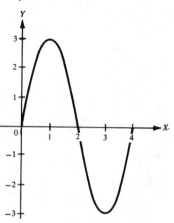

9. Amplitude 3, period $\dfrac{2\pi}{\frac{1}{3}} = 6\pi$,

x-intercepts: $\frac{1}{3}x + \pi/2 = \pi/2 + n\pi$

$$\frac{1}{3}x = n\pi$$
$$x = -3\pi, 0, 3\pi, \text{ etc.}$$

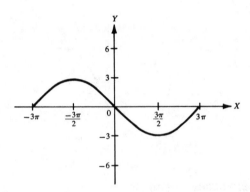

11. $y = \frac{1}{2}\cot 2x$, amplitude infinite, period $\pi/2$.

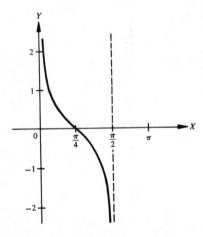

13. $y = 2 \csc x$, amplitude infinite, period 2π.

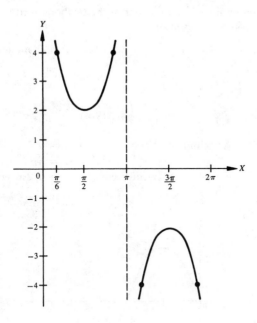

15. $y = \frac{1}{2}x + \cos 2x$. Amplitude and period not defined. Break the function into two parts, $y_1 = \frac{1}{2}x$ and $y_2 = \cos 2x$. Graph each part and then graph the composite curve by the addition of ordinates.

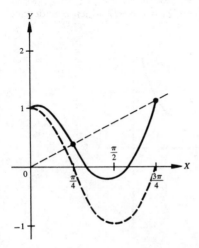

Chapter 9: Diagnostic Test   page 189

Following each problem number is the number of the textbook section (in parentheses) in which that kind of problem is discussed.

1. (9-2) Amplitude 4,

$$\text{period} = \frac{2\pi}{|b|} = \frac{2\pi}{\frac{2}{3}} = 3\pi.$$

2. (9-3) Amplitude infinite,

$$\text{period} = \frac{\pi}{|b|} = \pi/3.$$

3. (9-4) $2\pi/2 = \pi$.

4. (9-4) $\dfrac{2\pi}{|b|} = 2\pi/3$.

5. (9-2) $y = 4\sin\frac{1}{2}x$. Amplitude 4. When $y = 0$, $\sin\frac{1}{2}x = 0$

$$\text{period } \frac{2\pi}{\frac{1}{2}} = 4\pi$$

$$\frac{1}{2}x = n\pi$$

x-intercepts: $x = 0, 2\pi, 4\pi$, etc.

When $x = \pi$
$$y = 4\sin\frac{1}{2}\pi$$
$$= 4.$$

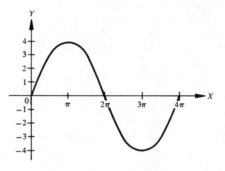

6. (9-3) $y = 2\tan\frac{1}{3}x$.
When $y = 0$, $\tan\frac{1}{3}x = 0$
$$\frac{1}{3}x = n\pi$$
x-intercepts: $x = 3n\pi = -3\pi, 0, 3\pi$, etc.
When $x = 3\pi/2$, $y = \tan\frac{1}{3}(3\pi/2) = \tan\pi/2$ which is undefined.

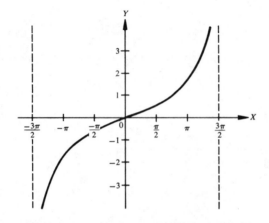

7. (9-4) $y = 2 \sec 2x$. First sketch $y = \cos 2x$.

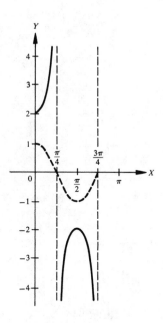

8. (9-5) $y = \cos x - x$. Graph $y_1 = -x$ and $y_2 = \cos x$, and then add ordinates.

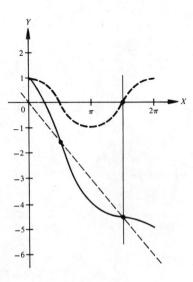

9. (9-3) $y = \cot \pi x$,   period $\dfrac{\pi}{\pi} = 1$.

When $y = 0$, $\cot \pi x = 0$

$$\pi x = \pm \pi/2, \pm 3\pi/2, \text{etc.}$$

x-intercepts: $x = \pm \frac{1}{2}, \pm \frac{3}{2}, \text{etc.}$

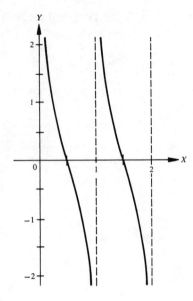

10. (9-4) $y = \csc \dfrac{\pi}{2} x$.

First graph $y_1 = \sin \dfrac{\pi}{2} x$ and take reciprocals of ordinates.

When $y_1 = 0$, $\sin \dfrac{\pi}{2} x = 0$

$$\dfrac{\pi}{2} x = n\pi$$

$$x = 2n$$

$$= -2, 0, 2, \text{etc.}$$

Therefore, $y = \csc \dfrac{\pi}{2} x$ is undefined when $x = -2, 0, 2$.

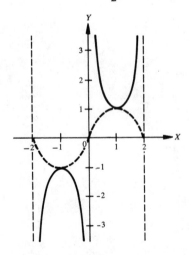

Exercises 10-3 page 193

1. $\log_2 8 = 3.$
3. $\log_{10} 100 = 2.$
5. $\log_{25} = \frac{1}{2}.$
7. $\log_8 \frac{1}{4} = -\frac{2}{3}.$
9. $\log_5 1 = 0.$
11. $\log_3 1 = 0.$
13. $\log_5 5 = 1.$
15. $\log_{10} 10 = 1.$
17. $4^3 = 64.$
19. $2^4 = 16.$
21. $16^{1/2} = 4.$
23. $10^1 = 10.$
25. $7^1 = 7.$
27. $b^1 = b.$
29. 3.
31. $\frac{3}{2}.$
33. 0.
35. $-\frac{1}{2}.$
37. $\frac{5}{2}.$
39. 9.
41. 4.
43. 9.
45. 2.

Exercises 10-4 page 196

1. 1.146.
3. .410.
5. .699.
7. .661.
9. 1.477.
11. 1.991.

13. .060.
15. 3.778.
17. $\log \dfrac{x}{y}.$

19. $\log x^3 y^2.$
21. $\log \dfrac{z\sqrt[4]{x}}{y}.$
23. $\log (x + y).$

25. $y = 9^x.$
27. $y = \dfrac{1}{x^2}.$
29. $y = x.$

31. $y = x^3.$
33. 11.
35. 99.

Exercises 10-7 page 202

1. 2.
3. 7.
5. $-2, 8 - 10.$
7. $-4, 6 - 10.$
9. 3.
11. $-3, 7 - 10.$
13. 3.
15. $-2, 8 - 10.$
17. $-1, 9 - 10.$
19. 12.
21. 127.8.
23. 506, 100.
25. 9.455.
27. .01008.
29. .00007599.

Exercises 10-9 page 204

1. 2.0969.
3. $9.7300 - 10.$
5. 1.8284.
7. $6.9031 - 10.$
9. 7.9685.
11. $8.7001 - 10.$
13. 6.150.
15. 4280.
17. .001200.
19. 17.62.

Exercises 10-10 page 207

1. 86.62.
3. 23.98.
5. .03492.
7. 2.101.
9. 2.949.
11. .9495.
13. .5144.
15. 6.995.
17. 8.618.
19. 2.896.
21. 87.70.
23. $-14.94.$
25. .07555.
27. 1.083.
29. .9144.
31. .2104.
33. 34.10.
35. $-2.694.$
37. 75.3.

Exercises 10-12 page 211

1. $8.3668 - 10.$
3. $9.8954 - 10.$
5. $9.9564 - 10.$
7. $9.5798 - 10.$
9. Not a real number.
11. $26° 00'$.
13. $58° 10'$.
15. $3° 10'$.
17. $9° 20'$.
19. $87° 26'$.
21. $.9622$.
23. 34.40.
25. -16.04.

Chapter 10: Review Exercises page 211

1. $\log_9 81 = 2$.
3. $4^2 = 16$.
5. $N = 3^2 = 9$.
7. $b^2 = \frac{9}{4}$
9. $10^x = 1000$
$b = \sqrt{\frac{9}{4}} = \frac{3}{2}$.
$x = 3$.

Note: In this book we take $b > 1$. For that reason $-\frac{3}{2}$ is not acceptable.

11. $2 \log x + 3 \log y$
$= \log x^2 + \log y^3$
$= \log x^2 y^3$.

13. $\log (x - y) - \log (x^2 - y^2)$
$= \log \dfrac{x - y}{x^2 - y^2} = \log \dfrac{\cancel{x - y}}{\cancel{(x - y)}(x + y)}$
$= \log \dfrac{1}{(x + y)}$
$= \log 1 - \log (x + y)$
$= 0 - \log (x + y) = -\log (x + y)$.

15. $\log x + \log (7 - x) = 1$
$\log x(7 - x) = 1$
$x(7 - x) = 10^1 = 10$
$7x - x^2 = 10$
$x^2 - 7x + 10 = 0$
$(x - 2)(x - 5) = 0$
$x - 2 = 0 \qquad x - 5 = 0$
$x = 2 \qquad x = 5$.
Therefore, $x = 2, 5$. (Both answers check in the original equation.)

17. $\log (2x + 1) = \log 1 + \log (x + 2)$
$\log (2x + 1) - \log (x + 2) = \log 1$
$$\log \frac{2x + 1}{x + 2} = 0$$
$$\frac{2x + 1}{x + 2} = 10^0 = 1$$
Therefore, $\qquad 2x + 1 = x + 2$
and $\qquad x = 1$.

19. $\sin 14° 30' = .2504 \Big\}9$
$\quad \sin \theta = .2513 \Big\}28$
$\sin 14° 40' = .2532 \Big\}$
Therefore, $\theta = 14° 33'$ (to the nearest minute).

21. $x = 2.567 \sin 25° 37'$

$\log x = \log 2.567 + \log \sin 25° 37'$

$\log 2.567 = \quad .4094$

$\log \sin 25° 37' = \quad 9.6358 - 10(+)$

$\overline{\log x = 10.0452 - 10}$

$x = \quad 1.110$ (rounded off to four significant digits).

23. $x = \dfrac{15.7 \sin 27° 15'}{\sin 102° 37'}$

$\log 15.7 = \quad 1.1959$

$\log \sin 27° 15' = \quad 9.6607 - 10(+)$

$\overline{\qquad\qquad\qquad 10.8566 - 10}$

$\log \sin 102° 37'$

$= \log \sin 77° 23' = 9.9894 - 10(-)$

$\log x = .8672$

$x = 7.365$ (rounded off to four significant digits).

Chapter 10: Diagnostic Test   page 212

Following each problem number is the number of the textbook section (in parentheses) in which that kind of problem is discussed.

1. (10-3) $\log_3 81 = 4$.   **2.** (10-3) $9^{1/2} = 3$.

3. (10-3) (a) 4, (b) −2, (c) 1, (d) 0.

4. (10-4) $4 \log x - 3 \log y$

$= \log x^4 - \log y^3$

$= \log \dfrac{x^4}{y^3}$.

5. (10-4) $2 \log (x + 3) = \log (7x + 1) + \log 2$. Rearrange terms and combine logarithms.

$2 \log (x + 3) = \log (7x + 1) + \log 2$

$\log (x + 3)^2 = \log 2(7x + 1)$

$(x + 3)^2 = 2(7x + 1)$

Solve for x.

$x^2 + 6x + 9 = 14x + 2,$

$x^2 - 8x + 7 = 0,$

$(x - 1)(x - 7) = 0.$

Therefore, $\qquad\qquad\qquad x = 1, \quad$ and $\quad 7$.

Both roots satisfy the original equation.

6. (10-12) $\log \tan 44° 40' = 9.9884 - 10$

$\qquad\quad \log \tan \theta = 9.9904 - 10 \Big\}\, {}^{20}_{58}$

$\quad \log \tan 44° 50' = 9.9942 - 10 \Big\}$

Therefore, $\theta = 44° 43'$ (to the nearest minute).

7. (10-12) $x = \log \sin 130° \, 45'$.

$$x = \log \sin 130° \, 45'$$
$$= \log \sin (180° - 130° \, 45') = \log \sin 49° \, 15'$$
$$= 9.8794 - 10.$$

8. (10-12) $x = \cos 23° \, 40$.

$$x = .9159.$$

9. (10-12) $x = 452 \tan 35° \, 10'$.

$$\begin{array}{rl}
\log 452 = & 2.6551 \\
\log \tan 35° \, 10' = & \underline{9.8479 - 10(+)} \\
\log x = & 12.5030 - 10 \\
x = & 318.4
\end{array}$$

10. (10-12) $x = \dfrac{27.4 \sin 54° \, 20'}{\cos 25° \, 40'}$.

$$\begin{array}{rl}
\log 27.4 = & 1.4378 \\
\log \sin 54° \, 20' = & \underline{9.9098 - 10(+)} \\
& 11.3476 - 10 \\
\log \cos 25° \, 40' = & \underline{9.9549} \\
\log x = & 1.3927 \\
x = & 24.70.
\end{array}$$

Solution for Question 10 on the calculator:

| C | 4 0 | ÷ | 6 0 | + | 2 5 | = | cos | STO | M1 |

| 2 0 | ÷ | 6 0 | + | 5 4 | = | sin | X | 2 7 | . | 4 | ÷ | RCL | M1 | = |

SEE THIS

| 24.69728914 |

Exercises 11-2   page 218

1. 664.8 square units.
5. 26.83 square units.
9. .05950 square units.
13. 25.78 square units.
17. 5268 square feet.
21. $2745.

3. 15.04 square units.
7. 9.354 square units.
11. 1.872 square units.
15. 62.70 square units.
19. 2.624 acres.
23. (a) 1.049 cubic feet, (b) 510.6 pounds.

Exercises 11-4   page 223

1. $B = 54° \, 04'$, $C = 90°$, $a = 16.85$.
3. $A = 42° \, 58'$, $C = 78° \, 42'$, $b = 5.182$.
5. $A = 77° \, 33'$, $B = 49° \, 47'$, $c = 53.00$.
7. $B = 38° \, 38'$, $C = 35° \, 42'$, $a = 283.8$.
9. $A = 30° \, 46'$, $B = 21° \, 54'$, $c = 1.344$.
11. $36° \, 13'$, $47° \, 17'$, 214.8 feet.

Exercises 11-5   page 228

1. $A = 61° 00'$, $B = 45° 45'$, $C = 73° 15'$.
3. $A = 27° 56'$, $B = 98° 48'$, $C = 53° 16'$ (to the nearest minute).
5. $A = 117° 28'$, $B = 35° 42'$, $C = 26° 52'$ (to the nearest minute).
7. $A = 33° 31'$, $B = 117° 11'$, $C = 29° 15'$ (to the nearest minute).
9. 66.53 yd.

Chapter 11: Review Exercises   page 229

1. $K = \frac{1}{2}bc \sin A$
    $= \frac{1}{2}(120)(100) \sin 39° 47'$
    $= 6000(.6399)$
    $= 3839.$

3. $K = \frac{1}{2}c^2 \dfrac{(\sin A)(\sin B)}{\sin C}$

    $= \frac{1}{2}(10^2)\dfrac{(\sin 31° 14')(\sin 84° 52')}{\sin 63° 54'}$

    $= 28.75.$

    The calculation using logarithms:
    $$\begin{aligned}
    \log 50 &= 1.6990 \\
    \log \sin 31° 14' &= 9.7147 - 10 \\
    \log \sin 84° 52' &= \underline{9.9982 - 10}(+) \\
    &\quad\; 21.4119 - 20 \\
    \log \sin 63° 54' &= \underline{9.9533 - 10}(-) \\
    \log K &= \overline{11.4586 - 10} \\
    K &= 28.75.
    \end{aligned}$$

5. $K = \sqrt{s(s - a)(s - b)(s - c)}$
    $= \sqrt{19(9)(6)(4)}$
    $= \sqrt{4104}.$
    $\log K = \frac{1}{2} \log 4104 = \frac{1}{2}(3.6132) = 1.8066$
    $K = 64.06.$

7. $\frac{1}{2}(A + C) = 63° 25'$
    $$\begin{aligned}
    a &= 161 \\
    c &= \underline{111} \\
    a + c &= \overline{272} \\
    a - c &= 50
    \end{aligned}$$

    $\dfrac{\tan \frac{1}{2}(A - C)}{\tan \frac{1}{2}(A + C)} = \dfrac{a - c}{a + c}$

    $\tan \frac{1}{2}(A - C) = \dfrac{50(\tan 63° 25')}{272}$

$$\log 50 = 1.6990$$
$$\log \tan 63° 25' = \underline{\ \ .3007\ \ } \quad (+)$$
$$11.9997 - 10$$
$$\log 272 = \underline{\ 2.4346\ } \quad (-)$$
$$\log \tan \tfrac{1}{2}(A - C) = 9.5651 - 10$$
$$\tfrac{1}{2}(A - C) = 20° 10'$$
$$\tfrac{1}{2}(A + C) = 63° 25'$$
$$\overline{\quad A = 83° 35'}$$
$$C = 43° 15'.$$

Use Law of Sines to solve for side b.

$$\frac{b}{\sin B} = \frac{a}{\sin A}$$

$$b = \frac{a \sin B}{\sin A} = \frac{161 \sin 53° 10'}{\sin 83° 35'} = 129.7.$$

9. $\sin \tfrac{1}{2}B = \sqrt{\dfrac{(s - a)(s - c)}{ac}}$ $\qquad s = \dfrac{a + b + c}{2} = \dfrac{25 + 18 + 21}{2} = 32.$

$$= \sqrt{\frac{(7)(11)}{(25)(21)}} \qquad\qquad \begin{array}{ll} s - a = 7 & a = 25. \\ s - c = 11 & c = 21. \end{array}$$

$$= \sqrt{\frac{11}{75}}$$

$$= .3830.$$
$$\tfrac{1}{2}B = 22° 31'$$
$$B = 44° 62' = 45° 02'.$$

11. $r = \sqrt{\dfrac{(s - a)(s - b)(s - c)}{s}}$ $\qquad \begin{array}{l} a = 150 \\ b = 125 \\ c = \underline{117} \\ 2s = 392 \\ s = 196 \end{array}$

$$= \sqrt{\frac{(46)(71)(79)}{196}} \qquad\qquad \begin{array}{l} s - a = 46 \\ s - b = 71 \\ s - c = 79. \end{array}$$

$$= 36.28.$$

$$\log 46 = 1.6628$$
$$\log 71 = 1.8513$$
$$\log 79 = \underline{1.8976}(+)$$
$$5.4117$$
$$\log 196 = 2.2923(-)$$
$$2\overline{|3.1194}$$
$$\log r = 1.5597$$
$$r = 36.28.$$

Chapter 11: Diagnostic Test   page 229

Following each problem number is the number of the textbook section (in parentheses) in which that kind of problem is discussed.

1. (11-2) $K = \frac{1}{2}ab \sin C$
   $= \frac{1}{2}(100)(140) \sin 81° 14'$
   $= 7000(.9883) = 6918.$

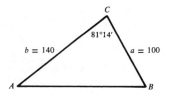

2. (11-2) $K = \sqrt{s(s-a)(s-b)(s-c)}$ $a = 15$
   $= \sqrt{18(3)(8)(7)}$ $b = 10$
   $= \sqrt{3024}$ $c = 11$
   $\log 3024 = 3.4806$ $2s = \overline{36}$
   $\log K = \frac{1}{2} \log 3024 = 1.7403$ $s = 18$
   $K = 54.99.$ $s - a = 3$
          $s - b = 8$
          $s - c = 7.$

3. (11-5) $\sin \frac{1}{2}A = \sqrt{\dfrac{(s-b)(s-c)}{bc}}$ $a = 21$
          $b = 32$
          $c = 27$
   $= \sqrt{\dfrac{(8)(13)}{(32)(27)}} = \sqrt{\dfrac{13}{108}}$ $2s = \overline{80}$
          $s = 40$
   $\log 13 = 11.1139 - 10$ $s - b = 8$
   $\log 108 = \underline{\quad 2.0334 \quad (-)}$ $s - c = 13.$
          $2\overline{|9.0805 - 10}$
   $\log \sin \frac{1}{2}A = \quad 4.5402 - 5 = 9.5402 - 10$
   $\frac{1}{2}A = 20° 18'$
   $A = 40° 36'.$

4. (11-2) $K = \frac{1}{2}a^2 \dfrac{\sin B \sin C}{\sin A}$

   $= \frac{1}{2}(12^2)\dfrac{(\sin 61° 53')(\sin 42° 53')}{\sin 75° 14'}$

   $= \dfrac{72 (\sin 61° 53')(\sin 42° 53')}{\sin 75° 14'}$

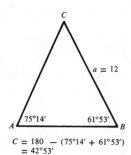

   $C = 180 - (75°14' + 61°53')$
   $= 42°53'$

   $\log 72 = \quad 1.8573$
   $\log \sin 61° 53' = \quad 9.9455 - 10$
   $\log \sin 42° 53' = \underline{\quad 9.8328 - 10(+)}$
   $\qquad\qquad\qquad 21.6356 - 20$
   $\log \sin 75° 14' = \underline{\quad 9.9854 - 10(-)}$
   $\log K = \overline{11.6502 - 10}$
   $K = 44.69.$

5. (11-3) $A + C = 180° - 53° 10'$
$\qquad\qquad\qquad = 126° 50'$
$\qquad \tfrac{1}{2}(A + C) = 63° 25'$
$\qquad\qquad a + c = 45$
$\qquad\qquad a - c = 5$

$$\frac{\tan \tfrac{1}{2}(A - C)}{\tan \tfrac{1}{2}(A + C)} = \frac{a - c}{a + c}$$

$$\tan \tfrac{1}{2}(A - C) = \frac{\tan 63° 25'}{9}$$

$$= \frac{1.9984}{9} = .2220$$

$\qquad \tfrac{1}{2}(A - C) = 12° 31'$
$\qquad \tfrac{1}{2}(A + C) = 63° 25'$
$\qquad\qquad\qquad \overline{A = 75° 56'}$
$\qquad\qquad\qquad\quad C = 50° 54'.$

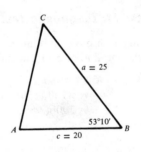

Use Law of Sines to find side b.

$$\frac{b}{\sin B} = \frac{a}{\sin A}$$

$$b = \frac{25 \sin 53° 10'}{\sin 75° 56'}$$

$\qquad\qquad \log 25 = \quad 1.3979$
$\quad \log \sin 53° 10' = \quad 9.9033 - 10(+)$
$\qquad\qquad\qquad\qquad \overline{11.3012 - 10}$
$\quad \log \sin 75° 56' = \quad 9.9868 - 10(-)$
$\qquad\qquad \log b = \quad \overline{1.3144}$
$\qquad\qquad\qquad b = 20.62$

Exercises 12-1   page 234

1. $\tfrac{1}{2}\pi$.

3. $\tfrac{1}{2}\pi, \dfrac{3\pi}{2}$.

5. $\tfrac{1}{4}\pi, \tfrac{5}{4}\pi$.

7. $\tfrac{7}{6}\pi, \tfrac{11}{6}\pi$.

9. π.

11. $\tfrac{1}{6}\pi, \tfrac{5}{6}\pi$.

13. $29° 00', 331° 00'$.

15. $21° 20', 201° 20'$.

17. $183° 20', 356° 40'$.

19. $168° 00', 348° 00'$.

21. $x = \tfrac{1}{3} \arcsin \tfrac{1}{2} y$.

23. $x = \tfrac{1}{2} \arccos 3y$.

25. $(1/\pi) \arctan y$.

27. $x = \tfrac{1}{3} \arccos (y - 1)$.

29. $x = \tfrac{1}{2} \sin y$.

31. $\tfrac{4}{3}$.

33. $\tfrac{3}{2}$.

35. $\tfrac{3}{7}$.

37. $\tfrac{1}{5}\sqrt{21}$.

39. 1.

Exercises 12-3 page 242

1. $\frac{1}{4}\pi$.
3. $\frac{1}{4}\pi$.
5. $\frac{2}{3}\pi$.
7. $-\frac{1}{3}\pi$.
9. $-\frac{1}{6}\pi$.
11. $\frac{1}{6}\pi$.
13. $\frac{2}{3}\sqrt{2}$.
15. $\frac{5}{12}$.
17. $\frac{5}{7}$.
19. $\frac{8}{17}$.
21. $\frac{7}{32}\sqrt{15}$.
23. $-\frac{24}{7}$.
25. 1.
27. $\frac{63}{65}$.
29. $\frac{156}{133}$.
31. $\frac{3}{4}$.
33. $\pi - \text{Arcsin}\,\frac{3}{5}$.
35. $\pi + \text{Arccos}\,\frac{5}{13}$.
37. $2\pi - \text{Arctan}\,\frac{3}{4}$.
39. $2\pi + \text{Arcsin}\,\frac{4}{5}$.
41. $\text{Arccos}\,(-\frac{3}{5})$.
53. 1.
55. $\frac{1}{5}\sqrt{5}$.
57. $\frac{1}{2}$.
59. $\frac{1}{2}$.
61. 0, 1.
63. 1.

Chapter 12: Review Exercises page 244

1. $\pi/3, 5\pi/3$.
3. $\pi/3, 4\pi/3$.
5. $-(\pi/2)$.
7. $2\pi/3$.
9. $\frac{2}{3}$.
11. $\frac{4}{5}$.

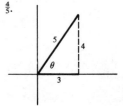

13. Let $\theta = \text{Arcsin}\,\frac{3}{5}$.

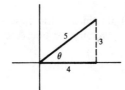

Then $\sin 2\theta = 2 \sin \theta \cos \theta$
$$= 2(\tfrac{3}{5})(\tfrac{4}{5})$$
$$= \tfrac{24}{25}.$$

15. Let $A = \text{Arctan}\,\frac{3}{2}$ and $B = \text{Arctan}\,\frac{1}{2}$.

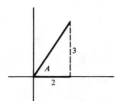 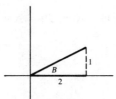

Then $\tan (A + B) = \dfrac{\tan A + \tan B}{1 - \tan A \tan B}$

$$= \frac{\frac{3}{2} + \frac{1}{2}}{1 - \frac{3}{2} \cdot \frac{1}{2}} = \frac{2}{1 - \frac{3}{4}} = \frac{2}{\frac{1}{4}} = 8.$$

17. $y = 3 \csc 2x$

$$\frac{y}{3} = \csc 2x$$

$$2x = \operatorname{arccsc} \frac{y}{3}$$

$$x = \frac{1}{2} \operatorname{arccsc} \frac{y}{3}.$$

19. $2y = \sec 3x - 1$
$2y + 1 = \sec 3x$
$3x = \operatorname{arcsec}(2y + 1)$

$$x = \frac{1}{3} \operatorname{arcsec}(2y + 1).$$

21. $\operatorname{Arctan} x + \operatorname{Arctan} 1 = \dfrac{\pi}{2}$

$$\operatorname{Arctan} x + \frac{\pi}{4} = \frac{\pi}{2}$$

$$\operatorname{Arctan} x = \frac{\pi}{4}$$

$$x = \tan \frac{\pi}{4} = 1.$$

23.   Take the tangent of both sides of the equation.
$\tan[\operatorname{Arctan}(x - 1) - \operatorname{Arctan} x] = \tan[\operatorname{Arctan}(-1)]$

$$\frac{x - 1 - x}{1 + (x - 1)(x)} = -1$$

$$\frac{-1}{1 + x^2 - x} = -1$$

$$x^2 - x + 1 = 1$$
$$x^2 - x = 0$$
$$x(x - 1) = 0$$
$$x = 0, 1.$$

Check for x = 0.
  $\operatorname{Arctan}(0 - 1) - \operatorname{Arctan} 0 = \operatorname{Arctan}(-1)$

$$-\frac{\pi}{2} \quad - \quad 0 \quad = \quad -\frac{\pi}{2}$$

$$-\frac{\pi}{2} = -\frac{\pi}{2}.$$

Check for x = 1.
  $\operatorname{Arctan}(1 - 1) - \operatorname{Arctan} 1 = \operatorname{Arctan}(-1)$

$$0 \quad - \quad \frac{\pi}{4} \quad = \quad -\frac{\pi}{4}.$$

Therefore, both 0 and 1 are roots of the equation.

25. For $0 \leq a \leq 1$

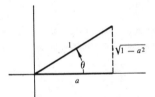

$\theta = \text{Arc cos } a$
$\theta = \text{Arc sin } \sqrt{1 - a^2}$
Therefore,
$\text{Arc cos } a = \text{Arc sin } \sqrt{1 - a^2}.$

Chapter 12: Diagnostic Test   page 244

Following each problem number is the number of the textbook section (in parentheses) in which that kind of problem is discussed.

1. (12-1)

(a) Let $\theta = \arcsin \dfrac{\sqrt{3}}{2}$.

Then $\sin \theta = \dfrac{\sqrt{3}}{2}$

and $\quad \theta = \dfrac{\pi}{3}, \dfrac{2\pi}{3}.$

(b) Let $\theta = \arccos(-1)$.
Then $\cos \theta = -1$
and $\quad \theta = \pi.$

(c) Let $\theta = \arctan(-1)$.
Then $\tan \theta = -1$

and $\quad \theta = \dfrac{3\pi}{4}, \dfrac{7\pi}{4}.$

2. (12-3)

(a) Let $\theta = \text{Arctan } 1$ where $-\dfrac{\pi}{2} < \theta < \dfrac{\pi}{2}.$
Then $\tan \theta = 1$

and $\quad \theta = \dfrac{\pi}{4}.$
(Principal value interval for Arctan numbers.)

(b) Let $\theta = \text{Arccos } \frac{1}{2}$ where $0 \leq \theta \leq \pi.$
Then $\cos \theta = \frac{1}{2}$
(Principal value interval for Arccos numbers.)

and $\quad \theta = \dfrac{\pi}{3}.$

(c) Let $\theta = \text{Arcsin} \left(-\frac{1}{2}\right)$ where $-\frac{\pi}{2} \leq \theta \leq \frac{\pi}{2}$.

Then $\sin \theta = -\frac{1}{2}$

and $\quad \theta = -\frac{\pi}{6}$.

(Principal value interval for Arcsin numbers.)

(d) Let $\theta = \text{Arcsin} \frac{4}{5}$ where $-\frac{\pi}{2} \leq \theta \leq \frac{\pi}{2}$.

Sketch θ.

Then $\tan \theta = \frac{4}{3}$.

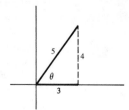

(e) Let $A = \text{Arcsin} \frac{5}{13}$ and $B = \text{Arccos} \frac{4}{5}$.

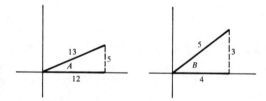

Then, $\sin \left(\text{Arcsin} \frac{5}{13} + \text{Arccos} \frac{4}{5}\right)$

$= \sin (A + B)$

$= \sin A \cos B + \cos A \sin B$

$= \left(\frac{5}{13}\right)\left(\frac{4}{5}\right) + \left(\frac{12}{13}\right)\left(\frac{3}{5}\right)$

$= \frac{20}{65} + \frac{36}{65}$

$= \frac{56}{65}$.

3. (12-1) $\quad \cot 2x = y + 3$

$2x = \text{arccot} (y + 3)$

$x = \frac{1}{2} \text{arccot} (y + 3)$.

4. (12-3) *Let* $\theta = \text{Arcsin} \, a$ *where* $-1 < a < 1$.

Sketch $0 < a < 1$ $\qquad\qquad -1 < a \leq 0$

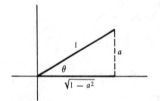
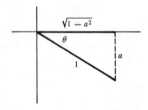

Then take the tangent of both sides of the equation.

$$\tan (\text{Arcsin } a) = \tan \left(\text{Arctan } \frac{a}{\sqrt{1 - a^2}} \right).$$

From figures we get

$$\frac{a}{\sqrt{1 - a^2}} = \frac{a}{\sqrt{1 - a^2}} \, .$$

which implies that $\text{Arcsin } a = \text{Arctan } \dfrac{a}{\sqrt{1 - a^2}}$ for $-1 < a < 1$.

5. (12-3) Take the tangent of both sides of the equation.

$$\tan [\text{Arctan } (x + 1) - \text{Arctan } (x - 1)] = \tan [\text{Arctan } 2]$$

$$\frac{x + 1 - (x - 1)}{1 + (x + 1)(x - 1)} = 2$$

$$\frac{2}{1 + x^2 - 1} = 2$$

$$\frac{2}{x^2} = 2$$

$$x^2 = 1$$

$$x = \pm 1 \text{ (Both roots check in the original equation.)}$$

Exercises 13-3 page 250

1. $7 + i.$

3. $-3 - i.$

5. $\frac{7}{3} + \frac{2}{3}i.$

7. $-5 + 12i.$

9. $10.$

11. $2 + 11i.$

13. $-i.$

15. $\frac{7}{9} + \frac{4}{9}\sqrt{2}i.$

17. $x = 2, y = 2.$

19. $x = 2, y = 16.$

21. $x = 3, y = 1.$

23. $x = -6, y = 6.$

25. $2 \pm i.$

27. $\dfrac{2 \pm 7i}{3}.$

Exercises 13-4 page 251

1.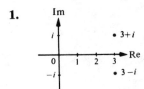

3.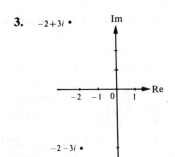

5.

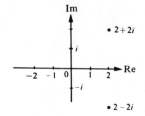

7.

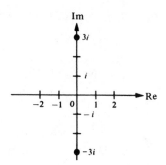

9.

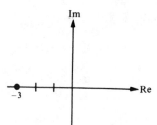

-3 is the conjugate of itself.

Exercises 13-5   page 254

1. $(2\sqrt{3}, 2)$.
3. $(3, 0)$.
5. $(0, -8)$.
7. $(-9.4, -3.4)$.
9. $(0, 6)$.
11. $(\sqrt{2}, \sqrt{2})$.
13. $(2\sqrt{2}, \frac{1}{4}\pi)$.
15. $(3, 0)$.
17. $(5, \pi)$.
19. $(2, \frac{7}{4}\pi)$.
21. $(2, 0)$.
23. $(-\sqrt{13}, 56° \, 20')$.

Exercises 13-6   page 258

1. $2\sqrt{2}, \frac{1}{4}\pi$.
3. $2, \frac{2}{3}\pi$.
5. $\sqrt{2}, \frac{7}{4}\pi$.
7. $\sqrt{3}, 234° \, 40'$.
9. $3, \frac{1}{2}\pi$.
11. $5, \pi$.
13. $6, 0$.
15. $\sqrt{13}, 303° \, 40'$.
17. $\frac{3}{2} + \frac{3}{2}\sqrt{3}i$.
19. $2\sqrt{2} + 2\sqrt{2}i$.
21. -1.
23. $-8i$.
25. $2\left(\cos\frac{1}{6}\pi + i\sin\frac{1}{6}\pi\right), \sqrt{3} + i$.
27. $5\left(\cos\pi + i\sin\pi\right), -5$.
29. $4\left(\cos\frac{3}{4}\pi + i\sin\frac{3}{4}\pi\right), -2\sqrt{2} + 2\sqrt{2}i$.

Exercises 13-8   page 263

1. $6\left(\cos 90° + i\sin 90°\right), 6i$.
3. $2\left(\cos 240° + i\sin 240°\right), -1 - \sqrt{3}i$.
5. $10\left(\cos 140° + i\sin 140°\right), -7.660 + 6.428i$.
7. $\cos 150° + i\sin 150°, -\frac{1}{2}\sqrt{3} + \frac{1}{2}i$.
9. $\frac{1}{32}\left(\cos 300° + i\sin 300°\right), \frac{1}{64} - \frac{1}{64}\sqrt{3}i$.
11. $324\left(\cos 180° + i\sin 180°\right), -324$.
13. $2\left(\cos 60° + i\sin 60°\right), 1 + \sqrt{3}i$.
15. $\frac{1}{13}\left(\cos 67° \, 50' + i\sin 67° \, 50'\right), 0.029 + 0.071i$.
17. $\cos 2\theta = \cos^2\theta - \sin^2\theta, \sin 2\theta = 2\sin\theta\cos\theta$.

Exercises 13-9 page 267

1. $\frac{1}{2}\sqrt{3} + \frac{1}{2}i$;
   $-\frac{1}{2}\sqrt{3} + \frac{1}{2}i$;
   $-i$.

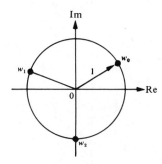

3. $\frac{1}{2}\sqrt{2} + \frac{1}{2}\sqrt{2}i$;
   $-\frac{1}{2}\sqrt{2} + \frac{1}{2}\sqrt{2}i$;
   $-\frac{1}{2}\sqrt{2} - \frac{1}{2}\sqrt{2}i$;
   $\frac{1}{2}\sqrt{2} - \frac{1}{2}\sqrt{2}i$.

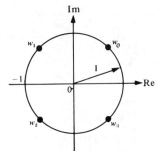

5. $1.397 + .221i$;
   $.221 + 1.397i$;
   $-1.260 + .642i$;
   $-1 - i$;
   $.642 - 1.260i$.

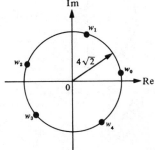

7. 1;
   $-\frac{1}{2} + \frac{1}{2}\sqrt{3}i$;
   $-\frac{1}{2} - \frac{1}{2}\sqrt{3}i$.

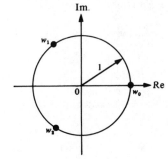

9. 2;
$1 + \sqrt{3}i$;
$-1 + \sqrt{3}i$;
-2;
$-1 - \sqrt{3}i$;
$1 - \sqrt{3}i$.

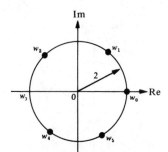

11. $1.932 - .518i$;
$.518 + 1.932i$;
$-1.932 + .518i$;
$-.518 - 1.932i$.

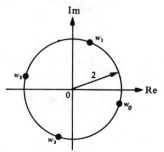

13. $2 + 3i$;
$-2 - 3i$.

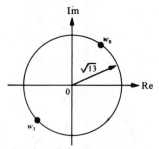

15. $.809 + .588i$; $-.309 + .951i$; -1; $-.309 - .951i$; $+.809 - .588i$.
17. $\frac{3}{2} + \frac{3}{2}\sqrt{3}i$; -3; $\frac{3}{2} - \frac{3}{2}\sqrt{3}i$.
19. $\frac{1}{2}\sqrt{6} + \frac{1}{2}\sqrt{2}i$; $-\frac{1}{2}\sqrt{6} - \frac{1}{2}\sqrt{2}i$;
$\frac{1}{2}\sqrt{6} - \frac{1}{2}\sqrt{2}i$; $-\frac{1}{2}\sqrt{6} + \frac{1}{2}\sqrt{2}i$.

Chapter 13: Review Exercises   page 268

1. $(3 + 4i) + (2 - 5i)$
$= (3 + 2) + (4 - 5)i$
$= 5 - i$.

3. $(5 - 2i) - (3 - 4i)$
$= (5 - 3) + (-2 + 4)i$
$= 2 + 2i$.

5. $(4 + i)(4 - i)$
$= 4^2 - i^2$
$= 16 - (-1)$
$= 17$.

7. $\dfrac{5+i}{5-i} \cdot \dfrac{\overline{|5+i|}}{|5+i|} = \dfrac{25+10i+i^2}{25-(i^2)} = \dfrac{25+10i-1}{25-(-1)}$

$\qquad\qquad = \dfrac{24+10i}{26} = \dfrac{12+5i}{13} = \dfrac{12}{13} + \dfrac{5}{13}i.$

9. $2(\cos 70° + i \sin 70°) \cdot 5(\cos 20° + i \sin 20°)$
$= 2 \cdot 5[\cos(70° + 20°) + i \sin(70° + 20°)]$
$= 10(\cos 90° + i \sin 90°)$
$= 10(0 + i)$
$= 0 + 10i = 10i.$

11. $[6(\cos 146° + i \sin 146°)] \div [3(\cos 86° + i \sin 86°)]$
$= \frac{6}{3}[\cos(146° - 86°) + i \sin(146° - 86°)]$
$= 2(\cos 60° + i \sin 60°)$
$= 2\left(\frac{1}{2} + i\dfrac{\sqrt{3}}{2}\right)$
$= 1 + \sqrt{3}i.$

13. $[2(\cos 15° + i \sin 15°)]^8$
$= 2^8(\cos 8 \cdot 15° + i \sin 8 \cdot 15°)$
$= 256(\cos 120° + i \sin 120°)$

$= 256\left(-\frac{1}{2} + i\dfrac{\sqrt{3}}{2}\right)$
$= 128(-1 + i\sqrt{3}) = -128 + 128\sqrt{3}i.$

15. $\dfrac{32}{(1 + i\sqrt{3})^4}$. Change $1 + i\sqrt{3}$ to polar form.

$\qquad\qquad 1 + i\sqrt{3} = 2(\cos 60° + i \sin 60°).$

Then $\dfrac{32}{(1 + i\sqrt{3})^4} = 32[2(\cos 60° + i \sin 60°)]^{-4}$

$\qquad\qquad = 32 \cdot 2^{-4}[\cos(-240°) + i \sin(-240°)]$

$\qquad\qquad = 2\left[-\frac{1}{2} + i\dfrac{\sqrt{3}}{2}\right]$

$\qquad\qquad = -1 + \sqrt{3}i.$

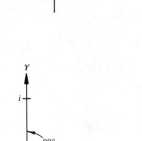

17. When we plot i we see that $\theta = 90°$ and $r = 1$.
$i = \cos(90° + k360°) + i(\sin 90° + k360°)$

$i^{1/5} = \cos\left(\dfrac{90°}{5} + \dfrac{k360°}{5}\right) + i \sin\left(\dfrac{90°}{5} + \dfrac{k360°}{5}\right)$

$\qquad = \cos(18° + k72°) + i \sin(18° + k72°).$
Let $k = 0, 1, 2, 3$, and 4; then
$w_0 = \cos 18° + i \sin 18°$
$w_1 = \cos 90° + i \sin 90°$
$w_2 = \cos 162° + i \sin 162°$
$w_3 = \cos 234° + i \sin 234°$
$w_4 = \cos 306° + i \sin 306°.$

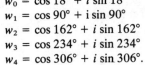

19. $2 + 2\sqrt{3}i$

$r = \sqrt{2^2 + (2\sqrt{3})^2}$

$= \sqrt{4 + 12} = 4.$

$\theta = \arctan \dfrac{2\sqrt{3}}{2} = 60°$

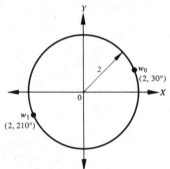

$2 + 2\sqrt{3}i = 4[\cos(60° + k360°) + i\sin(60° + k360°)]$

$w_k = \sqrt[2]{4}\left[\cos\left(\dfrac{60°}{2} + \dfrac{k360°}{2}\right) + i\sin\left(\dfrac{60°}{2} + \dfrac{k360°}{2}\right)\right]$

$w_0 = 2(\cos 30° + i\sin 30°)$

$w_1 = 2(\cos 210° + i\sin 210°).$

21. Use Definition II, Section 13-2.

$(x + 2y) + (x - 3y)i = 8 - 2i.$

$\begin{array}{r} x + 2y = 8 \\ x - 3y = -2 \\ \hline 5y = 10 \\ y = 2. \end{array}$

$x + 2y = 8$

$x + 2(2) = 8$

$x = 4.$

23. Use the quadratic formula.

$2x^2 + 6x + 13 = 0$

$x = \dfrac{-6 \pm \sqrt{36 - 104}}{4}$

$= \dfrac{-6 \pm \sqrt{-68}}{4}$

$= \dfrac{-6 \pm 2\sqrt{17}i}{4} = -\tfrac{3}{2} \pm \tfrac{1}{2}\sqrt{17}i.$

Chapter 13: Diagnostic Test   page 269

Following each problem number is the number of the textbook section (in parentheses) in which that kind of problem is discussed.

1. (13-3) $(7 - 2i) + (3 + 5i) = (7 + 3) + (-2 + 5)i$

$= 10 + 3i.$

2. (13-3) $(9 - 4i) - (6 - 2i) = (9 - 6) + (-4 + 2)i$

$= 3 - 2i.$

3. (13-3) $(3 + 2i)(3 - 2i) = 3^2 - (2i)^2$
$$= 9 - (-4)$$
$$= 13.$$

4. (13-3) $\dfrac{1 - 5i}{1 + 5i} \cdot \dfrac{\boxed{1 - 5i}}{\boxed{1 - 5i}} = \dfrac{1 - 10i - 25}{1 + 25} = \dfrac{-24 - 10i}{26}$

$$= -\frac{12}{13} - \frac{5}{13}i.$$

5. (13-8) $1 + i = \sqrt{2}(\cos 45° + i \sin 45°)$.
Then $(1 + i)^6 = [\sqrt{2}(\cos 45°) + i \sin 45°]^6$
$$= (\sqrt{2})^6(\cos 6 \cdot 45° + i \sin 6 \cdot 45°)$$
$$= 8(\cos 270° + i \sin 270°)$$
$$= 8(0 - i) = 0 - 8i = 8i.$$

6. (13-7) $12(\cos 80° + i \sin 80°) \div 3(\cos 20° + i \sin 20°)$
$$= \tfrac{12}{3}[\cos (80° - 20°) + i \sin (80° - 20°)]$$
$$= 4(\cos 60° + i \sin 60°)$$

$$= 4\left(\tfrac{1}{2} + i\frac{\sqrt{3}}{2}\right)$$

$$= 2 + 2\sqrt{3}i.$$

7. (13-8) $5(\cos 18° + i \sin 18°) \cdot 2(\cos 27° + i \sin 27°)$
$$= 5 \cdot 2[\cos (18° + 27°) + i \sin (18° + 27°)]$$
$$= 10(\cos 45° + i \sin 45°)$$

$$= 10\left(\frac{\sqrt{2}}{2} + i\frac{\sqrt{2}}{2}\right)$$

$$= 5\sqrt{2} + 5\sqrt{2}i.$$

8. (13-3) $\dfrac{\sqrt{3} - \sqrt{2}i}{\sqrt{3} + \sqrt{2}i} \cdot \dfrac{\boxed{\sqrt{3} - \sqrt{2}i}}{\boxed{\sqrt{3} - \sqrt{2}i}} = \dfrac{3 - 2\sqrt{6}i - 2}{3 + 2}$

$$= \frac{1 - 2\sqrt{6}i}{5} = \frac{1}{5} - \frac{2\sqrt{6}}{5}i.$$

9. (13-7) $\sqrt{3} + i = 2(\cos 30° + i \sin 30°)$

$$\frac{1}{(\sqrt{3} + i)^5} = [2(\cos 30° + i \sin 30°)]^{-5}$$

$$\frac{64}{(\sqrt{3} + i)^5} = \frac{64}{2^5}[\cos (-5)(30°) + i \sin (-5)(30°)]$$

$$= 2[\cos (-150°) + i \sin (-150°)]$$

$$= 2\left(\frac{\sqrt{3}}{2} - i\tfrac{1}{2}\right)$$

$$= \sqrt{3} - i.$$

10.   (13-9) $\theta = \arctan \dfrac{8\sqrt{3}}{-8} = 120°, \; r = \sqrt{(-8)^2 + (8\sqrt{3})^2} = 16$

$-8 + 8\sqrt{3}i = 16 \cos(120° + k360°) + i \sin(120° - k360°)$

$(-8 + 8\sqrt{3}i)^{1/4} = \sqrt[4]{16}\left[\cos\left(\dfrac{120°}{4} + \dfrac{k(360°)}{4} \right) + i \sin\left(\dfrac{120°}{4} + \dfrac{k(360°)}{4} \right) \right]$

$w_k = 2[\cos(30° + k90°) + i \sin(30° + k90°)]$
$w_0 = 2(\cos 30° + i \sin 30°)$
$w_1 = 2(\cos 120° + i \sin 120°)$
$w_2 = 2(\cos 210° + i \sin 210°)$
$w_3 = 2(\cos 300° + i \sin 300°).$

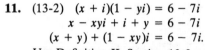

11.   (13-2) $(x + i)(1 - yi) = 6 - 7i$
             $x - xyi + i + y = 6 - 7i$
             $(x + y) + (1 - xy)i = 6 - 7i.$

Use Definition II, Section 13-2.

       $x + y = 6$ and $1 - xy = -7$ Solutions

       $x + \dfrac{8}{x} = 6$ $xy = 8$

x	y
2	4
4	2

     $x^2 - 6x + 8 = 0$ $y = \boxed{\dfrac{8}{x}}$

    $(x - 2)(x - 4) = 0$ $y = \dfrac{8}{2} = 4$

Therefore, $x = ②, ④.$ $y = \dfrac{8}{4} = 2.$

Both solutions satisfy the original equation.

12.   (13-1) $2x^2 - 2x + 3 = 0.$

Use the quadratic formula.

$x = \dfrac{-(-2) \pm \sqrt{(-2)^2 - 4(2)(3)}}{2(2)}$

   $= \dfrac{2 \pm \sqrt{-20}}{4} = \dfrac{2 \pm 2\sqrt{5}i}{4} = \tfrac{1}{2} \pm \dfrac{\sqrt{5}}{2}i.$

Index

TRIGONOMETRIC IDENTITIES

(1) $\csc \theta = \dfrac{1}{\sin \theta}.$ (2) $\sec \theta = \dfrac{1}{\cos \theta}.$ (3) $\cot \theta = \dfrac{1}{\tan \theta}.$

(4) $\tan \theta = \dfrac{\sin \theta}{\cos \theta}.$ (5) $\cot \theta = \dfrac{\cos \theta}{\sin \theta}.$

(6) $\sin^2 \theta + \cos^2 \theta = 1.$ (7) $1 + \tan^2 \theta = \sec^2 \theta.$ (8) $1 + \cot^2 \theta = \csc^2 \theta.$

(9) $\sin(-\theta) = -\sin \theta.$ (10) $\cos(-\theta) = \cos \theta.$ (11) $\tan(-\theta) = -\tan \theta.$

(12) $\cot(-\theta) = -\cot \theta.$ (13) $\sec(-\theta) = \sec \theta.$ (14) $\csc(-\theta) = -\csc \theta.$

(15) $\cos(A + B) = \cos A \cos B - \sin A \sin B.$

(16) $\cos(A - B) = \cos A \cos B + \sin A \sin B.$

(17) $\sin(A + B) = \sin A \cos B + \cos A \sin B.$

(18) $\sin(A - B) = \sin A \cos B - \cos A \sin B.$

(19) $\tan(A + B) = \dfrac{\tan A + \tan B}{1 - \tan A \tan B}.$

(20) $\tan(A - B) = \dfrac{\tan A - \tan B}{1 + \tan A \tan B}.$

(21) $\sin 2A = 2 \sin A \cos A.$

(22a) $\cos 2A = \cos^2 A - \sin^2 A$

(22b) $= 1 - 2 \sin^2 A$

(22c) $= 2 \cos^2 A - 1.$